The IMA Volumes
in Mathematics
and its Applications

Volume 87

Series Editors
Avner Friedman   Willard Miller, Jr.

Springer

New York
Berlin
Heidelberg
Barcelona
Budapest
Hong Kong
London
Milan
Paris
Santa Clara
Singapore
Tokyo

Institute for Mathematics and its Applications
IMA

The **Institute for Mathematics and its Applications** was established by a grant from the National Science Foundation to the University of Minnesota in 1982. The IMA seeks to encourage the development and study of fresh mathematical concepts and questions of concern to the other sciences by bringing together mathematicians and scientists from diverse fields in an atmosphere that will stimulate discussion and collaboration.

The IMA Volumes are intended to involve the broader scientific community in this process.

Avner Friedman, Director

Willard Miller, Jr., Associate Director

* * * * * * * * * *

IMA ANNUAL PROGRAMS

Continued at the back

Peter Donnelly   Simon Tavaré

Editors

Progress in Population Genetics and Human Evolution

With 59 Illustrations

Springer

Peter Donnelly
Departments of Statistics
  and Ecology and Evolution
University of Chicago
Chicago, IL 60637 USA

Simon Tavaré
Departments of Mathematics
  and Biological Sciences
University of Southern California
Los Angeles, CA 90089-1113 USA

Series Editors:
Avner Friedman
Willard Miller, Jr.
Institute for Mathematics and its
  Applications
University of Minnesota
Minneapolis, MN 55455 USA

Mathematics Subject Classifications (1991): 60J80, 60K35, 62M05, 62P10, 65U05, 92D10, 92D15, 92D20, 92D25

Library of Congress Cataloging-in-Publication Data
Progress in population genetics and human evolution / [edited by]
    Peter Donnelly, Simon Tavaré.
        p.   cm. — (IMA volumes in mathematics and its applications ;
    v. 87)
        ISBN 0-387-94944-5 (alk. paper)
        1. Population genetics—Mathematics—Congresses.   2. Human
    evolution—Mathematics—Congresses.   I. Donnelly, Peter J.
    II. Tavaré, Simon.   III. Series.
    QH455.P765   1997
    599.93´8—dc21   96-49167

Printed on acid-free paper.

Production managed by Allan Abrams; manufacturing supervised by Joe Quatela.
Camera-ready copy prepared by the IMA.
Printed and bound by Braun-Brumfield, Inc., Ann Arbor, MI.
Printed in the United States of America.

9 8 7 6 5 4 3 2 1

ISBN 0-387-94944-5   Springer-Verlag New York Berlin Heidelberg   SPIN 10557512

FOREWORD

This IMA Volume in Mathematics and its Applications

PROGRESS IN POPULATION GENETICS
AND HUMAN EVOLUTION

is based on the proceedings of a workshop "Mathematical Population Genetics" that was an integral part of the 1993–94 IMA program on "Emerging Applications of Probability." We would like to thank Peter Donnelly and Simon Tavaré for their hard work in organizing this meeting and in editing the proceedings. We also take this opportunity to thank the National Science Foundation, the Air Force Office of Scientific Research, the Army Research Office, and the National Security Agency, whose financial support made the workshop possible.

Avner Friedman

Willard Miller, Jr.

PREFACE

This volume is the Proceedings of the workshop on Mathematical Population Genetics, held at the IMA January 24 – 28, 1994 as part of the IMA Year on Emerging Areas of Probability.

For many years there has been a crucial interplay between the collection of population genetic data and the development of novel probabilistic and statistical methods for their interpretation. Recent advances in molecular biology have made the study of within-population molecular diversity a reality, and with these data comes the need for appropriate methods of analysis. While these vast amounts of molecular data have been accumulating, there has been a dramatic shift in the theoretical methodology of the field. From largely prospective approaches, it has changed focus to more retrospective ones; the development of 'ancestral methods' (or 'coalescent methods') has clearly revolutionized the field.

The time was right to have a workshop at which the interplay between new data collection methods and new methodology could be explored. This volume provides a summary of the 21 invited talks given at the workshop. They have been divided, somewhat arbitrarily, into three parts. The first focuses on issues relating to the use of mitochondrial DNA sequences to infer aspects of human evolutionary history, including discussion of problems relating to the time and location of 'Mitochondrial Eve.' The second part focuses on molecular techniques and some of the probabilistic, computational, and statistical techniques currently being developed to analyze such data. The third part addresses a variety of issues in the stochastic modeling of genealogy and population structure.

One of our central objectives in organizing the workshop was to encourage the close interaction between experimental molecular biologists and theoreticians. One consequence was a lot of spirited, not to say heated, discussion about how molecular population diversity could and should be studied. The level and extent of these interactions exceeded even our optimistic expectations. We hope that the participants' enthusiasm emerges from these pages.

We thank Avner Friedman, Willard Miller, Jr. and Michael Steele for the opportunity to organize a workshop at the IMA. The workshop was supported in part by grants from the National Science Foundation. Kathy Boyer, Ceil McAree, Kathi Polley, Pam Rech, and Mary Saunders made our job as organizers much easier by taking care of the local arrangements. Finally, this volume would never have appeared but for the extensive help of Melva Rogers, Stephan Skogerboe, Kaye Smith, and Patricia V. Brick in preparing the manuscripts for publication.

Peter Donnelly

Simon Tavaré

CONTENTS

Part III: Genealogy and population models

RECENT AFRICAN ORIGIN OF HUMAN MITOCHONDRIAL DNA: REVIEW OF THE EVIDENCE AND CURRENT STATUS OF THE HYPOTHESIS

MARK STONEKING*

Abstract. For the past seven years or so, much discussion and controversy in the field of human evolution has revolved around the application and interpretation of studies of human mitochondrial DNA (mtDNA) variation, particularly the hypothesis that all mtDNA types in contemporary populations can be traced back to a single African ancestor who lived about 200,000 years ago. In this review I describe the evidence that led to this hypothesis, subsequent work, and where things stand now, particularly with respect to recent criticisms concerning the adequacy of phylogenetic analyses of the mtDNA data.

Mitochondrial DNA (mtDNA) comprises only about 0.00006% of the total human genome, but the contribution of mtDNA to our understanding of human evolution far outweighs its minuscule contribution to our genome. The properties of mtDNA that make it so valuable for evolutionary studies include the high copy number, maternal mode of inheritance, and rapid rate of evolution (Wilson et al. 1985, Avise 1986, Stoneking and Wilson 1989). MtDNA is also perhaps the best-characterized eukaryotic genome, with the complete sequence and gene organization known for humans (Anderson et al. 1981) and several other organisms. The detailed knowledge of the molecular biology of the molecule, made possible by the complete sequence, has greatly facilitated evolutionary studies and offers a paradigm for what we can expect to gain from the effort to determine the complete sequence of the human genome.

In 1987, Rebecca Cann, the late Allan Wilson, and I proposed from our study of human mtDNA variation what has come to be popularly known as the "African Eve" hypothesis (Cann et al. 1987), although we prefer to call it the "recent African origin" hypothesis. In this review I will briefly describe the evidence that led to this hypothesis, as well as subsequent work (and criticism) relating to it; a more thorough review, from which much of this material is condensed, has recently appeared elsewhere (Stoneking 1993). While in many cases it should be obvious, I will also point out where new, creative insights into the analysis of the data are particularly needed.

1. The recent African origin hypothesis. There are three aspects to the recent African origin hypothesis: (1) all mtDNA types in contemporary human populations can be traced back to a single common ancestor; (2) this ancestor probably lived in Africa; and (3) this ancestor probably lived about 200,000 years ago. I shall consider each of these three aspects in turn.

---

* Department of Anthropology, Pennsylvania State University, University Park, PA 16802.

1.1. A single ancestor. What appeared to us to be a straightforward and logical conclusion from simple biological principles has in fact
been responsible for more consternation and misinterpretation, particularly (but not exclusively) by the popular press, than any other aspect of
the hypothesis. Simply put, given that there was a single origin of life on
this planet, and that all living things are descended from this single point
of origin, then it must be true that all of the variation in any segment of
DNA (mitochondrial or nuclear) must ultimately trace back to just one
ancestor. The only alternative would be to suppose multiple origins of life
on this planet, or perhaps an extra-terrestrial origin for mtDNA types in
some individuals!

Although this is a simple principle, there are additional implications
that have also led to misunderstanding. First, because mtDNA is maternally inherited, the mtDNA ancestor must have been female. However,
she was not the only female alive; she was a member of a population, but
the mtDNA types of her contemporaries eventually became extinct because
they or their maternal descendants either left no surviving offspring or left
only male offspring. This process of random extinction is illustrated in
Figure 1.1.

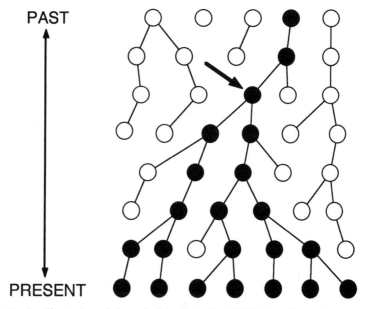

FIG. 1.1. *An illustration of the principle that all mtDNA types in contemporary populations must trace back to a single common ancestor. Solid circles indicate the path
of descent from the ancestor (arrow) to the present generation; empty circles represent
mtDNA types that went extinct (from Stoneking 1993).*

Second, she was not the first human to have appeared on this planet.
Like everyone else, she had ancestors, but she represents the point of co-

alescence of all contemporary mtDNA types. Indeed, she need not even have been human; in principle the coalescence to a single mtDNA ancestor could extend back to any point in our evolutionary history, going all the way back to the origin of mitochondria. For some genes, it is clearly the case that the ancestor was not human. For example, some alleles of the major histocompatibility complex are shared between humans and African apes (Figueroa *et al.* 1988, Lawlor *et al.* 1988, Gyllensten and Erlich 1989), indicating that the ancestor for these alleles predates the human-African ape divergence, some 5 million years ago.

Finally, the mtDNA ancestor need not have contributed any other genes or DNA segments to contemporary populations. While every gene or DNA segment must have a common ancestor, in principle each ancestor could be a different individual, living in different places and at different times. Clearly, then, the idea of a single common ancestor is not of particular interest or concern; what is interesting is determining where and when the mtDNA ancestor lived, and the implications this has for human evolution.

1.2. African origin. Two lines of evidence have been interpreted as supporting an African origin for the human mtDNA ancestor. The first of these is phylogenetic (tree) analysis of the data. The maternal, haploid inheritance of human mtDNA means that, with no recombination, the only source of new variation is mutation. Therefore, the number of mutations separating two mtDNA types reflects how closely related they are — the larger the number of mutations, the more distantly related the mtDNA types. Trees depicting the phylogenetic relationships of mtDNA types can therefore be readily constructed and interpreted as reflecting the maternal genealogical history of our species.

Cann *et al.* (1987) used the maximum parsimony method, which attempts to derive a tree that requires the fewest number of mutations, to construct a tree for 134 mtDNA types determined by high-resolution mapping of restriction site polymorphisms from 148 individuals. The resulting tree, shown schematically in Figure 1.2, had two primary branches, one consisting only of mtDNA types from Africa, the other consisting of all of the mtDNA types from everywhere else in the world, as well as some from Africa. This pattern was interpreted as indicating an African origin of the ancestor, with subsequent migrations out of Africa to the rest of the world.

The same type of tree as shown in Figure 1.2 was found in other RFLP analyses of human mtDNA variation (Johnson *et al.* 1983, Horai et al. 1987, Merriweather *et al.* 1991), as well as in studies that used the polymerase chain reaction (PCR) to amplify and sequence hypervariable segments of noncoding portions of the mtDNA genome (Vigilant *et al.* 1989, Horai and Hayasaka 1990, Kocher and Wilson 1991, Vigilant *et al.* 1991). However, the adequacy of the phylogenetic analysis performed by Cann *et*

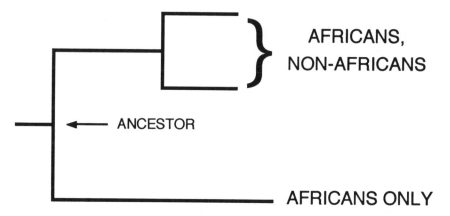

FIG. 1.2. *A simplified version of phylogenetic trees relating contemporary mtDNA types. Invariably, such trees consist of two primary branches, with only Africans represented on both branches (from Stoneking 1993).*

al. (1987) and Vigilant *et al.* (1991) has been called into question (Maddison 1991, Hedges *et al.* 1992, Maddison *et al.* 1992, Templeton 1992, Templeton 1993). These re-analyses have shown that maximum parsimony trees requiring fewer mutations can be constructed from the data, and that at least some of these trees do not conform to the pattern in Figure 1.2.

These re-analyses have been widely publicized, leading some to conclude that the geographic origin of the human mtDNA ancestor cannot be determined at all with existing data. A thorough discussion of the problem is outside the scope of this review; the interested reader should consult the above papers, as well as additional comments on these re-analyses (Wilson *et al.* 1991, Stoneking *et al.* 1992b, Stoneking 1993, Stoneking 1994). However, I believe this to be an overly pessimistic conclusion. Briefly, the crux of the problem lies specifically with present computer-based methods of maximum parsimony analysis of large datasets. For example, for the data of Vigilant *et al.* (1991), consisting of 135 distinct mtDNA types, there are about 8×10^{264} different bifurcating, unrooted trees, and there is thus no way to guarantee that the tree requiring the fewest number of mutations has been found. Furthermore, for these data there does not seem to be one single best parsimony tree; instead, there is an extremely large number of trees of equal length, and there is at present no way to guarantee that computer searches will recover all equally-parsimonious trees.

Since this problem specifically concerns maximum parsimony analysis, one way around it is to use alternative methods of tree construction, such as neighbor-joining (NJ) tree analysis (Saitou and Nei 1987). NJ analysis will produce a single "best" tree, even for very large datasets, and the NJ tree for the data of Vigilant et al. (1991) does conform to the pattern in, Figure 1.2 (Hedges et al. 1992). However, the drawback of NJ tree analysis is that there is no direct way of determining if the NJ tree provides a statistically significantly better fit to the data than alternative trees. An indirect method that addresses this issue, bootstrap resampling of the data, does not indicate strong statistical support for the NJ tree for the Vigilant et al. (1991) data (Hedges et al. 1992).

It should be kept in mind that while bootstrap analysis is widely used to assess the statistical significance of trees, it does not address the real question of interest. We would like to know if, for the data we have in hand, particular trees (or trees containing particular groups) are significantly better than alternative trees. Bootstrap analysis instead addresses the question of how often particular groups appear on trees derived from datasets artificially constructed (by sampling with replacement) to be similar to the original dataset. Clearly, there is a need for new methods for reconstructing and testing phylogenies, especially for large datasets.

The second line of evidence for an African origin of the human mtDNA ancestor is that all studies of worldwide human mtDNA variation have found that African populations, on average, exhibit more mtDNA sequence divergence than non-African populations (Johnson et al. 1983, Cann et al. 1987, Horai and Hayasaka 1990, Merriweather et al. 1991, Vigilant et al. 1991). Figure 1.3 illustrates this for one set of data, namely sequences of one of the hypervariable segments of the noncoding control region of the human mtDNA genome. It is important to note that the measure of mtDNA variation used is not a simple gene frequency-based measure of allelic variation, such as the average heterozygosity values typically calculated for blood group, serum protein, red cell enzyme, or other autosomal loci. Rather, it directly reflects the average number of mutations that have accumulated within populations, so the fact that African populations have the most mtDNA diversity means that they have accumulated the most mtDNA mutations. Since non-African populations appear to have a subset of the total mtDNA diversity found in Africa (also indicated by the phylogenetic analyses), it follows that the greater mtDNA mutational diversity in African populations indicates an African origin for contemporary human mtDNA diversity.

Thus, two lines of evidence appear to support an African origin for the human mtDNA ancestor: phylogenetic analysis indicating that the most divergent mtDNA types are in African populations, and more mtDNA diversity in African populations. To be sure, the adequacy of phylogenetic analysis for determining geographic origin remains debatable. And, there are other potential explanations for the greater mtDNA diversity in Africa, such as selection or a more rapid rate of mutation. However, it should

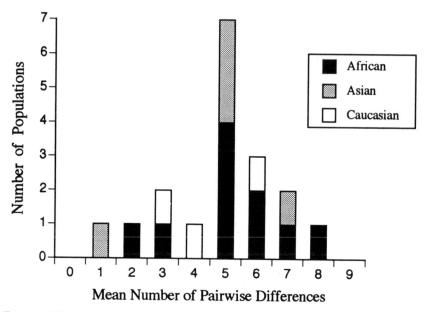

FIG. 1.3. *Histogram of the number of pairwise differences for 18 populations for which sequences of the first hypervariable segment of the noncoding mtDNA control region were obtained. Data include published studies (Horai and Hayasaka 1990, DiRienzo and Wilson 1991, Vigilant et al. 1991, Ward et al. 1991) and unpublished data from my laboratory. African populations clearly exhibit the most mtDNA diversity (from Stoneking 1993).*

be kept in mind that lack of statistical support does not mean that there is no information whatsoever regarding geographic origin (as some have claimed), but rather that the probability of a non-African origin is greater than 5%. Given that different methods of phylogenetic analysis on different datasets invariably lead to trees of the type depicted in Figure 1.2, and given that all worldwide studies of human mtDNA variation find the greatest divergence in African populations, an African origin of the human mtDNA ancestor still appears to provide the best explanation.

1.3. Age of the ancestor. The age of the human mtDNA ancestor can be inferred from the amount of sequence divergence that has arisen among contemporary human mtDNA types, if one knows the rate of mt-DNA evolution. The rate of human mtDNA evolution is typically estimated by calculating the average amount of sequence divergence between human and chimpanzee mtDNA (our closest non-human relative), and dividing by the divergence time for humans and chimpanzees. Thus, based on an average rate of mtDNA divergence of 2-4% per million years (Wilson *et al.* 1985) and an observed sequence divergence since the human ancestor of 0.56%, we previously calculated that the human ancestor lived between 140,000 and 280,000 years ago (Cann *et al.* 1987). This relatively recent date for the age of the human mtDNA ancestor was supported by the control region sequences of Vigilant et al. (1991), who obtained estimates of 11.5-17.3% per million years for the rate of divergence of these rapidly-evolving portions of the mtDNA genome, 2.87% for the amount of sequence divergence since the mtDNA ancestor, and hence an age for the mtDNA ancestor of 166,000 to 249,000 years ago.

However, there are several factors which complicate this seemingly straightforward procedure. First, the time of divergence between human and chimpanzee mtDNA must be known with reasonable accuracy. While most studies do support a date of about five million years ago for this divergence (Hasegawa and Kishino 1991, Horai et al. 1992), others have suggested that the divergence occurred as much as eight to ten million years ago (Gingerich 1985). Older divergence dates will lead to slower rates and hence older dates for the age of the human mtDNA ancestor.

Second, in estimating the amount of sequence divergence between human and chimpanzee mtDNA, correcting for multiple substitutions at the same nucleotide position becomes a factor. The peculiar dynamics of mtDNA sequence evolution (rapid rate of change, greatly elevated frequency of transitions vs. transversions, and existence of mutational "hot-spots") make this correction particularly troublesome; standard methods, which generally assume an equal probability of all types of mutations across all sites, are inadequate (Hasegawa and Horai 1991, Kocher and Wilson 1991).

Third, the above estimates of the age of the human mtDNA ancestor do not include standard errors. Thus, even though we estimate a relatively recent age for the human mtDNA ancestor, it could be that much older ages (even approaching a million years ago) would not be ruled out statistically. This is also not a trivial problem, as there are multiple sources of variance that must be considered in estimating a standard error, including variances related to the sampling of individuals, the sampling of mtDNA diversity, and the inherently stochastic nature of mtDNA evolution (Tajima 1983).

Nevertheless, a number of studies have recently attacked this problem, using a variety of methods and datasets. We have also developed an alternative approach that calibrates the rate of human mtDNA evolution not by comparing human mtDNA with chimpanzee mtDNA, but rather by estimating the amount of mtDNA sequence divergence that has accumulated within relatively isolated regions of the world (in particular, Papua New Guinea) since the time of first colonization, which is known from archaeological investigations (Stoneking et al. 1986, Stoneking et al. 1992a). The results of all of these studies are summarized in Figure 1.4 in the form of approximate 95% confidence intervals for the age of the human mtDNA ancestor. Despite the variety of methods and data used, the upper limits for the age of the human mtDNA ancestor are generally in good agreement, about 500,000 years ago.

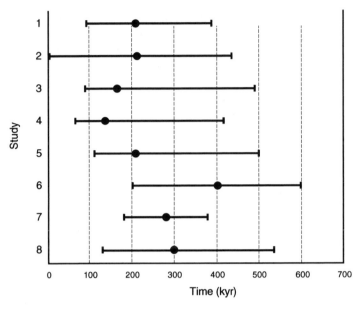

FIG. 1.4. *Estimates of the age and approximate 95% confidence interval for the human mtDNA ancestor from seven studies. Data are from: (1) Templeton (1993); (2) Hasegawa et al. (1993); (3) Tamura and Nei (1993); (4) Stoneking et al. (1992a); (5) Nei (1992); (6) Pesole et al. (1992); (7) Hasegawa and Horai (1991); and (8) Ruvolo et al. (1993). All ages and confidence intervals are based on a human-chimpanzee divergence date of four to six million years ago, with the exception of study (4), which used mtDNA divergence specific to Papua New Guinea and an initial colonization date for New Guinea of 60,000 years ago to calibrate the rate of human mtDNA evolution.*

2. Significance for human evolution. In summary, the available evidence would appear to indicate that the human mtDNA ancestor probably lived in Africa about 200,000 years ago (and not more than about 500,000 years ago). What, if anything, does this tell us about the origin of our species? There are currently two theories that dominate the study of recent human evolution. Both theories begin with the presumption that hominids (members of our genus, Homo) originated in Africa and first left Africa about one million years ago, dispersing throughout the old world. The multiregional evolution hypothesis holds that this was the only major dispersal in human evolution, and that there was no single origin of modern humans (our species, *Homo sapiens*). According to this hypothesis, old world populations are characterized by regional continuity in the fossil record, indicating genetic continuity over the past million years, and the mutations and biological traits that led to modern humans were spread in concert throughout the old world by gene flow (Wolpoff 1989, Wolpoff 1992). By contrast, the single origin hypothesis (also known as the "Garden of Eden") hypothesis holds that there was a single origin of our species in Africa about 100,000 years ago, and that modern humans subsequently

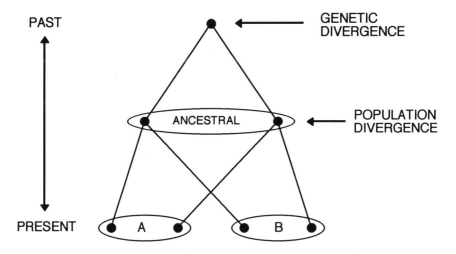

FIG. 2.1. *An illustration of the principle that genetic divergence tends to precede popu-*
lation divergence. An ancestral population is depicted that gave rise to two descendent
populations (A and B), each with two mtDNA types. The two mtDNA types in each
population did not diverge when the populations split, illustrating that a genetic di-
vergence need not correspond to any population event (cf. Nei 1987). However, the
genetic divergence predates and hence places an upper limit on the population diver-
gence, which is expected to be the case whenever the ancestral population is polymorphic
(from Stoneking 1993).

dispersed from Africa, mixing little or perhaps even not at all with the
non-modern old world populations (Stringer and Andrews 1988, Stringer
1992).

Clearly, the mtDNA evidence is most easily reconciled with the latter
hypothesis. In this regard, it is important to realize that a 200,000 year
old date for the age of the human mtDNA ancestor does not mean that
modern humans arose 200,000 years ago. The significance of the age of
the human mtDNA ancestor is that it marks a genetic divergence, which
need not correlate at all with a population divergence (Nei 1987). However, as Figure 2.1 illustrates, in general we expect genetic divergence to
precede population divergence. This will be the case whenever the ancestral population is polymorphic, containing more than one mtDNA type,
which certainly seems more reasonable than supposing that the ancestral
population had only one mtDNA type. Consequently, the coalescence for
the polymorphism must predate the population divergence. Therefore, if
the human mtDNA ancestor really lived no more than 500,000 years ago,
the corresponding inference is that human population differences are also
not more than 500,000 years old. Thus, an African mtDNA ancestor dated

to approximately 200,000 years ago is easily reconciled with an African origin of our species about 100,000 years ago. By contrast, the mtDNA evidence does not appear to be compatible with multiregional evolution, which holds that human population differences should be upwards of one million years old, and hence genetic divergences should be even older.

It should be pointed out that the above scenario (genetic divergence preceding population divergence) is strictly true only for species or populations that are reproductively isolated. In theory, it would be possible for human populations to have begun diverging a million years ago or more, but that an mtDNA type that arose in an African population 200,000 years ago would spread by gene flow, leading to a genetic replacement in the absence of any population replacement. Such genetic replacement could have been mediated by selection favoring this African mtDNA type, or it might have occurred simply by chance. The probability of a genetic replacement within the time period under consideration, across a species distributed from South Africa to northern Asia and Europe, does not seem very likely (at least, in the absence of strong selection). However, this question could and should be addressed by simulation studies, making use of demographic parameters estimated for early humans (Weiss 1984, Harpending *et al.* 1993).

Clearly, a recent African origin of our species is far from proven; much work remains to be done. For statistical resolution of the controversies surrounding the mtDNA story, more sequence information from more individuals is needed. We also need comprehensive genetic data from nuclear loci, in order to verify that the inferences based on patterns of mtDNA variation truly reflect the history of our species, and not just something peculiar about the evolutionary history of human mtDNA. After all, mtDNA is but a single locus (albeit a highly interesting and informative one), and even those of us who are enamored of mtDNA will grudgingly admit that, when it comes to understanding human evolution, it would help to know something about the other 99.99994% of the human genome! Methods of analysis that can then deal with the complexities of nuclear DNA inheritance, in particular segregation and recombination, are also urgently needed.

Gathering comprehensive information on patterns of mtDNA and nuclear DNA variation in human populations is the goal of the Human Genome Diversity initiative (Cavalli-Sforza *et al.* 1991), so if this project comes to fruition we can expect that our knowledge of patterns of genetic variation in contemporary human populations will become as good as it will need to be. In addition, other promising lines of research involving ancient DNA and the genetic basis of normal morphological variation should also shed light on modern human origins, as discussed in more detail elsewhere (Stoneking 1993). We can thus expect that applications of molecular genetics will continue to play a prominent role in furthering our understanding of the origin of our species, but it should also be apparent that new methods for analyzing and making sense of the ever-increasing molecular genetic data are also urgently needed. Our ability to obtain the data has far outpaced our ability to meaningfully analyze them, and it is hoped that this review and the other papers in this volume will stimulate statisticians and mathematicians to jump into the fray.

Acknowledgments. Research in my laboratory is supported by grants from the National Science Foundation and the National Institute of Justice.

REFERENCES

[1] Anderson, S., Bankier, A. T., Barrell, B. G., de Bruijn, M. H. L., Coulson, A. R., Drouin, J., Eperon, I. C., Nierlich, D. P., Roe, B. A., Sanger, F., Schreier, P. H., Smith, A. J. H., Staden, R. and Young, I. G. (1981). Sequence and organization of the human mitochondrial genome. Nature 290:457–465.

[2] Avise, J. C. (1986). Mitochondrial DNA and the evolutionary genetics of higher animals. Philosoph. Trans. Royal Soc. London B 312:325– 342.

[3] Cann, R. L., Stoneking, M. and Wilson, A. C. (1987). Mitochondrial DNA and human evolution. Nature 325:31–36.

[4] Cavalli-Sforza, L. L., Wilson, A. C., Cantor, C. R., Cook-Deegan, R. M. and King, M.-C. (1991). Call for a worldwide survey of human genetic diversity: a vanishing opportunity for the Human Genome Project. Genomics 11:490–491.

[5] Figueroa, F., Gunther, E. and Klein, J. (1988). MHC polymorphism pre- dating speciation. Nature 335:265–268.

[6] Gingerich, P. (1985). Nonlinear molecular clocks and ape-human divergence times. In: Hominid Evolution: Past, Present and Future (P. V. Tobias, ed.). Alan R. Liss, New York, pp. 411–416.

[7] Gyllensten, U. B. and Erlich, H. A. (1989). Ancient roots for polymorphism of the HLA-DQα locus in primates. Proc. Natl. Acad. Sci. USA 86:9986–9990.

[8] Harpending, H. C., Sherry, S. T., Rogers, A. R. and Stoneking, M. (1993). The genetic structure of ancient human populations. Cur. Anthropol. 34:483–496.

[9] Hasegawa, M. and Horai, S. (1991). Time of the deepest root for polymorphism in human mitochondrial DNA. J. Mol. Evol. 32:37–42.

[10] Hasegawa, M. and Kishino, H. (1991). DNA sequence analysis and evolution of Hominoidea. In: New Aspects of the Genetics of Molecular Evolution (M. Kimura and N. Takahata, eds.). Springer- Verlag, Berlin, pp. 303–317.

[11] Hasegawa, M., Di Rienzo, A., Kocher, T. D. and Wilson, A. C. (1993). Toward a more accurate time scale for the human mitochondrial DNA tree. J. Mol. Evol. 37:347–354.

[12] Hedges, S. B., Kumar, S., Tamura, K. and Stoneking, M. (1992). Human origins and the analysis of mitochondrial DNA sequences. Science 255:737–739.

[13] Horai, S. and Hayasaka, K. (1990). Intraspecific nucleotide sequence differences in the major noncoding region of human mitochondrial DNA. Am. J. Hum. Genet. 46:828–842.

[14] Horai, S., Hayasaka, K., Hirayama, K., Takenaka, S. and Pan, I. H. (1987). Evolutionary implications of mitochondrial DNA polymorphism in human populations. In: Human Genetics: Proceedings of the 7th International Congress (F. Vogel and K. Sperling, eds.). Springer-Verlag, Heidelberg, pp. 177–181.

[15] Horai, S., Satta, Y., Hayasaka, K., Kondo, R., Inoue, T., Ishida, T., Hayashi, S. and Takahata, N. (1992). Man's place in Hominoidea revealed by mitochondrial DNA genealogy. J. Mol. Evol. 35:32–43.

[16] Johnson, M. J., Wallace, D. C., Ferris, S. D., Rattazzi, M. C. and Cavalli- Sforza, L. L. (1983). Radiation of human mitochondria DNA types analyzed by re-

striction endonuclease cleavage patterns. J. Mol. Evol. 19:255–271.

[17] Kocher, T. D. and Wilson, A. C. (1991). Sequence evolution of mitochondrial DNA in humans and chimpanzees: control region and a protein-coding region. In: Evolution of Life: Fossils, Molecules, and Culture (S. Osawa and T. Honjo, eds.). Springer-Verlag, Tokyo, pp. 391–413.

[18] Lawlor, D. A., Zenmour, J., Ennis, P. P. and Parham, P. (1988). HLA-A and B polymorphisms predated the divergence of humans and chimpanzees. Nature 335:268–271.

[19] Maddison, D. R. (1991). African origin of human mitochondrial DNA reexamined. Syst. Zool. 40:355–363.

[20] Maddison, D. R., Ruvolo, M. and Swofford, D. L. (1992). Geographic origins of human mitochondrial DNA: phylogenetic evidence from control region sequences. Syst. Biol. 41:111–124.

[21] Merriweather, D. A., Clark, A. G., Ballinger, S. W., Schurr, T. G., Soodyall, H., Jenkins, T., Sherry, S. T. and Wallace, D. W. (1991). The structure of human mitochondrial DNA variation. J. Mol. Evol. 33:543–555.

[22] Nei, M. (1987). Molecular Evolutionary Genetics. Columbia University Press, New York.

[23] Nei, M. (1992). Age of the common ancestor of human mitochondrial DNA. Mol. Biol. Evol. 9:1176–1178.

[24] Pesole, G., Sbisa, E., Preparata, G. and Saccone, C. (1992). The evolution of the mitochondrial D-loop region and the origin of modern man. Mol. Biol. Evol. 9:587–598.

[25] Ruvolo, M., Zehr, S., von Dornum, M., Pan, D., Chang, B. and Lin, J. (1993). Mitochondrial COII sequences and modern human origins. Mol. Biol. Evol. 10:1115–1135.

[26] Saitou, N. and Nei, M. (1987). The neighbor-joining method: a new method for reconstructing phylogenetic trees. Mol. Biol. Evol. 4:406–425.

[27] Stoneking, M. (1993). DNA and recent human evolution. Ev. Anth. 2:60–73.

[28] Stoneking, M. (1994). In defense of REveS - a response to Templeton's critique. Am. Anth. 96:131–141.

[29] Stoneking, M., Bhatia, K. and Wilson, A. C. (1986). Rate of sequence divergence estimated from restriction maps of mitochondrial DNAs from Papua New Guinea. Cold Spring Harbor Symp. Quant. Biol. 51:433–439.

[30] Stoneking, M., Sherry, S. T., Redd, A. J. and Vigilant, L. (1992a). New approaches to dating suggest a recent age for the human mtDNA ancestor. Phil. Trans. R. Soc. Lond. B 337:167–175.

[31] Stoneking, M., Sherry, S. T. and Vigilant, L. (1992b). Geographic origin of human mitochondrial DNA revisited. Syst. Biol. 41:384–391.

[32] Stoneking, M. and Wilson, A. C. (1989). Mitochondrial DNA. In: The Colonization of the Pacific: A Genetic Trail (A. V. S. Hill and S. Serjeantson, eds.). Oxford University Press, Oxford, pp. 215–245.

[33] Stringer, C. B. (1992). Replacement, continuity and the origin of Homo sapiens. In: Continuity or Replacement: Controversies in Homo sapiens evolution (G. Brä Rotterdam, pp. 9–24.

[34] Stringer, C. B. and Andrews, P. (1988). Genetic and fossil evidence for the origin of modern humans. Science 239:1263–1268.

[35] Tajima, F. (1983). Evolutionary relationship of DNA sequences in finite populations. Genetics 105:437–460.

[36] Tamura, K. and Nei, M. (1993). Estimation of the number of nucleotide substitutions in the control region of mitochondrial DNA in humans and chimpanzees. Mol. Biol. Evol. 10:512–526.

[37] Templeton, A. R. (1992). Human origins and analysis of mitochondrial DNA sequences. Science 255:737.

[38] Templeton, A. R. (1993). The REveS hypotheses: a genetic critique and reanalysis. Am. Anth. 95:51–72.

[39] Vigilant, L., Pennington, R., Harpending, H., Kocher, T. D. and Wilson, A. C. (1989). Mitochondrial DNA sequences in single hairs from a southern African population. Proc. Natl. Acad. Sci. USA 86:9350– 9354.

[40] Vigilant, L., Stoneking, M., Harpending, H., Hawkes, K. and Wilson, A. C. (1991). African populations and the evolution of human mitochondrial DNA. Science 253:1503–1507.

[41] Weiss, K. M. (1984). On the number of members of the genus Homo who have ever lived, and some evolutionary implications. Human Biology 56:637–649.

[42] Wilson, A. C., Cann, R. L., Carr, S. M., George, M., Gyllensten, U. B., Helm-Bychowski, K. M., Higuchi, R. G., Palumbi, S. R., Prager, E. M., Sage, R. D. and Stoneking, M. (1985). Mitochondrial DNA and two perspectives on evolutionary genetics. Biol. J. Linn. Soc. 26:375–400.

[43] Wilson, A. C., Stoneking, M. and Cann, R. L. (1991). Ancestral geographic states and the peril of parsimony. Syst. Zool. 40:363–365.

[44] Wolpoff, M. H. (1989). Multiregional evolution: the fossil alternative to Eden. In: The Human Revolution: Behavioural and Biological Perspectives on the Origins of Modern Humans (P. Mellars and C. Stringer, eds.). Edinburgh University Press, Edinburgh, pp. 62–108.

[45] Wolpoff, M. H. (1992). Theories of modern human origins. In: Continuity or Replacement: Controversies in *Homo sapiens* evolution (G. Bräuer and F. H. Smith, eds.). Balkema, Rotterdam, pp. 25–63.

LINES OF DESCENT FROM MITOCHONDRIAL EVE: AN EVOLUTIONARY LOOK AT COALESCENCE

ROSALIND M. HARDING*

1. Introduction. Midway through the IMA workshop on *Mathematical Population Genetics* was an informal afternoon session to discuss the theoretical and statistical methods widely used to make inferences on molecular sequence data about human evolution. Such opportunities for dialogue between mathematicians and molecular biologists, who are divided by a gulf of mutual incomprehension for each other's field of expertise, are unusual. However, questions on human evolution, of interest to both sides, have motivated efforts to come together. What I learnt from the ensuing discussion is that the bridge-building efforts of population geneticists, among whom I count myself, lag about a decade behind the advances in the fields on either side. Here follows a presentation of what some mathematicians have to say to molecular biologists. Had more population geneticists attended, there would be more 'yes, but' rejoinders to report. The discussion and evaluation of the methods presented here may be remiss, therefore, in many details but are intended to challenge population geneticists to catch up or speak up. I admit that though I have tried to find reasons to continue with the simpler methods of popular population genetics, the mathematicians have won me over to their vision of coalescent theory.

In brief, the public debate over whether the ancestors of modern humans were localized within Africa in the late Pleistocene (the Noah's Ark model) or dispersed across Africa and Eurasia (the multiregional model) is being conducted mainly on the basis of evidence from analyses of mitochondrial DNA (mtDNA). The Noah's Ark model is preferred but controversial. Resolution of this debate requires more evidence and stronger inferences on available data. The potential for stronger inferences from DNA sequence data is provided by a powerful methodology mathematically based on coalescent theory. Yet little advantage has been taken of it. I think the reason for this lies in the credibility awarded to the popular methods for their origins in phylogenetic analysis, rather than mathematical population genetics. Biologists find phylogenetic concepts more intuititive than stochastic processes.

What are the weaknesses of the popular methods for mtDNA analysis? During discussion there emerged three areas of criticism. They were: pairwise difference estimates of coalescence time, the concept of effective population size, and the statistical use of theoretical expected values computed as means. For many at the workshop the mtDNA analyses motivating these criticisms are familiar. The workshop opened with Mark Stoneking's

---
* Institute of Molecular Medicine, John Radcliffe Hospital, Headington, Oxford OX3 9DU England, (fax: 44-865-222500; email: rharding@vax.ox.ac.uk).

comprehensive introduction to the debate about the evolutionary origins of modern humans and the contribution of mtDNA evidence. His chapter in this volume provides an important starting point. However, for a wider audience, more background seems warranted. Before presenting the discussion concerned with the criticisms of the popular methods, I will review them with reference to a small number of DNA (mainly mtDNA) studies, and summarize the conclusions in terms of the debate. I am dividing this discussion into two sections. In section 2, I will set out a personal view of the origins and use of pairwise difference estimates of coalescence times and follow with a presentation of the criticisms raised and alternatives recommended during the afternoon. Section 3 concerns estimation and models of evolutionary population size.

2. Coalescence time. Pairwise difference estimates of coalescence time indicate a recent common ancestor in the late Pleistocene

The analytical methods applied to population surveys of polymorphisms in mitochondrial DNA (mtDNA) combine techniques from phylogeny construction (trees) and population genetics. In a phylogenetic approach divergence between DNA lineages from different species is expected to be directly proportional to the mutation rate (μ) and assuming the molecular clock hypothesis of rate constancy, to time. In other words, the number of substituted mutations (k) between two lineages is expected to increase linearly with time, at least until the DNA sequence being scored is saturated by mutation and new mutations arise at previously mutated sites. At this point the expected relationship between sequence divergence and time flattens out. However, saturation should not be a problem for recent divergence, and accordingly, many analyses of mtDNA have calculated divergence time (T) within the human population using the phylogenetic relationship, $T = k/\mu$.

Cann, Stoneking and Wilson (1987) constructed a maximum parsimony tree from 133 human mtDNA types out of 147 sequences sampled, and defined a primary split. The different mtDNA types were assayed by restriction site polymorphisms sampling the whole mtDNA genome. The clades separated by the primary split indicated that the two oldest ancestral lineages had a pairwise divergence (k) of nearly 0.0057 per nucleotide site. Calibrating mtDNA sequence divergence between humans and chimpanzees from paleontological estimates of the time of the species' split, a mutation rate (μ) per million years of 0.04 was estimated assuming a 5 million year split, or alternatively, 0.02 for a 7 million year split. From this phylogenetic analysis ($T = k/\mu$) it was concluded that the common ancestor of all surviving modern human mtDNA lineages existed 142,500 to 285,000 years ago. The primary split defining the two oldest human mtDNA lineages in the tree constructed by Cann, Stoneking and Wilson (1987) separated an African clade and a clade containing all other humans including some Africans. A corollary of this analysis (if this particular tree is accepted) is that genetic diversity is greater within Africans than within any other population, and in conjunction with the recency of the common ancestor, this is taken as evidence for Africa as the geographical centre of modern human evolution.

The main focus of critical attention to date has concerned the construction of the tree for phylogenetic analysis (Templeton, 1993). However, another problem is recognized from the viewpoint of population genetics and that is the complication of genetic drift for the relationship between intra-species sequence divergence and time. Whereas divergence between species is scored from the accumulating number of mutations that have reached fixation, within-species diversity (θ) is scored from mutations in transit to fixation. Using population genetics theory, assuming strict neutrality, there is a very simple relationship between the expected number of sites (k) varying between two sequences, averaged over a sample, to give π on the one side, and the product of haploid female population size (N_f) and probability of mutation (μ) per allele per generation, on the other: $\pi = 2N_f\mu = \theta$. Unlike pairwise divergence of DNA lineages between species which continues to accumulate over time (up to a saturation level), pairwise divergence within a species is expected to equilibrate around an average value of $\pi = 2N_f\mu$ because of the finite size of the population.

Assuming neutrality, panmictic constant population size, (a demographic model such as that of Wright-Fisher), and the uniqueness of mutant substitutions, (the infinite-alleles or infinite-sites models), the expected time to fixation once a new mtDNA mutant has arisen is $2N_f$ generations (Ewens, 1979). In fact, the expected divergence time, T, for a sample of mtDNA sequences is $\theta/\mu = 2N_f$ generations. Continuing with the same assumptions, another important equality is between the expected divergence as a result of genetic drift going forwards in time, and the time back to a common ancestor for a sample of sequences tracing through their genealogical history. Adjusting the equations now for the diploid nuclear genome, $T = \theta/\mu = 4N$ where N is the number of individuals in the population. The expected time back to a common ancestor for a sample of sequences, n, from the nuclear genome, is $4N(1 - 1/n)$ which, for the number of sequences in the population, $n = 2N$, gives $4N - 2$ generations (i.e. total population divergence) and for $n = 2$ gives $2N$ generations (the average pairwise divergence).

A mathematical theory for the expected genealogy of a sample of nuclear DNA sequences was described by Tajima in 1983. For pairs of sequences (i.e. $n = 2$) Tajima (1983) formulated the expected times back to their respective common ancestors as a function of the distribution of k, the number of pairwise site differences: $E(T|k) = 2N(1 + k)/(1 + \theta)$. For the average of k (i.e. π) and $\theta = \pi$, $E(T|\pi)$ is $2N$ generations, the average pairwise divergence. The pair of sequences representing the maximum pairwise divergence for a sample of sequences ($k_{\max}$), estimates the expected total population divergence, $T = 4N$ generations. Because there is a limit on observable within-population divergence due to genetic drift $k_{max} = 1 + 2\pi = 4N$, approximately. Of course, the time that a nucleotide site remains polymorphic can be much shorter than $4N$ if the substituting variant arises in a family lineage where by chance offspring numbers over a number of generations are greater than two. Conversely, substitution by a mutational variant will take much longer if the new variant allele finds itself in a lineage with several generations of less than average reproductive success. The process of genetic drift at each polymorphic site in a sequence contributes to both the distribution of k and the variance in times back to a common ancestor for a sample of sequences, $Var(T|k) = 4N^2(1+k)/(1+\theta)^2$. In other words, genetic drift limits the total time depth of a gene tree, and as well, accounts for whether the time depth of a particular gene tree is deeper or shallower than $4N$ generations.

For the 147 individuals in the sample studied by Cann, Stoneking and Wilson (1987) π was reported as 0.0032 and there were 195 segregating sites in a sequence length of 4158.3 sites (approximated by Excoffier and Langaney, 1989). The largest pairwise difference observed was 30 sites. Substituting $\theta/2\mu$ for $2N$, as $T = [\theta/2\mu](1+k)/(1+\theta)$, and estimating θ from $S = s/(\sum 1/i)$ for $i = 1, \ldots, n-1$, where s is the number of segregating sites in the sample of n sequences ($\theta = 35.045$), Tajima's (1983) equation gives an estimate of the time back to the common ancestor of 90,602 years (μ=0.04 per million years per site) or 181,204 ($\mu = 0.02$ per million years per site). The proportion $(1 + k_{max})/(1 + \theta)$ is expected to equal 2 for equilibrium population diversity. It equals 0.86 for this example because the pairwise divergence is not as great as might be expected relative to the number of segregating sites. Tajima's (1983) equation yields revised estimates for the time back to the common ancestor which are considerably shorter than the original estimates by Cann, Stoneking and Wilson (1987) of 142,500 or 285,000 years.

With a new mtDNA data set, Ruvolo et al. (1993) found $\theta = 2.7$ by averaging diversities estimated from seven segregating sites in one particular mtDNA gene among six individuals and six segregating sites in another mtDNA gene among seven individuals. With relative divergence for the human mitochondrial ancestor as 1/27th of the human- chimpanzee divergence time (estimated from only third codon positions in these two genes) and assuming a 6 million year human-chimp split, a mutation rate, μ, of 0.0085 per site (over the two gene regions with 1580 sites) per million years was inferred. Using Tajima's (1983) equation and given $k = 7$ sites for the two most different sequences, the total divergence T for the sequence genealogy was estimated as 298,000 years with 95% confidence limits of 129,000-536,000 years, assuming T is approximately gamma distributed. In agreement with Cann, Stoneking and Wilson (1987), Ruvolo et al. (1993) concluded that the data were incompatible with a 1-Myr-old human mitochondrial ancestor and rejected the multiregional hypothesis for the emergence of modern humans.

2.1. Evaluating estimates from pairwise differences. Although pairwise difference distributions are appealing because it is not difficult to calculate what happens for two related individuals, they are remarkably complicated for samples greater than two. The patterns of distribution of pairwise differences, k, in a sample of sequences is unknown even for the simplest (infinite sites, constant N) model. Furthermore, k is not a consistent estimator of θ. An estimator is consistent if its sampling variance goes to zero as the sample size is increased, which is desirable. Finally, the use of pairwise differences as a summary statistic means that information in the full dataset is unnecessarily discarded.

Given a set of sequences from a region of the genome that can be appropriately represented by the infinite-sites model, an alternative statistic to average pairwise differences for estimating $\theta = 4N\mu$ is S (given above) based on the number of segregating sites, s (Watterson, 1975). The advantage of S is that θ can be estimated consistently. Admittedly, the variance of S converges logarithmically (at a rate of $\ln N$, where N is the sample size), which is a very slow rate of convergence and this behaviour kills the precision of the estimator. However, for the infinite-sites model, if not for any model, the sampling variance for S does get smaller with increasing sample size. In fact, the particular structure of the infinite-sites process which allows consistent estimators of θ, is quite atypical compared with other processes ie finite-alleles and recurrent mutation models.

As an example to illustrate the difference between estimates of θ based on k and s, I have made a set of computations to estimate the mean and sample variance in θ_k and θ_s from 50 draws with replacement of sample sizes from 2 to 30 out of 60 beta-globin sequences. Fullerton et al (1994, see figure 2) presented the original data showing 17 segregating sites in total among 61 sequences from a Melanesian population in Vanuatu. For this exercise I have excluded the B6 sequence, the sites in the 5′ flanking region except -541, -340 and -316, and additionally site 6 in exon 1, to make the data infinite-sites compatible. For these 60 sequences $\theta_k = 3.38$ and $\theta_s = 2.57$ (standard deviation for $\theta_s = 1.024$).

	θ_k		θ_s	
Sample size	mean	(variance)	mean	(variance)
2	3.38	(6.48)	3.38	(6.48)
3	3.49	(2.80)	3.49	(2.80)
4	3.31	(1.87)	3.25	(1.70)
5	2.99	(1.12)	2.88	(1.01)
6	3.43	(1.09)	3.20	(0.90)
7	3.37	(0.93)	3.11	(0.73)
10	3.22	(0.48)	2.96	(0.30)
15	3.13	(0.38)	2.71	(0.18)
20	3.14	(0.25)	2.55	(0.10)
30	3.28	(0.11)	2.48	(0.07)

Tajima (1983) noted that the variance in pairwise differences decreases to an asymptotic value as the number of sequences sampled from a population increases. That four sequences have approximately the same average pairwise difference as do 30 sequences, is illustrated above by the pattern in the mean values, justifying small sample sizes. (The trend in the sampling variances is not informative because they relate to the source sample of 60 not to the original population.) The average number of segregating sites, on the other hand, decreases with increasing sample size. While larger sample sizes yield more information for improved estimates of θ from s they do not using k. In this example θ is overestimated by k compared to the estimate based on s. The problem here is that the pairwise difference distribution gives an estimate of θ as a function of s and the square of the number of sequence haplotypes, which means each sequence in the sample, n, is not independently contributing information about the mutation process.

The statistics s and k generally give different estimates of θ, partly because of their different sampling properties and partly because they are not equally sensitive to departures from the underlying assumptions of the infinite-sites mutation model, constant N and neutrality. Clearly, Tajima's (1983) computation of T based on the ratio of the maximum pairwise difference to the average diversity based on the number of segregating sites, is sensitive to sampling error as well as to genetic drift. However, increasing sample size does reduce the sampling error around $k_{\max}$ more effectively than around the average value, π (Tajima, 1983). Furthermore, Tajima (1983) preferred k to s on the basis that heterozygosity should be a more robust estimate of neutral diversity if many new mutations are slightly deleterious and constrained from increasing in frequency by selection. Heterozygosity estimates are influenced more by common alleles than by rare alleles, and an allele that becomes common can be assumed neutral. If there is selective constraint, estimates of θ based on k may be lower than estimates based on s. Accordingly, estimates of θ based on k may be more appropriate given that mutation rates are estimated from substitutions, which under these same conditions would also be underestimated.

Cann, Stoneking and Wilson (1987) estimated per nucleotide mtDNA diversity, π, for Africans, Asians, and Europeans, respectively, as 0.47%, 0.35% and 0.23%. Excoffier and Langaney (1989) revised these estimates as 0.32%, 0.25% and 0.17% and also estimated diversity from the number of segregating sites, S, as 0.46%, 0.46% and 0.44%, respectively. Possibly the discrepancy between π and S is due to selective constraint. However, recent population expansion is expected to have the same impact on the frequency distribution of sequence haplotypes, generating lower estimates of θ from π than from S and this is the popular explanation from mtDNA analyses (discussed in greater detail below). Rather than invoking either selective constraint or population expansion, the discrepancy alternatively may be explained by sampling error. With sample sizes larger than 10 sequences the sampling error is expected to be lower for S than for π and accordingly, estimates of θ should be more reliable if based on S rather than on π.

Tajima (1989a,b) has determined the sampling distribution for $d = \pi - S$ under the usual assumptions — infinite-sites mutation, constant N and neutrality, and suggested that departures from these assumptions can be detected by significant values of $D = d/\text{st.dev.}(d)$. For the beta-globin dataset above, π is greater than S, flagging the possibilities of balancing selection, population structure, population collapse, or recombination as evolutionary forces acting additionally to genetic drift. However, for $d = 0.81$ and $D = 0.894$, the 90% confidence limits for 60 sequences are $-1.566 < D < 1.726$, indicating that the different estimates for θ based on π and S are not significant and can be explained by sampling error. These confidence limits can not be greatly reduced by increasing the sample size because the variance around π remains large.

2.2. Introducing the coalescent — historical background.

Tajima's (1983) presentation of a theory for the distribution of gene genealogies in a population evolving by genetic drift and mutation was an independent discovery of the coalescent process, already described by Kingman (1982 a,b) and subsequently reviewed in 1984 by Simon Tavaré. The coalescent is a stochastic model for the genealogical history of a sample of individuals in a population, which it describes independently of any mutation process. The coalescent is a more informative model than diffusion theory for describing genetic drift, and used in conjunction with the infinite-sites model for mutation, opens up a rich store of mathematical results and alternative formulations of classical problems in population genetics.

Coalescence as an idea has been around for at least a hundred years in the context of branching processes. The first problem which stimulated the development of branching process models was that of survival of family names over successive generations, investigated by the Rev. H. W. Watson and first discussed by Francis Galton. The next application of the theory (Fisher 1922, 1930a,b; Haldane 1927) was to the problem of the survival of rare alleles. Fisher looked at discrete branching process approximations to the initial growth phases of a mutant population, a situation in which genealogy has an important role. Later, the theory of branching processes was rediscovered and important ideas about coalescence were independently developed in the context of particle physics, in particular by Harris (1963), without knowledge that the same mathematics had been used in biology. For these genealogical problems of lineage survival or extinction, as for diffusion problems, a perspective of looking forwards in time was maintained.

When coalescence theory was constructed to analyze the ancestry of a population, the perspective was reversed. The idea of going backwards seems natural within the framework of population genetics. It is a relatively recent change in viewpoint in the history of stochastic processes. Perhaps what changed the perspective from looking forward to looking backward was the problem in population genetics that evolution turned out not to be predictable into the future. Data from a slice of time now, at best enables one to ask the question of whether it is concordant with hypotheses about evolutionary history. In the Darwinian tradition it is natural to try and account for past events rather than to speculate about the action of evolution and its future consequences.

In a population genetics setting, coalescents were first referred to in one of Joe Felsenstein's papers in 1971 and also by Bob Griffiths (1980). Earlier than this the idea of coalescence is incorporated in structured population models described in Malécot's (1948) book. However, it took a clear exposition by Kingman (1982 a,b) to describe what was really going on in this process. Subsequently, Tavaré (1984) paid particular attention to presenting the 'diffusion- time-scale' approximations of the coalescent, so that the estimation of θ and other population genetics results for strict neutrality and classical mutation models could be examined from this alternative viewpoint. In the meantime, Tajima (1983), had adjusted phylogenetic analysis from macroevolutionary theory to the study of microevolutionary divergence within a species. Whereas Tavaré described the sampling properties of alleles and segregating sites, Tajima had established an analytical approach to the analysis of DNA sequence data based on the distribution of pairwise differences.

2.3. A statistical framework using coalescent theory. The approach for studies of human evolution recommended by the mathematicians organizing this meeting is a statistical framework based on the coalescent. The conventional coalescent, as Kingman (1982 a,b) thought of it, is a more complicated version of the following. If one currently has n individuals then the process which counts backwards in time the number of ancestors at different times in the past is a death process and its rate goes down from n to $n - 1$ at rate $^nC_2$. The strengths of the coalescent can be summarized as follows. Firstly, coalescence theory provides an efficient way of estimating θ from DNA sequence data. Secondly, the coalescent is amenable for simulation modelling of large populations and the investigation of a wide range of alternative hypotheses, particularly concerning the effects of population expansion and structure.

Biologically, the coalescent describes the ancestral relationships between individuals in a population when there are random fluctuations in reproductive success. If mutations are superimposed on a coalescent tree, individuals, or their lineages, can be allocated into genetic categories. Condensing the coalescent tree in terms of its mutation history generates a phylogenetic tree. Time-scaling present in the coalescent tree is lost from the phylogenetic tree, and it has to be estimated from the mutation history. Depending on how many mutations have been scored, it may not be possible to deduce where in the original tree coalescences occurred. It is possible, however, given the special case of infinite-sites, to uniquely determine the phylogenetic tree from DNA sequence data. Conversely, a phylogenetic tree is the most complete way to summarize DNA sequence data. (The problem for mtDNA analyses that there is more than one way to construct the shortest tree is an indication that not all mtDNA variation is infinite- sites compatible. Recombination is not a problem, but recurrent mutation at hypervariable sites may be.)

The phylogenetic tree is not just the topology of the gene genealogy as described by Tajima (1983) but includes the frequencies in the sample of each of the sequence haplotypes. Compared with a pairwise difference distribution, it preserves not only the numbers of site differences between pairs of sequences, but higher order structure as well: the distribution of three-way and four-way differences, and so forth. If the ancestral bases are known from an outgroup, the root of the tree is also specified. Alternatively, all the rooted trees corresponding to an observed unrooted tree can be calculated and the root estimated probabilistically. The root of tree is the most recent common ancestor for the sample of sequences.

Estimating θ using the coalescent draws on sampling theory. Assuming infinite-sites mutation, neutrality, and the coalescent scaled for a particular constant population size, there is, for any value of θ, a sampling distribution for the rooted phylogenies given the data, just as there is a sampling distribution for the number of alleles given a sample of size n. For a given θ, the sampling distribution $p(\mathbf{Q}, \mathbf{f})$ is found from a recursion which relates the probability of observing a particular unrooted phylogeny, Q, with n sequences having frequencies $f = (f_1, \ldots, f_n)$, to the sum of the corresponding set of rooted tree probabilities. This problem is difficult to solve analytically. Instead $p(\mathbf{Q}, \mathbf{f})$ is represented as the expected value of a functional of a Markov chain which has transition probabilities derived from coefficients in the recursion. The probability is then evaluated by Markov chain Monte Carlo methods. The sampling distributions for the phylogenies are estimated for a range of θ values and a maximum-likelihood approach finds the θ value with the highest probability. Given this estimate of θ for the data, the total coalescence time, T, for the tree can also be estimated. For the 60 beta-globin sequences presented above, θ was estimated using the coalescent as 1.8, and T as 2.434 in units of $2N$ generations with a standard deviation for T of 0.756. Assuming 20 year generations, and estimating μ from human-chimp divergence as 1.6×10^{-9} per site per year, over 2500 nucleotides, $N = \theta/4\mu = 5,625$ and $T = 547,650$ years, with a standard deviation of 170,100 years.

2.4. Sampling issues for estimating θ. One of the options for further improving the estimates of parameters like θ and T is to reduce the sampling variance by increasing either the length of DNA sequenced, or the number of individuals sampled. It is difficult to evaluate for which the returns will be greater. Increasing the length of DNA sequenced increases the number of segregating sites observed. If there is a lot of clustering in the occurrence of segregating sites in DNA sequences, it will be important to increase sequence length to improve estimates of diversity. However, if long sequences are already available and the infinite sites assumption for them is appropriate, further increasing sequence lengths may not add much information. Assuming some degree of genealogical relationship, a sample of twenty individuals does not represent twenty independent lineages, and no matter how much DNA is sequenced, information is limited to that contained in the genealogy of the individuals sampled. More information is obtained by sampling additional individuals.

For populations with a lot of genealogical structure, sampling is critical. To represent say, twenty distinct lineages, a much larger number of individuals are needed in this case. If they are picked randomly, increasing the sample size may not return very many new 'unrelated' individuals. However, the assumption of random sampling is fundamental to all of the theory so far constructed. Ron Lundstrom has looked at what happens in a coalescent model if a number of individuals are closely related and apparently disastrous differences result between estimators from different samples. In particular, estimates of mutation rates, which are what calibrates the tree, vary wildly.

Another option for improving estimates of θ and T is to analyse multiple unlinked loci. Increasing the number of unlinked loci for which lengths of DNA are sequenced should reduce both the sampling variance around the estimators and the evolutionary variance around realizations of the coalescent tree. However, there are many homogeneities between loci to consider such as the nature of the mutation process, differences in selective constraints, and recombination. Also, the independence of unlinked loci is conditional on a population's genealogical history. For instance, polymorphisms at unlinked loci may become associated as a result of population bottlenecks. They are most likely to be independent if the population has been rapidly expanding.

3. Population size. Evidence for population expansion is consistent with a recent African origin and replacement of archaic hominids.

An origin for modern humans within the last few hundred thousand years from a small source population is one way to explain the rather low genetic diversity observed in mtDNA studies, as well as in multilocus surveys of both protein and nuclear DNA polymorphisms. Expansion to our current numbers of several billions apparently has been too recent to be reflected in the patterns of pairwise sequence differences or allelic heterozygosity popularly used to estimate θ. An estimate for effective population size (N_e) of 10^4 is sufficient to account for estimates of neutral nucleotide diversity, π, of 0.0011 at fourfold degenerate sites in nuclear coding DNA (Li and Sadler, 1991), 0.0014 for β-globin DNA (Fullerton et al, 1994) and 0.0032 for the mtDNA studied by Cann, Stoneking and Wilson (1987).

Takahata (1993) concluded that the effective hominid population size has been approximately 10^4 for the past 1 million years, perhaps a little smaller in the late Pleistocene, but to account for δ-globin polymorphism (assuming neutrality) or HLA polymorphism (assuming selection), more likely as large as 10^5.

Not only does the total genetic diversity seem low for a species currently of several billions, but there is very little genetic structure among populations. Interpopulation diversity as a proportion of total diversity, averaged over pairs of populations is given by Wright's fixation index F_{st}, which is approximately 10% for nuclear loci and 40% for mtDNA among the three major populations, i.e. negroids, caucasoids, and mongoloids. Takahata (1993) estimated the scaled effective migration rate $N_e m = (1 - F_{st})/4F_{st}$ for nuclear DNA and $N_f m = (1 - F_{st})/2F_{st}$ for mtDNA, where N_e is the effective population size (N_f is the effective number of females) and inferred a rate of gene flow of approximately 2 individuals (1 female) per generation. This rate is consistent with observed population differences and is sufficient to have prevented the isolation of ancestral subpopulations. Choosing between the Noah's Ark and multiregional hypotheses, therefore, depends on how far flung a Pleistocene population of $10^4 - 10^5$ can be and maintain gene flow of more than one individual per generation.

Population expansion during the last 200,000 years of human evolutionary history has been inferred from pairwise difference distributions. Rogers and Harpending (1992) fit a curve to the pairwise difference distribution presented by Cann, Stoneking and Wilson (1987) using three parameters: θ_0, θ_1 and τ, respectively to estimate $2N_0\mu$, $2N_1\mu$, and $2\mu\tau$. N_0 is the female population size before expansion, N_1 the size after, τ is the time in generations since the expansion and μ is the total mutation rate for the sequence per generation. In this model expansion is instantaneous. Using the method of nonlinear least squares, θ_0, θ_1 and τ, respectively, are estimated as 2.44, 410.69 and 7.18. Assuming 2-4% divergence, μ is estimated to be between 7.5×10^{-4} and 1.5×10^{-3}, giving N_0 between 800 and 1600, N_1 between 137,000 and 274,000 females, and τ between 60,000 and 120,000 years. Rogers and Harpending (1992) argue that the fit of this model for population expansion to the mtDNA data provides some support for the Noah's Ark hypothesis. Against the multiregional hypothesis is the evidence for a small founding population ($\leq$ 1600 females) which could hardly have continuously inhabited the vast regions of Africa, Europe and Asia known from the fossil record to have been occupied by *Homo erectus* and archaic *H. sapiens*. Rogers (1995) presents an extended account in this volume.

3.1. Evaluating the use of effective population size. In the established population genetics approach, based on diffusion theory and assuming neutrality, which is the basis of most of the computations done on DNA sequence data, N_e is the size of a theoretical population, not only constant but also panmictic and comprizing individuals having equal probabilities of reproductive success over some fixed time parameter. It is substituted for a realistic population that is neither constant nor panmictic because the assumption of constant N over evolutionary time is a necessary feature of Wright-Fisher demography. Breathtaking simplifications are not unusual in the construction of analytical mathematical models of the real world; nor should they be forgotten when attempts are made to translate N_e, estimated from genetic data, back into meaningful demographic numbers.

A population that has expanded may have the same effective population size as a population that was initially subdivided and has become randomly mixed. For expanding and subdivided populations with the same effective size as a constant population, not only the demographic histories, but the coalescent times also, may be very different. Strict neutrality is another assumption worth reappraisal in the estimate of demographic numbers from surveys of genetic diversity. Hitchhiking of neutral polymorphisms linked to DNA regions encoding genes or regulatory sequences and subject to directional selection, either as constraint or the substitutional sweep of an advantageous mutation, is expected to reduce genetic diversity in both mitochondrial and nuclear genomes (Hudson, 1994). Consequently, N_e may be underestimated from genetic diversity, and likewise the ancestral time depth for modern humans.

3.2. Large evolutionary variances and meaninglessness of comparing a single observation with an expectation. The estimation of parameters θ_0, θ_1, and τ by Rogers and Harpending (1992) for a population expansion model illustrates a classical approach for dealing with variability, which is to compute statistics, such as pairwise difference distributions, for many outcomes of a stochastic model and judge whether their average, or expected value, compares with some observed data. This procedure is easily implemented with a desktop computer and it is a very useful approach for handling variance due to sampling error. However, simulation studies of evolutionary models show that the variability between realizations, i.e. the evolutionary variance, particularly for constant-sized or structured populations, is extensive almost beyond credibility. For coalescent models, which have now been well studied, individual realizations look nothing like their expected value, calculated as the average over many realizations of evolution. But, in most cases data are available only for one realization of evolution and if this is very unlikely to look anything like the expected value, the classical statistical approach is not very helpful.

In defence of studies of pairwise difference distributions for dating population expansions, it should be pointed out that the variability around realizations for an expanding population is considerably reduced, and an empirical observation from simulation studies is that in this case the expected value is appropriate. The variability is reduced because there is more genealogical independence between lineages in a population which grows exponentially from a very small size. However, whether or not exponential expansion from a small initial value, without any fluctuation over time, is an appropriate assumption for population growth of modern humans has not been established.

This issue of variability and expectation also applies to calculations of the time taken for one mtDNA lineage, representing Eve, to become the ancestral mother of all mtDNA lineages in the current population, and has been presented by Krüger and Vogel (1989). They contrast the solutions suggested on the one hand by both Kimura and Ohta (1969) and Tajima (1983) with that, on the other, by Fisher (1930b). The former compute the 'mean' fixation time for a neutral mtDNA lineage in a female population of constant size N_f to be approximately $2N_f$ generations, and infer that the expected time until fixation of the maternal progeny of one primary mother as 200,000 years (for N_f of 5,000 and generations of 20 years). This computation is consistent with the phylogenetic analysis presented by Cann, Stoneking and Wilson (1987).

Fisher (1930b) proposed a model for computing the probability that only one of the mothers in the original population has maternal descendants after 'n' generations. This approach takes into account the variability in random walk models of fixation (looking forwards) or coalescence (looking backwards) and finds that the probability of observing the expected solution only to be 0.58. Assuming constant N_f of 5000, the probability that all contemporary mtDNA lineages descend from a single ancestor attains 0.9 only in 50,000 generations or 1 million years. Allowing expansion of N, which is a more realistic assumption, the probability that fixation (or coalescence) could occur in 200,000 years becomes very much lower than 0.5. However, the probability of fixation within 200,000 years is reasonably high if the number of primary mothers was much smaller than 5000, or in other words, if a small bottleneck in the founding population of modern humans is proposed.

3.3. Modelling N with coalescent theory. Mathematically, the coalescent is the limiting version of a stochastic process determined after a particular time-scaling and parameter-scaling to model the genealogical history of a population for which we have a DNA sequence phylogeny. The idea of scaling parameters in different ways and looking at the limiting process is well known to geneticists. Wright-Fisher diffusion processes and branching processes, which model an entire population of genes, assuming constant N, are obtained by using similar continuous time scalings as in coalescent theory. Kingman's continuous time approximation arises as a weak limit to many plausible reproduction processes, not only the Wright-Fisher model, but also others which have exchangeable offspring distributions (for modelling selection) or, generate variable or expanding population sizes. In fact, the coalescent is robust with respect to a broad class of haploid reproduction mechanisms, notwithstanding a number of simplifications including the assumptions of random mating and independence between generations.

With coalescent models the lineage history of a sample is simulated as a sufficient and efficient alternative to modelling the whole population through time. The realization that the reproductive contribution of individuals, other than those ancestral to a sample of interest from the current generation, can be ignored, is a great boon for population genetics and has enabled the simulation of large and expanding populations. It is a demonstration of the power of the coalescent that there is a new willingness to accept simulation as a viable way to calculate statistics for some very sophisticated models. Consider for example Hudson's (1983) work on recombination, conditioning on real frequencies in the sample to reduce the statistical variability. However, there are no closed form equations and the approach is very simulation intensive. It is probably because of this that most molecular biologists and many population geneticists have shied away from these methods.

The coalescent is embedded within the whole population, and work by Donnelly and Kurtz (1996 a,b) has characterized the population by considering the coalescent properties of samples of genes, where the samples are embedded in the larger constant-sized population represented by the diffusion process. The relationship of the coalescent to the genealogy in a branching process describing the whole population is currently being investigated by Donnelly and O'Connell (1995). It is for the whole population that variability in offspring numbers, due to both random factors and differential fitness, applies. For different assumptions about the whole population we need to know how to scale the coalescent which describes the genealogical history of the DNA sequences sampled from the current generation. Determination of the class of all possible random time changes of the coalescent will describe a general population genetic framework for making inferences from DNA phylogenies. Using this framework it will be feasible to make different assumptions about the aspect of biological reality being modelled; expanding or structured N as an alternative to the constant N assumption already has been studied by Marjoram and Donnelly (1994).

Marjoram and Donnelly (1994) investigated pairwise difference distributions in a coalescent model of modern human evolution, allowing population expansion and population structure. Using the coalescent to infer the demographic history implied by DNA sequence data is more informative for evaluating a range of more realistic hypotheses than fitting unimodal pairwise difference distribution to models of instantaneous population expansion. Marjoram and Donnelly (1994) concluded that exponential population expansion from N_0 of say 5000 does not consistently generate unimodal pairwise difference distributions and that multimodality is promoted by population structure. Unimodal pairwise difference distributions can be generated by assuming a bottleneck, say $N_0 = 500$, but this value needs to be further reduced, to say $N_0 = 50$, if gene flow was sufficiently restricted in the late Pleistocene to cause population structure. Accordingly, one interpretation of the mtDNA data is indeed population expansion from a bottleneck, in support of the Noah's Ark hypothesis. However, Marjoram and Donnelly (1994) also suggest alternative interpretations of the mtDNA data that should be investigated by simulation, including mutation at hypervariable sites and a selective sweep of an advantageous mtDNA genome.

Mutations superimposed on the coalescent can be generated by any process in line with the assumption that mutations in the sample lineages are independent of mutations in other lineages in the population. Markov processes, which have the probability of events in generation n dependent only on generation $n - 1$, fit this requirement. In the most popular model, 'infinite sites', each mutation occurs independently and in a site that has not previously mutated. It is this dual requirement of independence and uniqueness for the site of each mutation that implies 'infinite sites'. Other mutation models, including variations on the stepwise mutation model, are being investigated in studies of variable numbers of tandem repeats, trinucleotide repeats in particular (Roe, 1992; Harding, Marjoram and Sudbury, 1995).

4. Conclusions. There are three ways forward in studies of human evolution that are highly recommended. The first is to improve estimates of mutation rate, as is already appreciated by molecular biologists (Ruvolo et al., 1993). The second is to use data from loci additional to mtDNA, extending nuclear DNA analyses from surveys of random nucleotide variability to full sequence studies (Fullerton et al. 1994). The third recommendation is to use the coalescent both for estimation of θ and T for investigating different hypotheses about the evolutionary history of modern humans.

The statistical advantage of using coalescent theory is that phylogenies make available all of the information which is contained in DNA sequence data for estimates of θ and T assuming infinite-sites mutation. The use of pairwise differences, by comparison, causes information from the data to be thrown away. Sampling variances around estimates based on a phylogenetic tree are consequently lower than for estimates based on segregating sites alone or on average pairwise difference. However, even the best estimates of coalescence time will be insufficient to resolve the debate on human origins. The confidence limits for the estimate of T using the β-globin sequences from Vanuatu overlaps both the 200,000 year estimate consistent with the Noah's Ark model and the 1 million year estimate for the multiregional model. Much more progress will be made by using the coalescent to evaluate probabilistically the best-fitting demographic history that is implied by DNA sequence data. Current estimates of θ and T, using the coalescent assuming panmixia and constant size, must then be revised for alternative demographic hypotheses such as recent population expansion and ancient population structure. The complications of recombination for genealogies of nuclear DNA sequences are now being tackled by the mathematicians. To address questions of selection state-of-the-art methods are being developed using a more general mathematical framework based on branching processes. I suspect that these, and other advances I have neglected to mention, will do much in the next decade to give prominence to population genetics within the much larger arenas of evolutionary biology and the study of the human genome.

REFERENCES

Cann, R.L., Stoneking, M. and Wilson, A.C. 1987 Mitochondrial DNA and human evolution, *Nature*, 325: 31–36.

Donnelly, P. and Kurtz, T.G. 1996a A countable representation for the Flemming-Viot measure-valued diffusion, *Preprint*.

Donnelly, P. and Kurtz, T.G. 1996b Measure-valued population processes, *Preprint*.

Donnelly, P. and O'Connell, N. 1995 Population models directed by diffusions, *Preprint*.

Ewens, W. J. 1979 *Mathematical Population Genetics*, Springer, New York.

Excoffier, L. and Langaney, A. 1989 Origin and differentiation of human mitochondrial DNA, *Am. J. Hum. Genet.*, 44:73–85.

Felsenstein, J. 1971 The rate of loss of multiple alleles in finite haploid populations. *Theor. Pop. Biol.*, 2:39–403.

Fisher, R.A. 1922 On the dominance ratio, *Proc. R. Soc. Edin.*, 42:321- -341.

Fisher, R.A. 1930a The distribution of gene ratios for rare mutations. *Proc. R. Soc. Edin.*, 50:205–220.

Fisher, R.A. 1930b *The genetical theory of natural selection.* Oxford:Clarendon.

Fullerton, S.M., Harding R.M., Boyce, A.J. and Clegg, J.B. 1994 Molecular and population genetics analysis of allelic sequence diversity at the human β-globin locus, *Proc. Natl. Acad. Sci.USA.* 91:1805–1809.

Griffiths, R.C. 1980 Lines of descent in the diffusion approximation of neutral Wright-Fisher models. *Theor. Pop. Biol.*, 17:40–50.

Griffiths, R.C. and Tavaré, S. 1994 Ancestral inference in population genetics, *Statistical Science*, 9:307–319.

Haldane, J.B.S. 1927 A mathematical theory of natural and artificial selection. V. Selection and mutation, *Proc. Camb. Phil. Soc.*, 23:19–41.

Harding, R.M., Marjoram, P. and Sudbury, A. 1994 A coalescent model for the evolution of trinucleotide repeat variation in the Huntington's Disease gene. *Work in progress*.

Harris, T.E. 1963 *The theory of branching processes*, Berlin:Springer.

Hudson, R.R. 1983 Properties of a neutral allele model with intragenic recombination, *Theor. Pop. Biol.*, 23:183–201.

Hudson, R.R. 1994 How can the low levels of DNA sequence variation in regions of the *Drosophila* genome with low recombination rates be explained? *Proc. Natl. Acad. Sci.*, USA 91:6815–6818.

Kingman, J.F.C. 1982a On the genealogy of large populations, *J. Appl. Prob.*, 19A:27–43.

Kingman, J.F.C. 1982b The coalescent, *Stochastic Processes Appl.*, 13:235–248.

Kimura, M. and Ohta, T. 1969 The average number of generations until fixation of a mutant gene in a finite population, *Genetics*, 63:701–709.

Krüger, J. and Vogel, F. 1989 The problem of our common mitochondrial mother, *Hum. Genet.*, 82:308–312.

Li, W.-H. and Sadler, L.A. 1991 Low nucleotide diversity in Man, *Genetics*, 129:513–523.

Malécot, G. 1948 *Les mathématiques de l'hérédité* (1969,*The mathematics of heredity*, D.M. Yermanos (transl.) San Francisco: Freeman).

Marjoram, P. and Donnelly, P. 1994 Pairwise comparisons of mitochondrial DNA sequences in subdivided populations and implications for early human evolution,

Genetics 136:673–683.

Nei, M. and Tajima, F. 1981 DNA polymorphism detectable by restriction endonucleases, *Genetics* 97:145–163.

Nei, M. and Tajima, F. 1983 Maximum likelihood estimation of the number of nucleotide substitutions for restriction sites data, *Genetics* 105:207–216.

Roe, A. 1992 Correlations and interactions in random walks and population genetics, *Ph.D. thesis*, University of London, London, U.K.

Rogers, A.R. 1996 Population Structure and Modern Human Origins, this volume, pp 55–79.

Rogers, A.R. and Harpending, H. 1992 Population growth makes waves in the distribution of pairwise genetic differences, *Mol. Biol. Evol.*, 9:552–569.

Ruvolo, M., Zehr, S., Von Dornum, M. Pan, D., Chang, B. and Lin, J. 1993 Mitochondrial COII sequences and modern human origins, *Mol. Biol. Evol.*, 10:1115–1135.

Tajima, F. 1983 Evolutionary relationship of DNA sequences in finite populations, *Genetics*, 105:437–460.

Tajima, F. 1989a Statistical method for testing the neutral mutation hypothesis by DNA polymorphism, *Genetics*, 123:585–595.

Tajima, F. 1989b The effect of change in population size on DNA polymorphism, *Genetics*, 123:597–601.

Takahata, N. 1993 Allelic genealogy and human evolution, *Mol. Biol. Evol.*, 10:2–22.

Tavaré, S. 1984 Line-of-descent and genealogical processes, and their applications in population genetic models, *Theor. Pop. Biol.*, 26:119–164.

Templeton, A.R. 1993 The "Eve" hypotheses: a genetic critique and reanalysis, *Am. Anthrop.*, 95:51–72.

Watterson, G.A. 1975 On the number of segregating sites in genetical models without recombination, *Theor. Pop. Biol.*, 7:256–276

PHYLOGEOGRAPHY OF HUMAN MTDNA: AN AMERINDIAN PERSPECTIVE

R.H. WARD*

Abstract. Recent advances in molecular biology have increased both the number and diversity of DNA-based assays that are available to evolutionary biologists. While larger numbers of independent assays result in increased precision for estimates of genetic kinship within and among populations, the real impact of molecular biology lies in the creation of assays that facilitate the construction of intraspecific genomic phylogenies. Phylogeographic principles, which emphasize the distribution of gene genalogies through space and time, provide a powerful tool for evaluating how evolutionary processes have influenced the comtemporary distribution of genetic variety. However, application of this new paradigm to traditional small-scale human societies remains an essentially untried venture. This paper examines the way in which mtDNA sequence variability has been applied to evaluate the phylogeographic structure of a series of Amerindian tribes, and discusses some of the benefits and problems associated with this new toolkit for evolutionary biology.

1. Introduction. Recent advances in molecular biology are in the process of revolutionizing evolutionary biology. In the field of population genetics, the most obvious effect is the exponential increase in the number of polymorphic markers that can be assessed by assaying genetic variability at the level of DNA. The ability to utilize hundreds, if not thousands, of independent polymorphic markers for population studies promises a substantial increase in the precision of estimating population parameters, and a corresponding increase in the specificity of discriminating between competing hypotheses. The increased precision offered by a direct assessment of DNA variability also allows a more precise definition of the genomic ancestry of segregating alleles. The ability to interpret genetic variability in terms of identity by descent, rather than identity by state, as was the case for classical polymorphic markers, now makes it possible to discriminate between the effects of drift and admixture [8]. It also gives the ability for a more precise reconstruction of the evolutionary relationships between populations [14].

A second advance, less obvious but arguably of greater importance, is the advent of a relatively new paradigm for evolutionary population genetics termed phylogeography [3]. This new strategy, represents an outgrowth of the recent emphasis of using molecular phylogenies to infer the evolutionary relationships between a set of taxa. In phylogeography, the molecular phylogenies are intraspecific and used to infer the evolutionary relationships between populations, rather than between taxa. Just as the comparative method uses molecular phylogenies to test hypotheses about the evolution of attributes that characterize taxa, phylogeography uses the spatial distribution of molecular phylogenies to infer the action of evolu-

---

* 2100 EIHG, Human Genetics, University of Utah Salt Lake City, UT 84112.

tionary processes. The geographic distribution of intra-specific phylogenies can be used to determine the probable influence of such deterministic evolutionary parameters as migration and selection on the distribution of allelic variability among a set of sub-populations [5, 79]. When applied in a more general sense, phylogeographic analyses can be used to test inferences about population histories, including the influence of demographic fluctuations, and similar phenomena, which will have had an impact on the magnitude of stochastic drift.

ONE POPULATION TREE: MANY GENE TREES

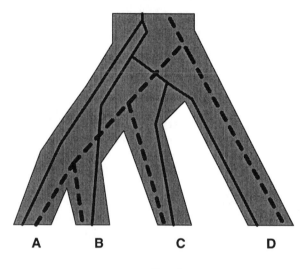

A   B   C   D

FIG. 1.

However, stochastic factors can result in a considerable lack of congruity between the genealogy of a single genomic region and the evolutionary phylogeny of a set of populations. Figure 1 illustrates this problem by depicting a phylogeny for four sub populations (A, B, C and D), within which two genomic phylogenies are contained. One genomic phylogeny, illustrated by the dashed line, is a faithful representation of the true underlying population phylogeny: the lineages found in populations A and B are most closely related, while the lineage found in population D is the most divergent. By contrast, the second genomic region, represented by the solid line, has a phylogeny that is the reverse of the population phylogeny: the most ancient lineage is found in population A, while the two most closely related lineages are in populations C and D. This simple figure emphasizes the potential discordance between a molecular genealogy and a population phylogeny. Such lack of congruence is not due to error in phylogenetic reconstruction, nor to the action of selection or migration, but results from the fact that every molecular genealogy contained within a population phylogeny is drawn from a distribution of all possible phylogenies.

Thus, choosing a specific genomic region to sequence, in order to elucidate the evolutionary history of a set of populations, represents a sample of size one from the underlying distribution of gene genealogies contained within the population phylogeny. Hence, consistent estimates of population history can only be achieved by using the phylogenetic information from multiple genomic regions. While molecular phylogenies for a single genomic region are more likely to resemble the first example (dashed line) than the second (solid line), the sampling distribution of gene trees contained within a population phylogeny is essentially unknown. The action of selection and migration only complicates the situation. Although the difference between "gene trees" and "species trees" has long been recognized in the field of molecular evolution [49], the relatively large evolutionary time span separating the genomes of species means the problem is less acute. In the phylogeographic situation, where the molecular phylogenies are contained within species, the short time span required for population divergence increases the probability of discordance between molecular phylogenies and population history. As molecular advances result in the increasing generation of molecular phylogenies from independent genomic regions, it will become even more important to assess the distribution of congruence between gene trees and the population phylogenies within which they are contained.

There is another distinction between the idealized molecular phylogenies used to define taxonomic relationships and the set of molecular phylogenies obtained in a phylogeographic study of a series of subpopulations. In a phylogeographic study, each subpopulation will tend to contain a set of descendent molecular lineages, rather than just a single lineage. Again, stochastic variability can lead to discrepancies between the observed distribution of molecular phylogenies and the distribution expected on the basis of the population phylogeny. Discrepancies may occur with respect to: i) the degree of divergence between constituent lineages within a single subpopulation; ii) the number of lineages contained within a subpopulation; iii) the degree of divergence between lineages among subpopulations; iv) the amount of lineage sharing among subpopulations.

In Figure 2, subpopulations A and B are sister populations of equal demographic size, subpopulation C is a smaller isolated population, and subpopulations D and E are small sister populations, recently diverged from a very isolated population. The coalescence time scale for the ancestry of the *populations* is at the left of the diagram. Given this representation of population ancestry, the distribution of lineages within and among the five subpopulations differs from the predicted distribution in a number of important respects. While the sister subpopulations A and B each contain the same number of lineages (in this case 5), the lineages in subpopulation A have diverged from a common ancestor much more recently than have the lineages in subpopulation B. Hence, subpopulation A will ex-

GENE TREES vs POPULATION TREES

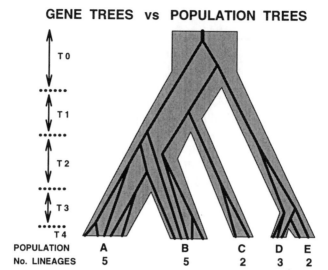

FIG. 2.

hibit a substantially smaller mean pairwise difference than subpopulation B. The difference between the mean pairwise differences would usually be interpreted to infer that subpopulation B is either larger, or older, than subpopulation A — at variance with the true state of affairs.

Also, while subpopulations C, D, and E are all the same size (smaller than A and B), they contain different numbers of lineages. In addition, the levels of evolutionary divergence exhibited by the molecular lineages contained within these three subpopulations is at variance with their phylogenetic relationships: the mean pairwise differences in D will be greater than those in either C, or E. As a consequence, the interpretation of the distribution of pairwise sequence differences within and among subpopulations requires a considerable element of caution. Lacking data on the probable distribution of sequence diversity within and among subpopulations that have a defined history, caution is the only appropriate defense against misinterpretation of phylogeographic data derived from a single genomic region.

Notwithstanding these caveats, the phylogeographic strategy represents a powerful extension to the repertoire of the evolutionary biologist's toolkit. Accordingly, we have started to use these molecular and conceptual advances as part of our goal of developing a deeper understanding of the evolutionary history of contemporary Amerindian populations. In the context of human molecular evolution, Amerindian tribal populations represent a rare opportunity to evaluate the relationship between population ancestry and the distribution of molecular variability [9, 15]. Unlike most major ethnic groups, the geographic and temporal aspects of origin can be identified with reasonable certainty. There is little doubt that the affinities of the Amerindians lie with the populations of Northeastern Asia [20, 61, 62]. Similarly, despite persistent arguments about the exact dates of entry [17, 43], Amerindians are a relatively recent branch of the human tree, with a probable entry into the Americas some 12,000–14,000 years ago [32, 43]. In addition, apart from the catastrophic disruptions of the post-Columbian period, Amerindian populations have mostly evolved without experiencing the mass population movements that occurred on the Eurasian and African continents. Consequently, Amerindian populations not only offer an exceptional opportunity to study the effects of evolution in situ, but may also provide a paradigm for much of the evolutionary history of modern *Homo sapiens*.

2. Mitochondrial DNA and human evolution. As detailed elsewhere in this volume (Stoneking chapter), analysis of mtDNA has provided many important, and provocative, insights about the pattern and process of human evolution [12, 18, 33, 71]. There are a number of reasons why mtDNA has proved exceptionally informative. The characterization of the entire nucleotide sequence [1] allows the evolution of the molecule to be understood in terms of specific nucleotide substitutions [10]. The rapid accumulation of nucleotide substitutions [10] makes mtDNA exceptionally informative for studying short range evolution [5, 79], especially since most of the observed sequence variability appears to be neutral [80]. Also, due to its essentially uniparental inheritance, mtDNA can be treated as a haploid entity. This not only validates the application of such statistical techniques as the coalescent [36, 40], but allows molecular phylogenies to be estimated without the ambiguities caused by recombination [5, 25].

The initial studies of mtDNA, which focussed on variable restriction sites scattered around the genome [11], suggested a relatively recent origin of modern *Homo sapiens* and a rapid dispersal from Africa [11, 12]. While initially contentious, this interpretation is now supported by a great deal of archeological data, as well as by extensive data on nuclear genetic polymorphisms [8, 14]. Hence, the molecular and archeological data suggest that contemporary human populations owe their origin to successive waves of dispersal. The first, originating in Africa some 100,000 to 120,000 years ago, led to the establishment of anatomically modern populations in the Middle East, then Europe and Asia. Though still a hotly debated issue, it appears that the establishment of modern populations throughout much of Eurasia may have resulted in the extinction of earlier human populations that had inhabited these regions for millennia. Subsequent waves of dispersal led to the colonization of formerly uninhabited geographic areas: the colonization of Australasia some 40,000 to 50,000 years ago, the entry into the New World 12,000 to 14,000 years ago and finally, only 4,000 years ago, the dispersion of Malayo-Polynesian peoples into the Pacific and elsewhere

— the last great human expansion into empty territory.

Although the first mtDNA studies indicated that mtDNA restriction types clustered in major ethnic groups [6, 29, 120], they had insufficient resolution to provide information about local evolutionary processes [53, 59]. However, once PCR allowed routine sequencing of the rapidly evolving mtDNA control region [10], it became possible to evaluate more recent evolutionary events [38, 79]. The first analyses of sequences from the highly variable mtDNA control region duplicated the global phylogeny, albeit at higher resolution [33, 34, 71]. Further, studies of sequence variation in an Amerindian tribe, confirmed the utility of sequence data for studying micro-differentiation by identifying a high degree of molecular variability *within* tribal populations [75, 76]. Substantial variability was also found between Amerindian tribes [35, 56], suggesting that previous conclusions about the lack of molecular diversity in Amerindian populations [53, 68] were ill-founded. These latter data indicated that the distribution of mtDNA sequence data within and among human populations could lead to highly informative phylogeographic analyses — as had previously been predicted for other species [3, 5].

3. Amerindian evolution: a molecular perspective. The analysis of mtDNA unambiguously shows that the ancestral populations of contemporary Amerindian populations originated from northeastern Asia [35, 53, 56, 68, 75]. It is now clear that multiple mitochondrial lineages have contributed to Amerindian populations, whether revealed by restriction site analysis [53, 66, 68], or by high resolution sequence analysis [35, 75, 76]. However, mtDNA represents only a small fraction of the human genome and a phylogeny based only on mtDNA is an imperfect estimate of the population phylogeny [42, 49]. Earlier studies used nuclear polymorphisms to good effect to investigate the processes underlying Amerindian evolution: "classical" genetic markers emphasized the importance of intra-tribal population structure [16, 46, 74], with intra-tribal micro-differentiation accounting for a substantial portion of the continent wide variability [45]. Analysis of nuclear polymorphisms demonstrated broad continental trends of genetic diversity [13, 58, 60], with apparent stasis in Central America [7]. This latter interpretation is now supported by recent mtDNA data [67]. Overall, despite the occasional furor about the finer points of linguistic classification [17, 23, 48, 51] and time of genetic divergence [67, 75], there are a number of consistent relationships between language, culture and biological affinities throughout most of the Americas [19, 24, 60]. Hence, an intensive investigation of molecular diversity at the sequence level in Amerindian populations is extremely likely to reveal information about the tempo and mode of molecular evolution in this important chapter of human history.

In the context of Amerindian origins, the Beringian area holds a special importance: this is the region through which the earliest migrants passed, and the evolutionary relationships of contemporary peoples of this region are likely to reflect early affinities. While the archeological record for this area was initially shallow [20], recent work has extended the time depth considerably [21, 32, 39]: Beringia was occupied at least 12,000 years ago by more than one culture [32], this early settlement preceding the subsequent rapid colonization of the Americas [32]. This agrees with our conclusion, based on mtDNA [75], that the first migrants to the New World had already undergone considerable genetic differentiation, but is at variance with the "genetic bottleneck" hypothesis [53, 73]. The archeological evidence also suggests that contemporary Eskimo populations have a relatively recent origin [21], with the ancestors of the Eskimo-Aleut peoples appearing in the New World only some 4,500 years ago [21] — in remarkable agreement with our mtDNA estimates [56, 76]. A competing interpretation, which presumes a identity between linguistic affinities and migratory waves, posits three successive waves of migrants entering the New World between 12,000 and 9,000 years ago [24, 78]. Such an interpretation would equate the deep divisions between these three language phyla [23, 51] with an equally deep genetic divergence. However, as indicated below, this prediction is not borne out by the mtDNA sequence data. Hence, although more data is clearly needed, it seems likely that most contemporary Amerindian populations derive their ancestry from a single colonizing event some 12,000 years ago, with a second, more recent, migration contributing to the ancestry of contemporary Na-Dene and Eskimo-Aleut speakers.

4. Mitochondrial diversity within a single tribe: the Nuu-Chah-Nulth. Before embarking on a general survey of mitochondrial variability throughout the New World, we first analyzed the distribution of molecular diversity within a single tribal population. Our much earlier work with "classical" polymorphisms indicated that the magnitude of genetic micro-differentiation within an Amerindian tribe represented a substantial portion of continent-wide genetic differentiation [45]. In depth studies of micro-differentiation within a single tribe helps identify the predominant socio-demographic processes that influence the magnitude and distribution of genetic diversity [74]. It also helps define the appropriate sample size required to obtain valid estimates of molecular diversity, and indicates which level of population subdivision represents the most appropriate sampling unit.

Our genealogical database for the Nuu-Chah-Nulth identified 401 four generation matrilines [70]. We obtained sequence data for 128 individuals, sampled to represent a random selection of these independent matrilines. All 128 individuals were sequenced for the first 360 nucleotides of the control region (5' end), where 35 variable sites defined 33 lineages. A subset of 65 individuals (containing 31 of these lineages) were also sequenced for 250 nucleotides at the 3' end, and were also evaluated for 12 informative restriction sites [53], plus the 9 bp deletion [29, 81]. Combining the 250 3' nucleotides with the 360 5' nucleotides increased the number of lineages from 31 to 40 — a 30% increase in phylogenetic information. One of the 40 lineages was observed in 5 individuals, one in 4 individuals, five occurred in three individuals each, four occurred in two individuals each and 28 lineages occurred only once [70]. This frequency spectrum for the 40 mtDNA lineages observed in 65 randomly sampled Nuu-Chah-Nulth, is remarkably similar to the distribution observed in the continent-wide sample of 707 Amerindians, and reflects the consequences of a relatively short evolutionary history, coupled with a rapid substitution rate.

Our initial analysis of 63 maternally unrelated Nuu-Chah- Nulth revealed 28 lineages defined by 26 variable sites [75]. This level of sequence diversity was much higher than predicted by previous analyses [73], and supported the contention that sequence analysis will be more informative for micro-evolutionary studies than restriction site analysis [75]. The magnitude of sequence diversity in the tribe was also surprisingly large: the average pairwise difference of 1.5% is approximately 80% of the mean sequence difference observed in a large Japanese sample [34] and approximately 60% of the mean value for a heterogeneous sample of sub-Saharan Africans [135]. The amount of molecular diversity within a single Amerind speaking tribe is therefore a substantial fraction of the continent-wide diversity — in agreement with our earlier assessment of the importance of studying tribal populations [45]. Moreover, when the divergence between lineages is translated into a temporal scale, the major clades observed within this tribal population appear to have originated over 25,000 years ago — well *before* humans are thought to have migrated to the New World. If one accepts the more reliable archeological dates [21, 32, 43], the Nuu-Chah-Nulth data imply that the original migrants to the Americas were probably an offshoot of a relatively large, genetically heterogeneous, population [75].

5. Population Subdivision and Distribution of lineages. The

Nuu-Chah-Nulth data set also allowed us to evaluate the effect of population subdivision upon the distribution of sequence diversity and also allowed us to examine the evolutionary relationship between mates [69]. This latter analysis indicates that the distribution of pairwise sequence differences between mates is strongly bimodal: a minor mode (40%) where mates have identical sequences, and a major mode, where mates are essentially a random sample from the entire tribe [69]. The distribution of sequence divergence *within* a band, compared to the sequence divergence *between* bands was then evaluated. Unlike the situation for classical polymorphisms [45, 74], the average sequence divergence between bands is *no greater* than the mean divergence observed within bands. Further, the distribution of lineages amongst Nuu-Chah-Nulth bands is essentially random and fails to reflect the geographic, linguistic and socio-political affinities between bands [70]. In conjunction with the distribution of sequence differences between mates, the failure of mtDNA sequence analysis to reveal intra-tribal population structure suggests that migration between bands is sufficient to obliterate the effects of population subdivision. Intra-tribal population subdivision influences neither the spatial distribution of mtDNA sequence divergence, nor the spatial distribution of mtDNA lineages. The tribe, rather than the band/village, is the most appropriate sampling unit for mtDNA.

6. Linguistic versus Molecular Diversity in Pacific Northwest Tribes. Since the Pacific Northwest exhibits the greatest degree of linguistic diversity within the Americas [27, 48], we studied two additional tribes from this area in order to assess the congruence between molecular diversity and the differentiation of New World languages. We studied the Haida (Na-Dene language phylum) and the Salishan speaking Bella Coola (Amerind phylum). Two unexpected results emerged:
1) the Haida exhibited a substantial reduction in sequence divergence compared to the two Amerind tribes;
2) The average sequence divergence between the Na-Dene speaking Haida and either of the two Amerind groups was no greater than the divergence between the two Amerind tribes. This led to the following conclusions:
1) The Haida appear to be relatively recent arrivals to the New World, with insufficient time for an appreciable number of substitutions to have accumulated;
2) based on the distribution of sequence divergence, the linguistic differences that distinguish the Na-Dene phylum from the Amerind phylum appear to have arisen in the same time span that was required for the differentiation of two closely related Amerind languages — Wakashan and Salishan [76]. Neither conclusion is in agreement with the majority view concerning the peopling of the Americas [24, 82], nor with the current interpretation of the linguistic data [23, 48].

While the reduced sequence diversity within the Haida was partly due to lineage identity, a high proportion of distinct Haida lineages differed by only 1–2 substitutions — in contrast to the two Amerind tribes where the majority of lineages differ by 3–5 substitutions. In addition, the Haida sample lacked representatives of all four clades [75], which were present in both Amerind tribes. Instead the majority of Haida lineages were located within a single clade (cluster II). Hence, a short evolutionary history for the Haida, rather than a recent bottleneck (or population subdivision) appears the most likely cause of these observations [76].

Besides failing to find increased sequence divergence between the two language phyla, we found no evidence that lineages clustered on the molecular phylogeny as a function of linguistic affiliation. Instead, lineages of all three tribes were intermingled throughout the phylogeny (although the Haida lineages were largely constrained to cluster II). Admixture could be discounted, as there was very little lineage sharing between tribes. Further, with one exception, lineages that occurred in more than one tribe occupied nodal positions on the phylogeny — suggesting the perpetuation of an ancestral lineage, rather than recent admixture [36, 63]. This initial study suggested that Na-Dene tribes might have arisen relatively recently, and might contain less molecular diversity than Amerind tribes of comparable size. It also raised the possibility that the rate of linguistic evolution could be quite uneven, with some major linguistic differences arising due to "saltations." A more intensive evaluation of the evolutionary divergence of northern Native American populations was required — preferably one that analyzed molecular data for all three language phyla.

7. Homogeneity of mitochondrial lineages in "Circumarctic" populations. The genetic relationships between Beringian populations holds the key to understanding of the peopling of the New World [55, 61]. In this context, it is significant that Inuit populations (thought to be late arrivals [21]) exhibit little genetic diversity throughout their range [21, 58, 62]. However, there is less agreement whether the Northern Na-Dene (Athapaskans) have a similar reduction in diversity [54, 62]. Since our initial results suggested reduced mitochondrial sequence diversity among the Na-Dene speaking Haida [76], a more detailed analysis of "Circumarctic" populations, and comparison with Amerind populations, was warranted. Accordingly, mtDNA sequences were generated for six additional "Circumarctic" groups ranging from Siberia (Altai, Chukchi, and Siberian Yup'ik) through Alaska (Athapaskan, Inupiaq) to Greenland (Greenlandic Eskimo). Our initial conclusions [56] relied mainly on populations represented by an adequate sample size: Altai, Athapaskans, Greenlandic Eskimo, and Haida, plus the Amerind speaking Nuu-Chah-Nulth, Bella Coola and Yakima. However, small data sets from the Chukchi and Inupiaq were also included. We have recently obtained more adequate sample sizes for the Chukchi, Siberian Yup'ik and Yukaghir — all from Siberia, and find these revised data corroborate the conclusions outlined below. As indicated in Figure 3, which depicts the distribution of pairwise sequence differences (equivalent to the standard measure of nucleotide diversity, π_{ij}, [47, 77]), Circumarctic populations as a whole have a significant reduction in molecular diversity, compared to a "representative" Amerind tribe. Overall, the level of mitochondrial diversity within the entire sample of 90 individuals (35 lineages) from the six Circumarctic populations (solid line) is less than 50% of diversity within a *single* Amerind tribe (dotted line). Further, while there is a slight increase in lineage identity in the Circumarctic data, the underlying reason for the reduction is the fact that nearly 45% of pairwise lineage comparisons in the Circumarctic data involve 1–2 substitutions, compared to only 15% in an Amerind tribe. Conversely, while the majority of Amerind comparisons (54%) involve 5–8 substitutions, only 14% of the Circumarctic comparison fall in this range. Hence, both Eskimo-Aleut and Na-Dene populations appear to be of recent origin, compared to Amerind populations, with a concomitant reduction in molecular diversity. (This finding agrees with the reduced level of restriction site diversity observed in the Na-Dene

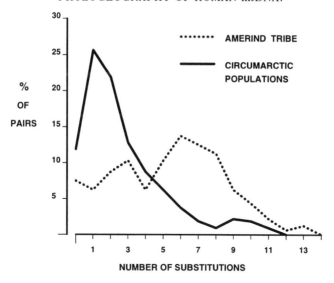

FIG. 3.

speaking Dogrib [68]). Not only is there a reduction in molecular diversity within each Circumarctic group, but widely separated groups speaking languages belonging to different phyla (e.g. Haida and Greenlandic Inuit) have an average sequence divergence of only 2 substitutions (compared to the modal value of 5–9 within an Amerind tribe).

In addition, the molecular phylogeny for the 35 mtDNA lineages found in the six Circumarctic populations (Figure 4) is characterized by a virtual absence of major clades. This is a stark contrast to the situation within a single Amerind speaking tribe, such as the Nuu-Chah-Nulth, where there are four distinct clades representing divergences that predate entry into the Americas. Further, as seen by the distribution of mtDNA lineages in Figure 4, lineages that are unique to a specific group are scattered indiscriminately within the phylogeny, with no aggregation by linguistic affiliation, nor by geographic location. With only one exception (lineage #58), lineages found in more than one population (starred) occur at nodes within the phylogeny, suggesting that lineage sharing in these populations is due to incomplete lineage sorting following population divergence, rather than recent migration. Thus despite the caveats exemplified by the distribution of lineages in Figure 2, it is difficult to postulate recent admixture as the cause of the genetic homogeneity of these six Circumarctic groups [56].

While the reduction in intra-population diversity is compatible with small effective population size, the relatively small values of lineage identity within populations plus the high proportion of pairwise differences involving 1–2 substitutions, is more suggestive of a young evolutionary age. The similarity of mitochondrial sequences found among the widely scattered

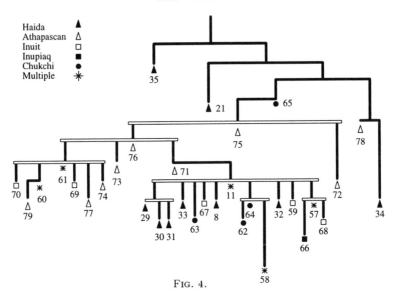

Fig. 4.

Circumarctic populations, the small number of substitutions separating sequences, their intermingling within the molecular phylogeny and the lack
of major clades, all combine to suggest these populations arose by a recent
and rapid evolutionary radiation. As indicated by the population phylogeny, our results suggest a very recent origin for the linguistic distinction
between Eskimo-Aleut and Na-Dene, and a recent arrival of both groups
in the New World. By contrast, Amerind groups display a much deeper
ancestry — though neither group displays evolutionary affinities with the
Siberian Altai. If the mitochondrial data accurately reflects the population
history of these groups (our working hypothesis), it implies that the biological affinities of the early archeological deposits were with contemporary
Amerind speakers, rather than Eskimo-Aleut/Na-Dene. It also calls into
question the presumption that rates of linguistic diversification are similar
across phyla as is commonly believed [23, 48, 51].

**8. A multitude of "Amerindian" mitochondrial lineages indi-
cates many original colonists.** Our mtDNA data set now encompasses
control region sequences for some 707 Amerindians and Siberians, distributed among 27 tribal populations, representing all three major language
phyla, Eskimo-Aleut, Na-Dene, and Amerind [23, 48, 51]. The **Eskimo-
Aleut** language phylum is represented by four populations: Greenlandic
Eskimos; Inupiaq speakers and Yup'ik speakers from Alaska; and Siberian
Yup'ik speakers. The **Na-Dene** phylum is represented by only two groups:
the Haida and Alaskan Athapaskans. Lastly, the more numerous and diverse **Amerind** language phylum [48, 51] is represented by:

a) North America — three Pacific North west groups (Nuu-Chah- Nulth, Bella Coola and Yakima); three South West groups (Pai, Pima and Zuni); plus the Creeks from the southeast.

b) Central America — seven Chibchan groups (Bribri, Buglé, Cabecar, Guatuso, Huetar, Kuna, and Ngöbé) plus the Embera.

c) South America — three Brazilian groups (Xavante, Gaviõ and Zoró). We also have sequence data from three additional groups in Siberia: Chukchi, Yukaghir and Altai. As detailed elsewhere, each group, or set of groups, has been selected to answer a specific question about Amerindian tribal relationships. However, the entire set of data can also be used to provide an overview of the distribution of mtDNA sequence variability throughout Amerindians.

Within the first 360 nucleotides of the control region (5' end), this collection of sequences data includes 112 variable sites. Hence, during the formation of contemporary Amerindian populations, approximately one out of every three nucleotides in this region has undergone at least one mutation. Further, these 112 variable sites define a minimum of 181 distinct mitochondrial lineages, of which 142 are unique to a single tribe. The 39 lineages which are found in more than one tribe not only tend to be more numerous than the tribally unique lineages, but also occupy "nodal" positions in the molecular phylogeny. Thus, we interpret the majority of these 39 lineages to be ancestral variants that have survived in the Amerindian population since the arrival of the first colonists. By analogy with the diversity observed within the contemporary Nuu-Chah-Nulth, this implies that the effective population size of the original migrants was no less than 600 (i.e. a standing population of 4,000 [75]). Since a proportion of ancestral lineages must have been lost due to drift and since our sample by no means includes all major groups of Amerindians, the actual number of migrants was probably much greater. Since we observe a significant north-south gradient of sequence diversity within Amerind speakers (the diversity being greatest in the Pacific Northwest, and least in Central America and South America tribes), it is likely that migration occurred over a considerable time span, with the later migrants failing to penetrate very far, once the Central American Isthmus became extensively populated.

9. Mitochondrial diversity in "Amerindians" compared to other ethnic groups.
Based on a small number of restriction site polymorphisms, the initial study of mtDNA variability in Amerindians concluded that the founding population had passed through a "dramatic" bottleneck [73]. While use of an expanded set of restriction site polymorphisms identified somewhat more variability, it was still believed that Amerindians had a substantial reduction in molecular diversity, compared to other populations [6, 53, 68, 138]. Our evidence of extensive sequence diversity challenged this view [75, 76], and suggested it would be informative to compare the distribution of sequence divergence in Amerindian samples

with that observed in African [135], Caucasian [18] and Asian (Japanese, Chinese, and Papua New Guinean) populations [34, 71]. While the African sample exhibited the greatest sequence divergence, the sequence divergence within Amerindian tribes is similar to the sequence divergence in large, regional, Asian populations. While the greater differentiation observed in African populations agrees with the restriction site data [11, 12], the relatively high degree of diversity within Amerindian tribes was unexpected.

The distribution of sequence divergence within and between tribes, suggests a possible explanation for these observations. As noted above, Amerindian tribes tend to be characterized by a unique set of lineages, with very little lineage sharing between tribes (only 39 out of 181 lineages). Although the data is less extensive, the same appears true for African tribes [135]. However, the average sequence divergence between Amerindian tribes (6.5 substitutions) is only marginally greater than the divergence within tribes (5.8 substitutions). By contrast, in the African data the inter-tribal divergence is 50% greater (9.95 substitutions) than average intra-tribal sequence divergence (6.1 substitutions). The phylogenetic relationships between the Amerindian lineages reveal the reason for this result. Although each Amerindian tribe tends to have a unique set of lineages, these lineages are all derived from the same set of four clades that occur throughout the New World [35, 75]. Since the average sequence divergence within clades is only a third of the divergence between clades, the unique lineages within a tribe adds little to the diversity between tribes. By contrast, there appear to be more clades on the African continent, with tribally unique lineages drawn from different clades. Compared to African tribes, Amerindian tribes may only have been isolated for a relatively short time period — sufficient for no more than one to two substitutions (i.e. about 10,000 years). Hence, molecular diversity in "Amerindian" populations appears to have been influenced by two processes — extensive genetic differentiation that occurred *before* entry to the New World, resulting in sequence divergence of 5–6 substitutions (i.e. a time depth of 25,000 years, or more), and evolution in situ resulting in only 1–2 additional substitutions.

10. Using the coalescent model to estimate substitution rates. In molecular evolution, the estimation of substitution rates requires not only an estimate of sequence divergence, but also an *a priori* estimate of the date of species divergence [10, 79]. There are two reasons why this is a questionable strategy for intraspecific data: i) dates of population divergence are usually unknown, ii) the depth of molecular divergence is usually substantially greater than the depth of population divergence [47, 49]. By contrast, the coalescent model that defines the ancestry of a random sample from a finite population [36, 37], incorporates the statistical distribution of time to the common ancestor. Hence, an estimation procedure based on the coalescent would eliminate the need for estimates of divergence time between populations. It also eliminates the potential for error if substitu-

TABLE 1

Type of Transition	$C \to T$	$T \to C$	$A \to G$	$G \to A$	Pyri-midine	Purine	Total
Probability	17	25	2	12	20	6	14
10% Lower Bound	8	17	2	3	11	2	7
90% Upper Bound	33	50	13	42	40	18	30

tion rates vary significantly between species. Further, with the increased resolving power provided by a large population based sample of sequences, an estimation strategy based on the coalescent would allow individual estimates for different categories of nucleotides, as well as an estimate of the "overall" rate [40].

The coalescent model for DNA sequences has two components. The first generates the genealogy of the sequences by giving the distribution of time between nodes in the ancestral tree. Under the coalescent model, the amount of time that the sample has j distinct ancestors is an exponential random variable, with parameter $j(j-1)/2$, where time is measured in units of N generations, and N is the population size. The second component of the coalescent superimposes nucleotide substitutions on the branches of the ancestral tree. Substitutions occur at rate $\theta/2$ along each branch, where θ is the scaled substitution parameter. For haploids, $\theta = 2N\mu$ where μ is the probability of a neutral substitution per site per generation. When a substitution occurs at a site containing nucleotide i, the probability that it is changed to nucleotide j is the i, j entry of the substitution matrix P. The off-diagonal entries of the product, $\theta.P$, define the substitution rates of the process.

By summarizing the sequence data in terms of the fraction of sites with a given base composition, we developed two methods to estimate θ: A least square method which minimizes the sum of the squared differences between observed and expected fractions, and a maximum-likelihood method which assumes sites are independent and maximizes the likelihood of the observed fractions [40]. Both methods gave nearly identical estimates when applied to the distribution of transitions observed in the sample of 63 Nuu-Chah-Nulth sequences. Although the estimates were initially defined in terms of their relative magnitude, estimating the effective population size (600 females for the Nuu- Chah-Nulth) gave the absolute estimates shown in Table 1.

Statistical properties of these methods, determined via simulation, showed that the methods were nearly unbiased and had acceptable standard errors. The resulting 90% confidence bounds are also given in Table 1 (all substitution rates are multiplied by 10^6). Using this method, the estimated substitution rates for pyrimidines (2.5×10^{-5} to 1.7×10^{-5}) were significantly larger than the rates for purines (1.2×10^{-5} to 3.8×10^{-6}) [40]. Since the relatively few purine transitions in this data set have correspondingly large standard errors, there is concern that the observed difference may not reflect biological reality. However, a more comprehensive analysis, using the "Monte Carlo" approach [26], also indicates a substantially lower rate for the purine transitions, suggesting the existence of a real substitutional bias [26]. Also, when the "overall" substitution rate is estimated by comparing the sequence divergence between the first 360 nucleotides of the human control region with that of the chimpanzee, the result (4×10^{-6} [75]) is approximately three times lower than the overall rate of 14×10^{-6} estimated by the coalescent method. This suggests that using divergence between rapidly evolving regions will lead to underestimation of substitution rates.

11. Sequence diversity is influenced by demographic factors and site specific variability. The ability to generate sequence variability at the population level also led to new theoretical and analytical techniques. One notable development was the recognition that the distribution of sequence differences could reflect the genealogical ancestry of the mtDNA molecule — in turn a reflection of the demographic history of the population [4, 5, 30, 36, 65]. Hence, the characteristic "unimodal" distribution of sequence differences seen in most populations [18], including ours [56, 75], is very unlikely to occur in demographically stable populations [50, 57], but expected in populations that experienced marked fluctuations in the past. This led to estimations of the magnitude and timing of past demographic fluctuations [30, 50], with the conclusion that ancestral human populations experienced a major demographic expansion 80,000 to 30,000 years ago [30]. However, our analytical work shows that heterogeneity of site specific mutation rates can also lead to unimodal distributions [41]. In addition, since sequences are sampled from an existing gene genealogy, the expectation for a pair of individuals differs appreciably from sample expectations [42, 57]: With only a single evolutionary realization of a molecular genealogy (a sample of size one), analysis is problematic at best [42]. Overall, the interaction between population structure, demographic fluctuations, and heterogeneity in mutation rates remains unresolved and requires both an appropriately designed simulation study [42], as well as a series of extensive analyses of comparative data sets.

As noted earlier, the characteristic "unimodal" distribution of pairwise sequence differences observed in the Nuu-Chah-Nulth (as in most other populations) is inherently unlikely in a demographically stable population [30, 50, 57]. Overall, there are three evolutionary factors which could influence the distribution of sequence differences:

    1) Population growth and/or past demographic fluctuations,

    2) Admixture between genetically distinct tribes or a polyphyletic

origin of the initial founding population, and

3) Site-specific variability in mutation rates.

While the first factor has received considerable attention, few analyses have evaluated the potential impact of the latter two biological factors. We used the coalescent model to investigate the consequence of each factor with respect to two issues:

1) the impact on the estimated substitution rates, and

2) the form of the distribution of pairwise sequence differences.

We found that each factor considered alone exerts only a relatively minor effect on the magnitude of the estimated substitution rates, and certainly cannot explain the difference between the "coalescent" estimates and estimates based on estimated divergence times [40, 41]. For example, to investigate the effects of admixture, we separately analyzed the two largest mtDNA clades observed in the Nuu-Chah-Nulth [75], and also analyzed pooled data from three tribes (Nuu-Chah-Nulth, Bella Coola, and Haida). After accounting for the changes in effective population size, the estimates ranged from only 25% higher to 35% lower, indicating that even appreciable admixture rates (e.g. 25% admixture) have only minimal influence on estimates of substitution rates [40].

To explore how different factors might interact, we have begun a more systematic analysis. As others have predicted [18, 30, 50, 57], we find demographic fluctuations have a marked influence on the distribution of sequence differences. The exact nature of the perturbation is a function of the type of demographic fluctuation, whether the population is exponentially growing, and by the overall substitution rate. We also find that existence of hypervariable sites can have a profound effect on the distribution of sequence divergence [41]. As a consequence the importance of demographic fluctuations may be less important in a finite sites model than formerly supposed [41]. However, we regard these findings to be only tentative. Before the relative influence of each type of evolutionary factor can be assessed, many more analyses will need to be accomplished

REFERENCES

[1] Anderson S, Bankier AT, Barrell BG, de Bruijn MHL, Coulson AR, Drouin J, Eperon IC, Nierlich DP, Roe BA, Sanger F, Schrier PH, Smith AJH, Staden R, Young IG: Sequence and organization of the human mitochondrial genome. Nature 290:457–465, 1981.

[2] Avise JC, Neigel JE, Arnold J: Demographic influences on mitochondrial DNA lineage survivorship in animal populations. J. Mol. Evol. 20:99–105, 1984.

[3] Avise JC: Mitochondrial DNA and the evolutionary genetics of higher animals. Phil. Trans. Roy. Soc. Lond. (B) 312:325–342, 1988.

[4] Avise JC, Ball RM, Arnold J: Current versus historical population size in vertebrate species with high gene flow. Mol. Biol. Evol. 5:331–344, 1988.

[5] Ball RM, Neigel JE, Avise JC: Gene genealogies within the organismal pedigrees of random mating populations. Evolution 44:360–370, 1990.

[6] Ballinger SW, Schurr TG, Torroni A, Gan YY, Hodge JA, Hassan K, Chen K-H,

Wallace DC: Southeast Asian mitochondrial DNA analysis reveals continuity of ancient Mongoloid migrations. Genetics 130:139–152, 1992.

[7] Barrantes R, Smouse PE, Mohrenweiser HW, Gershowitz H, Azofeifa J, Arias TD, Neel JV: Microevolution in lower Central America: genetic characterization of the Chibcha speaking groups of Costa Rica and Panama, and a consensus taxonomy based on genetic and linguistic affinity. Amer. J. Hum. Genet 46:63–84, 1990.

[8] Bowcock AM, Kidd JR, Mountain JR, Herbert JM, Carotenuto, L, Kidd KK, Cavalli-Sforza, LL: Drift, admixture and selection in human evolution: A study with DNA polymorphisms. Proc. Natl. Acad. Sci. (US) 88:839–843, 1991.

[9] Bowcock AM, Cavalli-Sforza LL: The study of variation in the human genome. Genomics 11:491–498, 1991.

[10] Brown WM, George M, Wilson AC: Rapid evolution of animal mitochondrial DNA. Proc. Natl. Acad. Sci. (U.S.) 76:1967–1971, 1979.

[11] Cann RL: Human dispersal and divergence. Trends Ecol. Evol. 8:27–31, 1993.

[12] Cann RL, Stoneking M, Wilson AC: Mitochondrial DNA and Human Evolution. Nature 325:31–36, 1987.

[13] Cavalli-Sforza LL, Menozzi P, Piazza A: Demic expansion and human evolution. Science 259:639–646, 1993.

[14] Cavalli-Sforza LL, Piazza A, Menozzi P, Mountain J: Reconstruction of human evolution: Bringing together genetic, archaeological and linguistic data. Proc. Natl. Acad. Sci. (U.S.) 85:6002–6006, 1988.

[15] Cavalli-Sforza LL, Wilson AC, Cantor CR, Cook-Deegan RM, King M-C: Call for a world-wide survey of human genetic diversity. Genomics, 11:490–491, 1991.

[16] Chakraborty R, Smouse PE, Neel JV: Population amalgamation and genetic variation: observations on artificially agglomerated populations of Central and South America. Amer. J. Hum. Genet. 43:709–725, 1988.

[17] Dillehay TD, Collins MB: Early cultural evidence from Monte Verde in Chile. Nature, 332:150–152, 1988.

[18] DiRienzo A, Wilson AC: Branching pattern in the evolutionary tree for human mitochondrial DNA. Proc. Natl. Acad. Sci. (US), 88:1597–1601, 1991.

[19] Driver HE, Coffin JL: Classification and development of North American Indian cultures: A statistical analysis of the Driver-Massey sample. Trans. Am. Phil. Soc. 65:1–32, 1975.

[20] Dumond DE: The archeology of Alaska and the peopling of America. Science 209:984–991, 1980.

[21] Dumond DE: A reexamination of Eskimo-Aleut prehistory. Amer. Anthrop. 89:32–56, 1987.

[22] Foran DR, Hixson JE, Brown WM: Comparisons of ape and human sequences that regulate mitochondrial DNA transcription and D-loop DNA synthesis. Nucleic Acids Res 16:5841–5861, 1988.

[23] Greenberg JH: Language in the Americas Stanford: Stanford University Press, 1987.

[24] Greenberg JH, Turner CG III, Zegura SL: The settlement of the Americas: A comparison of linguistic, dental and genetic evidence. Curr. Anthrop. 27: 477–497, 1986.

[25] Griffiths RC: An algorithm for constructing genealogical trees. Stat. Res. Rep. Monash, Australia, 163, 1987.

[26] Griffiths RC, Tavaré S: Ancestral inference in population genetics. Stat Sci 9:307-319, 1994.

[27] Gruhn R: Linguistic evidence in support of the coastal route of earliest entry into the New World. Man 23:77–100, 1988.

[28] Gyllensten U, Erlich HA: Generation of single stranded DNA by the polymerase chain reaction and its application to the direct sequencing of the HLA- DQA locus. Proc. Natl. Acad. Sci. (US), 85:7652–7656, 1988.

[29] Harihari S, Hirai M, Saitou Y, Shimizuk K, Omoto K: Frequency of a 9-bp deletion

in the mitochondrial DNA among Asian populations. Hum. Biol. 64:161–166, 1992.

[30] Harpending HC, Sherry ST, Rogers AR, Stoneking M: The genetic structure of ancient human populations. Current Anthrop 34:483–496, 1993.

[31] Hill AVS, Serjeantson SR (Eds): The Colonization of the Pacific: A Genetic Trail. Oxford, Oxford University Press, 1989. 67.

[32] Hoffecker JF, Powers WR, Goebel T: The colonization of Beringia and the peopling of the New World. Science 259:46–53, 1993.

[33] Horai S: Molecular phylogeny and evolution of human mitochondrial DNA. In: M Kimura, N Takahata (Eds) New Aspects of the Genetics of Molecular Evolution. Tokyo: Japan Science Press, pp 135–152, 1991.

[34] Horai S, Hayasaka K: Intraspecific nucleotide sequence differences in the major noncoding region of the human mitochondrial DNA. Amer. J. Hum. Genet., 46:828–842, 1990.

[35] Horai S, Kondo R, Nakagawa–Hattori Y, Hayashi S, Sonoda S, Tajima K: Peopling of the Americas founded by four major lineages of mitochondrial DNA. Mol. Biol. Evol. 10:23–47, 1993.

[36] Hudson RR: Gene genealogies and the coalescent process. Oxford Surveys of Evol. Biol. 7:1–44, 1990.

[37] Kingman JFC: On the genealogy of large populations. J. Appl. Prob., 19A:27–43, 1982.

[38] Kocher TD, Wilson AC: Sequence evolution of mitochondrial DNA in humans and chimpanzees: the control region and a protein coding region. In: S Osawa, T Honjo (eds) Evolution of Life: Fossils, Molecules and Culture Berlin: Springer Verlag, pp. 391–413, 1991.

[39] Laughlin WS: From Ammassalik to Attu: 10,000 years of divergent evolution. Obj. et Mond., 25:141–148, 1987.

[40] Lundstrom R, Tavaré S, Ward RH: Estimating substitution rates from molecular data using the coalescent. Proc. Nat. Acad. Sci. 89:5961–5965, 1992.

[41] Lundstrom R, Tavaré S, Ward RH: Modeling the evolution of the human mitochondrial genome. Math. Biosci. 112:319–335, 1992.

[42] Marjoram P, Donnelly P: Pairwise comparison of mitochondrial DNA sequences in subdivided populations and implications for early human evolution. Genetics 136:673–683, 1994.

[43] Meltzer DJ: Pleistocene peopling of the Americas. Evol Anthrop 1:157–169, 1993.

[44] Merriwether DA, Clark AG, Ballinger SW, Schurr TG, Goodyall H, Jenkins T, Sherry ST, Wallace DC: The structure of human mitochondrial DNA variation. J. Mol. Evol 33:543–555, 1991.

[45] Neel JV, Ward RH: Village and tribal genetic distances among American Indians and the possible implications for human evolution. Proc. Natl. Acad. Sci. (U.S.) 65:323i–330, 1970.

[46] Neel JV, Thompson EA: Founder effect and the number of private polymorphisms observed in Amerindian tribes. Proc. Natl. Acad. Sci. (US), 75:1904–1908, 1978.

[47] Nei M: Molecular Evolutionary Genetics New York: Columbia University Press, 1987.

[48] Nichols J: Linguistic diversity and the first settlement of the new world. Language 66:475–521, 1990.

[49] Pamilo P, Nei M: Relationships between gene trees and species trees. Mol. Biol. Evol., 5:568–583, 1988.

[50] Rogers A, Harpending H: Population growth makes waves in the distribution of pairwise differences, Mol. Biol. Evol., 1992.

[51] Ruhlen M: The Amerind phylum and prehistory of the New World. In SL Lamp, ED Mitchell (eds), Sprung from some common source: The prehistory of Languages. Stanford: Stanford University Press, pp 330–350, 1992.

[52] Saccone C, Pesole G, Sbisa E: The main regulatory region of mammalian mito-

chondrial DNA: Structure-function model and evolutionary pattern. J. Mol. Evol. 33:83–91, 1991.

[53] Schurr TG, Ballinger SW, Gan YY, Hodge JA, Merriwether DA, Lawrence DN, Knowler WC, Weiss KM, Wallace DC: Amerindian mitochondrial DNAs have rare Asian mutations at high frequencies, suggesting they derived from four primary maternal lineages. Amer. J. Hum. Genet., 46:613–623, 1990.

[54] Scott EM: Genetic diversity of Athabascan Indians. Ann. Hum. Biol. 6:241–247, 1979.

[55] Shields GF, Hecker K, Voevoda MI, Reed JK: Absence of the Asian-specific region V mitochondrial marker in Native Beringians, Amer. J. Hum. Genet. 50:758–765, 1992.

[56] Shields GF, Schmiechen AM, Frazier BL, Redd A, Voevoda MI, Reed JK, Ward RH: Mitochondrial DNA sequences suggest a recent evolutionary divergence for Beringian and Northern North American populations. Amer. J. Hum. Genet. 53:549–562, 1993.

[57] Slatkin M, Hudson RR: Pairwise comparisons of mitochondrial DNA sequences in stable and exponentially growing populations. Genetics, 29:555–562, 1991.

[58] Spuhler JS: Genetic distances, trees and maps of North American Indians. In: WS Laughlin, AB Harper (eds.) The First Americans: Origins, Affinities and Adaptation. New York: Fischer, pp 135–183, 1979.

[59] Stoneking M, Jorde LB, Bhatia K, Wilson AC: Geographic variation in human mitochondrial DNA from Papua New Guinea. Genetics, 124:717–733, 1990.

[60] Suarez BK, Crouse JD, O'Rourke DH: Genetic variation in North American populations: The geography of gene frequencies. Am. J. Phys. Anthrop. 67:217–232, 1985.

[61] Szathmary EJE: Peopling of North America: Clues from genetic studies. Acta. Anthro. 8:79–110, 1984.

[62] Szathmary EJE: Genetics of aboriginal North Americans. Evol. Anthrop. 1(6):202–220, 1993.

[63] Takahata N: Genealogy of neutral genes and spreading of selected mutations in a geographically structured population. Genetics, 29:585–595, 1991.

[64] Tamura K, Nei M: Estimation of the number of nucleotide substitutions in the control region of mitochondrial DNA in humans and chimpanzees. Mol. Biol. Evol. 10:512–526, 1993.

[65] Tavaré S: Lines of descent and genealogical processes, and their application in population genetics models. Theor. Pop. Biol., 26:119–144, 1984.

[66] Torroni A, Chen Y-S, Semino O, Santachiara-Beneceretti AS, Scott CR, Lott MT, Winter M, Wallace DC: mtDNA and Y-chromosome polymorphisms in four Native American populations from Southern Mexico. Amer J Hum Genet, 54:303–318, 1994.

[67] Torroni A, Neel JV, Barrantes R, Schurr TG, Wallace DC: Mitochondrial DNA "clock" for the Amerinds and its implications for timing their entry into North America. Proc Nat Acad Sci (US) 91:1158–1162, 1994.

[68] Torroni A, Schurr TG, Yang C-C, Szathmary EJE, Williams RC, Schanfield MS, Troup GA, Knowler WC, Lawrence DN, Weiss KM, Wallace DC: Native American mitochondrial DNA analysis indicates the Amerind and the Na-Dene populations were founded by two independent migrations. Genetics 130:153–162, 1992.

[69] Valencia D, Ward RH: Using mitochondrial sequence data to estimate evolutionary divergence between mates. Amer. J. Phys. Anthrop. 1991.

[70] Valencia D: "Mitochondrial DNA evolution in the Nuu-Chah- Nulth population", MS thesis, Univ. of Utah, 1992.

[71] Vigilant L, Stoneking M, Harpending H, Hawkes K, Wilson AC: African populations and the evolution of human mitochondrial DNA. Science, 253:1503–1507, 1991.

[72] Waits LP, O'Brien SJ, Ward RH: Molecular phylogeny for bears: multiple region

mtDNA data indicates recent paraphyly. Mol. Biol. Evol (submitted), 1995.

[73] Wallace DC, Garrison K, Knowler WC: Dramatic founder effects in Amerindian mitochondrial DNAs. Amer. J. Phys. Anthrop. 68:149–155, 1985.

[74] Ward RH: The genetic structure of a tribal population, the Yanomama Indians. V. Comparison of a series of genetic networks. Ann. Hum. Genet. 36:21–43, 1972.

[75] Ward RH, Frazier BL, Dew-Jaeger K, Pääbo S: Extensive mitochondrial diversity within a single Amerindian tribe. Proc. Natl. Acad. Sci. (US), 88:8720–8724, 1991.

[76] Ward RH, Redd A, Valencia D, Frazier BL, Pääbo S: Genetic and linguistic differentiation in the Americas. Proc. Nat. Acad. Sci. (US), 90:10663–10667, 1993.

[77] Weir BS: Genetic Data Analysis. Sinauer, 1990.

[78] Williams RC, Steinberg AC, Gershowitz H, et al.: Gm allotypes in Native Americans: Evidence for three distinct migrations across the Baring Land bridge. Am. J. Phys. Anthrop. 66:1–19, 1985.

[79] Wilson AC, Cann RL, Carr SM, George M, Gyllensten UB, Helm- Bychowski KM, Higuchi RG, Palumbi SR, Prager EM, Sage RD, Stoneking M: Mitochondrial DNA and two perspectives on evolutionary genetics. Biol J Linn Soc 26:375–400, 1985.

[80] Woodbury AC: Eskimos and Aleut Languages. Handbook of North American Indians 5:49–63, 1984.

[81] Wrischnik LA, Higuchi RG, Stoneking M, Erlich HA, Arnheim N, Wilson AC: Length mutations in human mitochondrial DNA: direct sequencing of enzymatically amplified DNA. Nuc. Acid Res., 15:529–542, 1987.

POPULATION STRUCTURE AND MODERN HUMAN ORIGINS*

ALAN R. ROGERS[†]

Abstract. This paper reviews statistical methods for inferring population history from mitochondrial mismatch distributions and extends them to the case of geographically structured populations. Inference is based on a geographically structured version of the coalescent algorithm that allows for temporal variation in population size, in the number of subdivisions, and in the rate of migration between subdivisions. Confidence regions are inferred under several models of population history. If the pattern in mitochondrial DNA reflects population growth rather than selection, then the confidence regions reject the multiregional hypothesis of modern human origins more strongly than has previously been possible. They do not reject the replacement hypothesis.

Key words. coalescent, mitochondrial DNA, modern human origins, mismatch distribution, population structure.

1. Introduction. A mitochondrial mismatch distribution is a histogram that describes variation in the amount of genetic difference between pairs of individuals in a sample. In several recent articles, my coauthors and I have suggested that the mismatch distribution is rich in information about population history [19,17,6,21,5,18]. This work suggests that the human population experienced a population explosion during the late Pleistocene, some 30,000 to 130,000 years ago.

This work has encountered two kinds of criticism. Some have objected to our use of a theory describing pairs of individuals when our data consist not of pairs but of much larger samples. Others have objected that when populations are subdivided, the mismatch distribution may not contain much information about population history, and methods such as ours should not work [15].

In response to the first criticism, sections 1–5 of this paper will emphasize that the theory in question has never been used as a basis for inference but is used instead as a basis for intuition. It will review the reasons why this approach seems plausible and what it has accomplished. Later sections will emphasize that statistical inference has been based on computer simulation, not on the theoretical mismatch distribution.

The second criticism would not appear relevant to the work of Harpending et al [6], whose model incorporates the effect of genetic population structure. It may however bear on other work that assumes a randomly mating population. The final sections of the paper will therefore explore the effect of population structure on statistical methods that I develop elsewhere [18].

* This research was supported in part by National Science Foundation grant DBS–9310105.

† Dept. of Anthropology, University of Utah, Salt Lake City, UT 84112.

2. What is a mismatch distribution, and how can it inform us about history? Genetic data provide a record of population history that stretches back tens or even hundreds of thousands of years. This record exists for two reasons. First, genetic differences between individuals measure the length of the genealogy that connects them. Second, genealogical distances tend to be longer in large populations than in small ones. For example, a random pair of individuals are more likely to be brothers, and thus connected by a short genealogy, in a population of 100 than in one of 100 million.

To get a feel for this effect, consider Figure 1. The upper panel there shows the history of a hypothetical population, with time measured in units of $1/(2u)$ generations before the present. Here, u is the aggregate mutation rate over the region of DNA under study. This scale of measurement is useful because it makes the time separating two individuals equal to the expected genetic difference between them. [1] For concreteness, I assume that $u = 0.0015$.[2] If each generation lasts 25 years, this mutation rate makes each unit of mutational time equal to 8333 years. In the figure, N_F denotes the effective female population size. The hypothetical population expanded by 500-fold at 7 units of mutational time (58,000 years) before the present.

The middle panel shows a simulated mitochondrial genealogy of 50 individuals, which was generated from this population history using the "coalescent" algorithm that I describe below. The 50 individuals in this sample are represented by 50 horizontal lines at the left edge of the genealogy, which corresponds to the present. The vertical lines in the genealogy mark places where two lineages have a common ancestor and "coalesce" into a single lineage. In this genealogy, coalescent events occur only rarely during the period from the expansion to the present. 35 of the 49 coalescent events are compressed into a relatively brief interval just prior to (to the right of) the expansion. This reflects the history of population size. After the expansion, the population was large and a random pair of individuals was unlikely to share the same mother. Therefore, coalescent events were rare. But prior to the expansion the population was small, and coalescent events were common. The result is that coalescent events are concentrated in a relatively brief interval prior to the expansion. This pattern is characteristic of expanded populations (see Rogers and Jorde [20] for another hypothetical example). It also appears in many gene genealogies estimated from human mtDNA [3].

A wave such as that in Figure 1 might also have been produced by selection rather than population growth. Under this interpretation, Figure 1 tells a different story: A favorable mitochondrial allele appears by

---

[1] For example, if two lineages have been separate for $7/(2u)$ generations, the expected number of mutations separating the them is $2u \times 7/(2u) = 7$.

[2] This estimate was obtained by Rogers and Harpending [19] for the data of Cann, Stoneking, and Wilson [2], which I discuss further below.

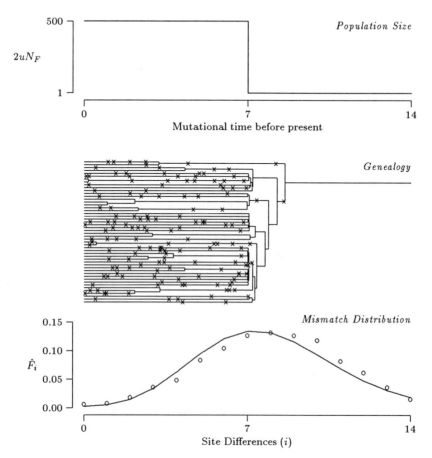

FIG. 1. *Mitochondrial genealogy and mismatch distribution of a hypothetical population*
The top panel shows $2uN_F$ as a function of time before present, with time measured in units of $1/(2u)$ generations. Here, N_F is the effective female population
size and u the aggregate mutation rate over the region of DNA under study. The
population was small prior to time 7. The middle panel shows the genealogy
of a sample of 50 individuals drawn from this population. The crosses represent mutations. The open circles in the bottom panel show these same data as
a mismatch distribution. The solid line there shows the theoretical mismatch
distribution for the parameters of the hypothetical population.

mutation at 7 units of mutational time before the present and then spreads rapidly to fixation. In the upper panel of the figure, N_F now refers not to the female population size but to the number of female descendants (or maternal ancestors) of the mutant female. There were few individuals in this lineage before the mutation (in fact there was only one: the mutant female has a single maternal ancestor in each generation) but many in the generations that followed the mutation. Although natural selection is responsible for the increase in the size of this lineage, the individuals within the lineage are selectively neutral with respect to each other, since each carries a copy of the same mitochondrial allele. Thus, it is appropriate to assume that variation within the lineage (and within our modern sample) is selectively neutral. This selective interpretation therefore leads to a genealogy and a mismatch distribution that are indistinguishable from that produced by population growth. Since the mismatch distribution is consistent with two interpretations (population growth and natural selection), the choice between these interpretations must be based on other data [20].

The crosses on the genealogy in Figure 1 represent mutations, which occur randomly along each branch. If this were a real population, we could count the mutational differences between pairs of individuals,[3] but we could not know either the true genealogy or the population history. These could only be estimated.

But any effort to infer the genealogy in Figure 1 from genetic data would be doomed to failure. In these data, nearly all of the 157 mutations occur after the expansion, in the part of the genealogy with few coalescent events. There are 35 coalescent events prior to the expansion but only 7 mutations. Consequently, no statistical method could succeed in telling us much about the topology of this genealogy—the data are essentially devoid of phylogenetic information. This example shows how a population expansion (or the selective sweep of a favorable allele) can lead to data with low phylogenetic resolution. With such data there is little point in trying to reconstruct the gene genealogy.

This is not to say that methods of phylogenetic inference are useless. Even when these methods cannot tell us the topology of the tree, they might still tell us that coalescent events were clustered in a narrow interval of time [3,8]. This would imply a small effective population size during this interval. Thus, the data may tell us about population history even if they are devoid of phylogenetic information.

But when the sample of individuals is large, phylogenetic inference is a formidable business. There is no efficient way to search the immense set of possible genealogies for those which best describe the data, and computer runs take many hours. The method to be described below is a short-cut that

---

[3] Strictly speaking, this is not so. We can only count nucleotide (or restriction) site differences between pairs of individuals, and such a difference may reflect more than one mutation [13]. This issue is discussed further in the "Discussion" section.

avoids this problem. To introduce it, I turn once again to the hypothetical data in Figure 1.

Let us assume that each mutation produces a detectable nucleotide site difference (the so-called model of "infinite sites" [12] that is discussed in footnote 3). The number of site differences between each pair of individuals in Figure 1 is then equal to the number of crosses along the path connecting them. For example, there are six site differences between the top-most pair of individuals in the genealogy. With 50 individuals in the sample, there are 1225 pairs of individuals, and we can count the differences between each pair. The open circles in the lower panel represent these 1225 differences as a scatter plot. Such plots are sometimes called "distributions of pairwise differences," and sometimes "mismatch distributions" [7,6]. For simplicity, I will use the latter term. Notice that the mismatch distribution in Figure 1 peaks just to the right of the point $i = 7$. This is because, as the genealogy shows, many pairs of individuals are separated by a little more than 7 units of mutational time and are therefore expected to differ by a little more than 7 mutations (see footnote 1). Thus, the mismatch distribution peaks just prior to the expansion at a point corresponding to the part of the genealogy at which coalescent events are concentrated.

Had the expansion happened earlier, the peak would have been farther to the right. As time passes the peak will move from left to right, traversing one unit of the horizontal axis in $1/(2u)$ generations [19]. Thus, the distribution looks and acts like a wave moving very slowly from left to right. The horizontal position of the wave measures time since the expansion in mutational time units.

This example suggests that the mismatch distribution might provide information about the history of population size and/or natural selection. Unfortunately, there is no statistical theory to tell us how this information can best be extracted. We can of course define ad hoc statistics and explore their behavior through computer simulation, but it is difficult to know in advance which statistics are likely to prove useful. To use simulations effectively, we need some basis for intuition about the behavior of the mismatch distribution. To gain such intuition, it is useful to study the "theoretical mismatch distribution."

3. The theoretical mismatch distribution. There are no simple theoretical formulas for subdivided populations, so I shall rely on results for a population that mates at random. Even there, we have no explicit formulas for samples of arbitrary size and must make do with formulas for samples of only two individuals. Watterson [23] showed how to calculate the probability that two individuals would differ by i nucleotide sites in a population of constant size. His model is compared to simulated data in Figure 2. Watterson's theoretical mismatch distribution is drawn as a solid line in the lower panel. The contrast between it and the simulated mismatch distribution—shown by the open circles—could not be greater. Whereas

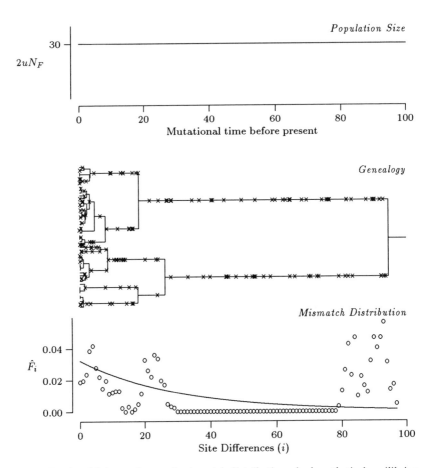

FIG. 2. *Mitochondrial genealogy and mismatch distribution of a hypothetical equilibrium population*
This hypothetical population has always had a constant size, $N_F = 30/(2u)$. Definitions are as in figure 1.

the theoretical distribution declines smoothly from a maximum at $i = 0$, the simulated distribution is ragged, with multiple peaks and a maximum value at $i = 93$. To recover the theoretical curve, we would need to average a large number of simulated mismatch distributions with the same population history. With real data, this would require the impossible—averaging mismatch distributions from a series of parallel worlds [22].[4] Since we have only one world to study, there are grounds for skepticism about the utility of the theoretical mismatch distribution.

But let us press on, nonetheless, to the case in which population size has not been constant. Li [14, Equation 5] developed the relevant formula, which was used by Rogers and Harpending [19] to study population histories such as that in Figure 1. Their Equation (4) was used to draw the solid line representing the theoretical distribution in the lower panel of that figure. The agreement between theory and simulated data is much better in the non-equilibrium case (Figure 1) than in the equilibrium case (Figure 2). This is no fluke: theory and simulated data often agree in expanded populations, provided that the initial population was fairly small [19]. Thus, the theoretical formula may be useful after all as a basis for intuition about the empirical distributions of expanded populations.

This argument is not rigorous, but it doesn't need to be. I am trying to justify using the theory as a basis for intuition, not as a basis for inference. Below, the theory will suggest which statistics should be calculated, and how they might be related to parameters describing population history. But these suggestions are only tentative, and are therefore checked by computer simulation. Thus, statistical inference will be justified by computer simulation, not by appeal to the theoretical mismatch distribution.

4. What should be estimated? We cannot hope for a complete description of the population's history. That would require one parameter— the population's size—for each time period. If the population were subdivided, we would need additional parameters for migration rates in each time period. Yet it is never possible to estimate more than a few parameters

---

[4] One reviewer disagreed with this claim, so I will provide a proof: Let

$$\delta_{ij}(k) \equiv \begin{cases} 1 & \text{if the } i\text{th and } j\text{th DNA sequences in the sample differ by } k \text{ sites} \\ 0 & \text{otherwise} \end{cases}$$

If the sample was drawn at random, then for any distinct i and j the expectation $E[\delta_{ij}(k)]$ is by definition equal to F_k, the kth term of the theoretical mismatch distribution. The empirical mismatch distribution is $\hat{F}_k \equiv \binom{n}{2}^{-1} \sum_{i<j} \delta_{ij}(k)$, where n is the number of DNA sequences in the sample. Its expectation is therefore

$$E[\hat{F}_k] = \binom{n}{2}^{-1} \sum_{i<j} E[\delta_{ij}(k)] = F_k$$

Thus, the theoretical mismatch distribution is the expectation of the empirical distribution. Were it possible to average independent realizations of $\hat{F}_k$, the law of large numbers would guarantee that this average would converge to F_k as the number of cases became large.

at once. We must content ourselves with some simplified representation of population history.

Fortunately, the theoretical mismatch distribution suggests that a simple model may be useful. The population history in Figure 1 has just three parameters: N_0 (the female population size before expansion), N_1 (the post-expansion size), and t (the time in generations since the expansion). Unfortunately, these parameters are all confounded with the mutation rate so that the mismatch distribution depends only on

(1) $$\theta_0 \equiv 2uN_0$$
(2) $$\theta_1 \equiv 2uN_1$$
(3) $$\tau \equiv 2ut$$

where as before u is the aggregate mutation rate over the region of DNA under study. Because N_0, N_1, and t are confounded with u, it is not possible to estimate any of these parameters directly from genetic data. Consequently, I address myself to the problem of estimating θ_0, θ_1, and τ.

This three-parameter model is not complex enough to describe the history of any real population, but it may nonetheless provide a fair description of the data. Analysis of the theoretical mismatch distribution [19] shows that the three-parameter model is robust in several ways: (1) When a population's size is small, convergence to the equilibrium is rapid. This implies that "bottlenecks," or temporary reductions in population size, amount to growth from an equilibrium population unless the bottleneck is very brief. Thus, it is not unreasonable to assume that the pre-expansion population was at equilibrium. (2) Instantaneous growth has an effect on the mismatch distribution that is indistinguishable from exponential growth over thousands of years. (3) After the population has grown large, subsequent episodes of growth and minor bottlenecks have little effect on the mismatch distribution. Because of these properties, the three-parameter model should prove useful even in populations whose histories are more complex than that in Figure 1.

This conclusion is based on the theoretical mismatch distribution, and should therefore be regarded with caution. It may not hold in circumstances where empirical and theoretical distributions tend to differ. Consider therefore the hypothetical population whose history, genealogy, and mismatch distribution are shown in Figure 3. This population is similar to that in Figure 1 in that it too experienced a burst of growth at mutational time 7, was small for a long while before, and was generally large thereafter. But there the similarity ends. The population in Figure 3 has seen several growth spurts and minor bottlenecks. Even the spurt at time 7 is different, being exponential rather than instantaneous. Yet none of this has any important effect. The genealogies of the two populations both show the same pattern—coalescent events are concentrated in a brief interval just prior to the population expansion. The lower panel of Figure 3 includes both

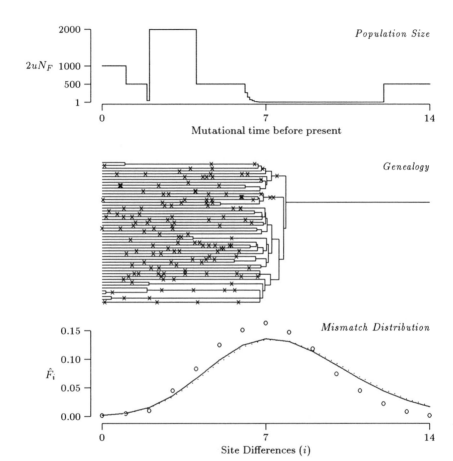

FIG. 3. *The minimal effect of a complex population history*
The dotted line shows for comparison the theoretical mismatch distribution from
Figure 1. Everything else is as defined in Figure 1.

the theoretical distribution of the complex population history (shown as a solid line) and of the simple population history (the dotted line). There is hardly any difference between them. This illustrates the robustness of the theoretical mismatch distribution [19]. The empirical distributions shown in Figures 1 and 3 are also very similar; the difference between them is no larger than that typically seen in different distributions simulated with the same population history. Were we to analyze the data in Figure 3 using the simpler population history of Figure 1, we would not be led astray. We would conclude correctly that a substantial episode of growth had occurred at around $\tau = 7$. Thus, the simple three-parameter model of population history can be useful even with populations whose histories are far more complex.

It would be easy to over-interpret these examples. While they suggest that the empirical mismatch distribution is insensitive to many details of population history, they don't amount to a proof. Furthermore, they deal only with the history of population size and therefore tell us nothing about the effect of other assumptions that may be violated. It is not even true that the difference between the population histories of Figures 1 and 3 has no effect on the mismatch distributions, for the second history very occasionally produces distributions with a second peak very far to the right. I found 2 such distributions in 100 trials with the complex history but none with the simpler history. All the other simulated distributions looked similar to those in Figures 1 and 3. Since the two histories produce similar distributions 98 times in 100, it is reasonable to estimate the parameters of the simple history in either population. The rare outliers of the complex history will however affect the statistical properties of these estimates. For example, the mean depth of genealogies from the complex history is about twice that of genealogies from the simpler history. This will require further discussion below in the section on confidence intervals.

5. Estimators. Statisticians have developed various methods for finding statistics to estimate particular parameters. The simplest of these, called the *method of moments*, proceeds by equating theoretical moments (the mean, the variance, and so on) with observed moments and solving the resulting equations. There is no good reason for confidence that this method will yield well-behaved estimators. The problem is that mismatch distributions do not offer a set of independent, identically distributed observations. Each pair of individuals in the data is correlated to a greater or lesser degree with many others. Thus, computer simulations will be needed not only to determine the statistical properties of the estimators proposed below but also to verify that they behave as estimators at all.

Given three parameters, θ_0, θ_1, and τ, a straightforward application of the method of moments would use three equations, obtained from the theoretical formulas for the mean, the variance, and the skewness. This approach works poorly here, because it is often impossible to solve these

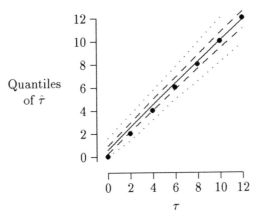

FIG. 4. *Quantiles of $\hat{\tau}$*

At least 1000 data sets were simulated at each of several values of τ, and each was used to estimate the model's parameters. The bold dots indicate points at which $\hat{\tau} = \tau$. The solid line is the median, the dashed lines enclose the central 50% of the distribution, and the dotted lines the central 95%. Each simulated data set was generated using the coalescent algorithm with $\theta_0 = 1$, $\theta_1 = 500$, and 147 subjects.

three equations.

However, the theoretical mismatch distribution tells us that there is usually very little information about θ_1 anyway. A large value of θ_1 has essentially the same effect as an infinite value. Therefore, we may hope to estimate θ_0 and τ from a model in which $\theta_1 \to \infty$. In this case, there are only two equations, and these have a simple solution [18]:

$$(4) \qquad \hat{\theta}_0 = \sqrt{v - m}$$
$$(5) \qquad \hat{\tau} = m - \hat{\theta}_0$$

where m is the mean and v the variance of the empirical mismatch distribution. I propose to use these statistics as estimators.

This proposal, however, is only based on intuition—the intuition provided by the theoretical mismatch distribution. To justify this intuition it is necessary to show that $\hat{\theta}_0$ and $\hat{\tau}$ do in fact behave as estimators. For each set of parameter values, the sampling distributions of the estimators should ideally be narrow, and centered around the true parameter values.

I have verified this behavior elsewhere [18] and will here include only a single figure. Figure 4 was obtained by simulating 1000 data sets at each of several values of τ and using each of these to calculate $\hat{\tau}$. At each value of τ, the 1000 estimates were used to estimate the quantiles of $\hat{\tau}$, and these quantiles are shown in the figure. At each value of τ, the median (shown as a solid line) is close to the bold dot that marks the true value of τ, and the distribution is relatively narrow. Thus, not only does $\hat{\tau}$ behave as an estimator, it is an estimator with admirable statistical properties. A similar analysis [18] shows that $\hat{\theta}_0$ is also well behaved when $\theta_0 \geq 1$ but is incapable of discriminating values in the range $0 < \theta_0 \leq 1$. This is not a serious limitation. It means only that when $\theta_0 \approx 1$, the confidence interval will reach all the way to zero.

6. Simulating data sets with population structure.

Thus far, I have merely summarized previous work, which assumed a randomly mating population. What happens when these methods are applied in a subdivided population? To find out, I have implemented the geographically structured coalescent algorithm described by Hudson [9]. My implementation breaks the population history into an arbitrary number of "epochs," within each of which all parameters are constant. Within epoch i, the population is described by four parameters,

$$\theta_i \;=\; 2uN_i, \text{ where } N_i \text{ is the effective female population size during epoch } i;$$

$$M_i \;=\; \text{the number of migrants per generation between each pair of groups during epoch } i;$$

$$\tau_i \;=\; 2ut_i, \text{ where } t_i \text{ is the length of epoch } i \text{ in generations;}$$

$$K_i \;=\; \text{the number of subdivisions during epoch } i.$$

If $K_i = 1$, then M_i is undefined and the entire population mates at random. The earliest epoch is epoch 0 and has infinite duration, i.e. $\tau_0 = \infty$.

The algorithm begins with the last epoch, which I denote as epoch L. The n individuals of the sample are at first divided evenly among the K_L groups of epoch L. Thus, the algorithm requires that n be evenly divisible by K_L.[5]

As the algorithm moves backward into the past, two types of event occur. Migrations occur when an individual moves from one group to another, and "coalescent events" occur when two individuals have a common ancestor and therefore coalesce to become a single individual.

The hazard h at time τ is defined so that $h\,d\tau$ is the probability that an event of either type will occur between τ and $\tau + d\tau$, where τ measures mutational time looking backwards into the past. The hazard depends on prevailing values of the population history parameters, on the number of individuals, and on how these are distributed among groups. At any given time, let s_j denote the number of individuals within group j, $S \equiv \sum_j s_j$ (the total number of individuals), and $R \equiv \sum_j s_j^2$ (the sum of these

---

[5] The allocation of individuals among groups in the simulation should match that in the data under study. Thus, the allocation used here is most appropriate when the real data include samples of equal size, drawn from several groups.

numbers squared). Then the hazard of an event is[6]

$$(6) \qquad h = [SM_i + (R - S)/2]/\gamma_i$$

where $\gamma_i \equiv \theta_i/K_i$, and measures group size in epoch i.

The algorithm first sets $S = n$, $R = K_L(n/K_L)^2$, and then sets h using these values together with the parameters of the final epoch, L. It then enters a loop that is executed repeatedly. I describe the steps of this loop briefly before describing each step in detail.

Overview of coalescent loop.

1. Find the time of the next event, changing epochs and recalculating h as necessary.
2. Determine whether the next event is a migration or a coalescent event.
3. Carry out the next event.

These steps are repeated until $S = 1$. Mutations are then added along each branch.

Step 1. Let T_i denote the amount of time that we have already traveled (backwards) into epoch i. To find the time of the next event, draw a random number x from an exponential distribution whose parameter equals unity. In a constant world, the time of the next event would be $T_i + x/h$. If this time lies within epoch i (i.e. if $T_i + x/h < \tau_i$), then we have found the time of the next event. Otherwise, change epochs as follows:

    a Subtract off the portion of x that is "used up" by epoch i, i.e. subtract $h \cdot (\tau_i - T_i)$ from the value of x.

    b Reset population history parameters to those of epoch $i-1$ and set T_i to zero. If $K_{i-1} < K_i$, join groups at random to diminish the number of groups. If $K_{i-1} > K_i$, increase the number of groups, but allocate no individuals to the new groups. Individuals will enter the new groups only through migration.[7]

    c Reset R and h. Subtract 1 from the value of i.

This process repeats until $T_i + x/h < \tau_i$.

---

[6] Let m denote the migration rate per generation, g the group size, and $M \equiv mg$. The hazard per generation is

$$h^* \equiv \sum_j [s_j m + s_j(s_j - 1)/(2g)] = (1/g)[SM + (R - S)/2]$$

The cumulative hazard in t generations is

$$h^* t = \frac{2ut}{2ug}[SM + (R - S)/2] \equiv \frac{\tau}{\gamma}[SM + (R - S)/2],$$

where $\tau \equiv 2ut$ and $\gamma \equiv 2ug$. Equation 6 follows from the observation that, by definition, the hazard h in mutational time obeys $h\tau \equiv h^* t$.

[7] The assumption for $K_{i-1} > K_i$ implies that, in forward time, the number of groups has decreased because some groups have died out. Other assumptions are possible and the present one was chosen only for computational convenience.

Step 2. Once the time of the next event has been established, step 2 classifies the event as either a migration or a coalescent event. Equation 6 implies that the event is a migration with probability

$$P_M = \frac{SM_i}{SM_i + (R - S)/2}$$

Thus, step 2 calls the next event a migration with probability P_M and a coalescent event with probability $1 - P_M$.

Step 3. If the next event is a migration, then move a random individual into a new, randomly chosen group. Then reset R and h.

Otherwise, we have a coalescent event and the procedure is as follows. First choose a group at random, weighting each group by the number of pairs of individuals within it. Then choose a random pair of individuals from within the chosen group, replace the two individuals with a single individual (their common ancestor), reduce S by 1, and reset R and h.

Mutation. I use the infinite sites model of mutation, which implies that the number of mutations along each branch is a Poisson random variable with parameter ut, where u is the mutation rate and t the length of the branch in generations [12]. In mutational time, branch lengths equal $\tau \equiv 2ut$ and the Poisson distribution has parameter $\tau/2$.

To execute this algorithm, it is necessary to specify the sample size n and the parameters $(\theta_i, M_i, \tau_i,$ and $K_i)$ that describe the population's history. There is no need to specify the mutation rate, the number of individuals in the population, or the number of generations in each epoch.

7. Confidence regions. A 95% confidence region is a set of parameter values constructed by any procedure that guarantees the following property: *If, each time we construct a 95% confidence region, we assert that it includes the true parameter value, we will in the long run be correct 95% of the time (and incorrect 5% of the time).* One way to construct such a region is to define some statistical test whose outcome depends only on the data and the parameters of interest. The set of parameter values that cannot be rejected at significance level α will constitute a $100 \times (1 - \alpha)\%$ confidence region [11, p. 110].

I shall apply this method to hypotheses about population history. A hypothesis is rejected at significance level α if it implies that data sets "at least as extreme" as the real data occur with a frequency less than α. Given a precise definition of "at least as extreme," it is easy to estimate the frequency of such events. I generate a large number of simulated data sets using the coalescent algorithm described above, and estimate α by the fraction of the simulated data sets that are at least as extreme as the observed data.

It remains to decide when a simulated data set will be deemed at least as extreme as the observed data. This decision might be made in any

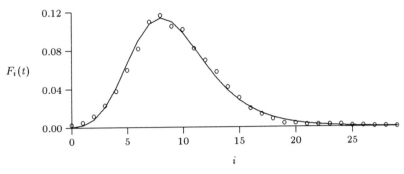

FIG. 5. *The model of sudden expansion fit to the data of Cann, Stoneking and Wilson* The open circles show the mismatch distribution (the relative frequencies of pairs of individuals whose mtDNA samples differ by i sites) of Cann, Stoneking, and Wilson [2, fig. 1]. The solid line shows the fit of the non-equilibrium distribution [19] obtained using the two-parameter method of moments estimator [18]. This figure summarizes comparisons among $n = 147$ subjects.

number of ways, all of which would yield valid confidence regions. Yet these definitions would not all be equally useful. I have tried to design a test that would yield small confidence regions in expanded populations. The test is described in full elsewhere [18] and is summarized only briefly here: A simulated data set is deemed to be at least as extreme as that observed if (a) the Mahalanobis distance [[10],pp. 423–424] between the vector $(\hat{\theta}_0, \hat{\tau})$ and the simulation mean is at least as large for the simulated data as for the real data, and (b) the mean squared error (MSE) between empirical and theoretical mismatch distributions is at least as large for the simulated data as for the observed data. Occasionally, the covariance matrix used in calculating the Mahalanobis distance turns out to be singular and this test fails to provide an answer.

This method yields information not only about θ_0 and τ, but also about θ_1, because for given values of θ_0 and τ, the MSE tends to decrease with θ_1. Simulations show that this method yields confidence regions that usually enclose the true parameters values and are reasonably small in expanded populations under random mating [18].

In structured populations, these confidence intervals could presumably be made smaller by using a test that distinguishes within-group from between-group variation [6]. Nonetheless, I will work instead with distributions calculated from the population as a whole, using the test developed for my earlier random-mating model [18]. This will make it possible to assess the bias that is introduced when that model is applied to data from a structured population. Thus, I take no account of population structure when estimating parameters. The statistical properties of these estimates, however, will be investigated under various assumptions about population structure. To make comparisons simple, I also analyze the same data set that was used in that earlier publication. These data were published by Cann, Stoneking, and Wilson [2] and are shown in Figure 5.[8]

8. Population structure. In this section I consider three models of population history. The first—that of world-wide random mating—is not realistic. I include it because it is simple, because it has been used before [18], and because I want to find out whether it yields useful answers even when reality is more complex. The second model assumes that the human population is been subdivided as far as we can see back into the past, while the third assumes that subdivision appeared more recently. These models represent the multiregional and the replacement models of modern human origins, respectively.

8.1. Random mating. My earlier confidence intervals [18] made the simplest possible assumption about population structure: that of an undivided randomly mating population. With the present model, this case corresponds to a population history of the following form:

Epoch	θ_i	M_i	τ_i	K_i
1	θ_1	0	τ	1
0	θ_0	0	∞	1

In each epoch, there is only a single subdivision ($K_i = 1$) and there can of course be no migration between subdivisions ($M_i = 0$). In an earlier publication [18] I used this population history together with the data in Figure 5 to infer the confidence region shown in Figure 6. The open circles there represent hypotheses that were rejected at the 0.05 significance level. The filled circles represent hypotheses that could not be rejected. Thus, the filled circles delimit a 95% confidence region for the parameters defined in Equations 1–3. The confidence region implies that $\theta_0 < 10$, $\theta_1/\theta_0 > 100$, and that $4 < \tau < 9$. Yet these results rely on an assumption that has surely been violated: They assume random mating, whereas the human population is geographically structured.

The history of this structure is a matter of debate. There are two competing views, which I discuss below.

8.2. The multiregional hypothesis. The multiregional hypothesis of modern human origins [24,4] holds that our species evolved within a widespread population that has inhabited much of Europe, Africa, and Asia for the past million years. Favorable mutations arising in one location spread to others by gene flow rather than by replacement of entire populations.

This hypothesis implies that the geographic structure of our species

---

[8] Do not read too much into the close fit between the observed and the theoretical distributions. This fit provides no strong support either for the theory or for the statistical methods. See my earlier paper [18] and the discussion above dealing with Figure 4.

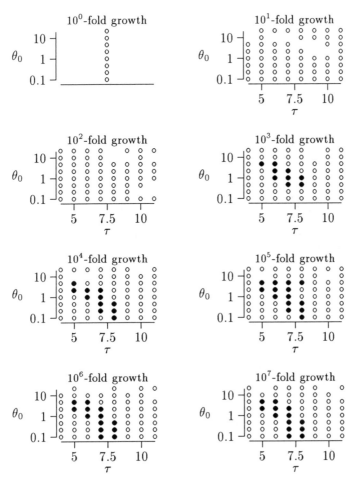

FIG. 6. *A 95% confidence region for the CSW data under the assumption of random mating*
Large filled circles (•) indicate points within the 95% confidence region, and open circles (○) indicate points outside of the confidence region. 10^x-fold growth means that $\theta_1/\theta_0 = 10^x$. Missing circles indicate parameter values for which no test was possible because the covariance matrix of $(\hat{\theta}_0, \hat{\tau})$ was singular. Data are from Cann, Stoneking, and Wilson [2]. Reproduced from Rogers [18].

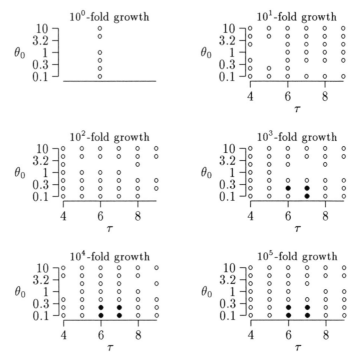

FIG. 7. *Multiregional hypothesis, M = 0.1*

goes back at least a million years, originating well before the common mitochondrial ancestor. For our purposes, this is equivalent to assuming that the structure has existed forever. Thus, I use a population history of form

Epoch	θ_i	M_i	τ_i	K_i
1	θ_1	M	τ	3
0	θ_0	M	∞	3

with $K_0 = K_1 = 3$ to represent the three major races, and migration measured by the parameter M. In words, this history assumes that the population has always been divided into three groups, which have always exchanged M migrants per generation. The history allows for a change in population size from θ_0 to θ_1 at τ units of mutational time before the present.

This history was used to generate the confidence regions shown in Figures 7–9. The three confidence regions differ in their assumptions regarding the level M of migration. Figure 7 assumes that migration is weak ($M = 0.1$), Figure 8 that it is moderate ($M = 1$), and Figure 9 that it is strong ($M = 10$). In all three figures, the confidence region is smaller than that under random mating and includes no parameters values that are not also included within the random-mating confidence region. There is one

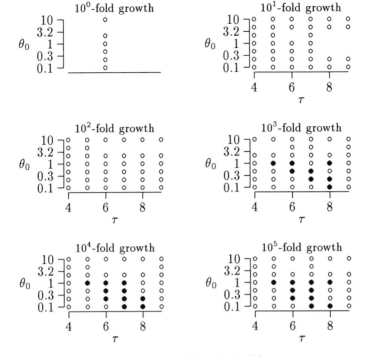

FIG. 8. *Multiregional hypothesis, $M = 1$*

important difference: The multiregional hypothesis requires that $\theta_0 < 2.15$, whereas the model of random mating requires only that $\theta_0 < 10$.

Using published estimates of u, my previous paper found that the random-mating result requires that $N_0 < 7000$ breeding females [18].[9] Since this number seemed too small to populate the continents of Europe, Africa, and Asia, I viewed this estimate as evidence against the multiregional hypothesis. Yet now it is clear that my earlier estimate was too generous. The reduced upper bound on θ_0 implies that $N_0 < 1500$ breeding females. Thus, if the wave in the mitochondrial data reflects population growth rather than selection, then the analysis *with* population structure rejects the multiregional hypothesis even more strongly than the one *without*.

It seems clear, moreover, that this conclusion would not be altered much by other assumptions about M. The upper bound on θ_0 increases with M (compare Figures 7–9), so smaller values of M would not lead to any favorable assessment of the multiregional hypothesis. On the other hand, larger values of M might do so. However, as M grows large, the confidence

---

[9] I used the smaller of the two published estimates of u ($\hat{u} = 7.5 \times 10^{-4}$) in order to make the upper bound on N_0 as large as possible [19].

ALAN R. ROGERS

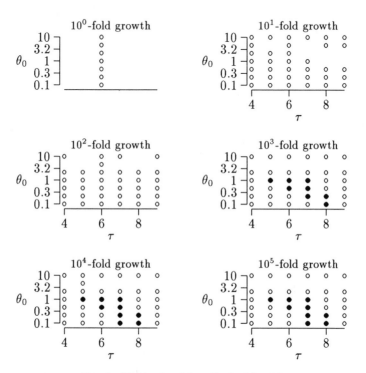

FIG. 9. *Multiregional hypothesis, M = 10*

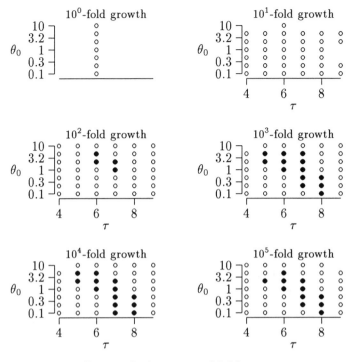

FIG. 10. *Replacement model, M = 0.1*

region will approach the random-mating result, which still allows only 7000 breeding females. Thus, it is not possible to salvage the multiregional hypothesis by a judicious choice of M.

8.3. The replacement hypothesis. The replacement hypothesis holds that modern humans evolved in one region and spread from there throughout the world some 50,000–100,000 years ago, replacing earlier peoples as they went. This view implies that the geographic structure of the human population is less ancient, having developed after the initial expansion of modern humans. Thus, the population history becomes

Epoch	θ_i	M_i	τ_i	K_i
1	θ_1	M	τ	3
0	θ_0	0	∞	1

The difference here is that $K_0 = 1$, rather than 3. In words, this says that prior to the expansion that occurred τ time units ago, there was only 1 ancestral population rather than 3.

The confidence interval generated under this hypothesis is shown in Figure 10 and is similar to that in Figure 6. The main difference seems to be that the hypothesis of random mating rejects 100-fold growth, while the replacement hypothesis does not. This makes sense, since subdivision increases a population's effective size [16]. Thus, a 100-fold increase that is combined with subdivision is equivalent to a larger increase without subdivision. Apart from this amendment, the replacement hypothesis is well approximated by a model of random mating and allows the initial population to be nearly five-fold larger than does the multiregional hypothesis. This paradoxical result says that if our ancestors were all in one place then their population may have been of moderate size ($N_0 < 7000$), yet if their population was subdivided then it must have been extremely small ($N_0 < 1500$).

9. Discussion. Two studies [15,6] have shown that when the initial population is relatively large, and especially if it is geographically structured, mismatch distributions are often rough and ragged like that in Figure 2. Smooth waves such as that in Figure 1 occur only when the initial population is extremely small. These results have been interpreted as evidence that statistical inference from mismatch distributions is a perilous business [15], but I would argue otherwise. Indeed, the raggedness of these distributions makes my confidence regions smaller. According to the confidence regions, the upper bound on θ_0 is 2.15 when the initial population is strongly structured but 10 when it mates at random. This is because intermediate values such as $\theta_0 = 5$ produce ragged mismatch distributions if the initial population is structured but not if it mates at random. Since ragged distributions look nothing like the observed distribution (Figure 5), the statistical method rejects the parameter values that generate them. Far from being a problem, raggedness made the present estimates more accurate [5].

On the other hand, raggedness is not always a blessing. When the observed mismatch distribution is ragged, the present statistical methods yield large confidence regions [18]. But this is as it should be. Large confidence regions may be disappointing, but they are unlikely to lead us astray. They demand no more caution than is normal in statistical analysis.

The results presented here suggest that geographic structure affects the mismatch distribution primarily by way of effective population size. Effective size is larger if a population is structured than if it mates at random [16]. Consequently, a population with a structured initial population behaves like one with a large initial population—its mismatch distribution tends to be ragged. By the same token, a population that expands and at the same time subdivides behaves like a population that has undergone a much larger expansion. This accounts for the fact that a 100-fold expansion is rejected under the model of random mating but not under the replacement hypothesis. A smaller expansion is allowed in the latter case because geographic structure makes the effective expansion larger than the expansion in numbers.

The expansion that is inferred here can be interpreted in two ways. It may have been either an expansion in population size or the expansion in frequency of an advantageous mitochondrial allele. Harpending et al. [6] and Rogers and Jorde [20] present arguments in favor of the former interpretation, but additional data will be needed to settle the issue.

There is also cause for concern about my use of the infinite sites model of mutation. R. Lundstrom (unpublished data) has shown that with a finite number of sites and mutation rates that vary from site to site, waves can be generated in the theoretical distributions even of equilibrium populations. Fortunately, my own calculations indicate that this effect is unlikely to be important in the theoretical distributions considered here [17]. But even if this effect is negligible in theoretical distributions, its effect on the statistical distribution of my estimates may be important [1].

Finally, it has been argued that mismatch distributions should be interpreted with special caution because we have only one world to study [15,1]. Because of this limitation, the empirical mismatch distribution amounts to a single observation from a distribution that (depending on parameter values) may be highly variable. This problem is real, for it makes parameter estimates less precise—confidence regions would surely be smaller with data from parallel worlds. Yet it is not a fatal problem, since the confidence regions are small enough to be useful even without parallel worlds.

10. Summary. This paper began with a review showing how the theoretical mismatch distribution has been useful in previous research as a basis for intuition. It then introduced a method for inferring confidence regions with data from subdivided populations, emphasizing that these methods are based on computer simulation, not on the theoretical mismatch distribution. The statistical methods show that if the pattern in the mitochondrial data reflects population growth rather than selection, then (1) the multiregional hypothesis of modern human origins is rejected more strongly than before, (2) the replacement hypothesis of modern human origins is not rejected but allows the expansion of population size to be smaller than did my earlier random-mating model, (3) the model of random mating yields a confidence region that encompasses that of the multiregional hypothesis and differs only slightly from that of the replacement hypothesis. (4) Population structure does not make confidence intervals larger, at least for the data considered here. Consequently, these results provide no support for the view that population structure reduces the value of mismatch distributions for statistical inference.

Acknowledgements. I thank Peter Donnelly, Henry Harpending, Lynn Jorde, and two anonymous reviewers for comments. This work was supported in part by a grant from NSF (DBS–9310105).

REFERENCES

[1] Giorgio Bertorelle and Montgomery Slatkin. The number of segregating sites in expanding human populations, with implications for estimates of demographic parameters. manuscript, 1994.

[2] Rebecca L. Cann, Mark Stoneking, and Allan C. Wilson. Mitochondrial DNA and human evolution. *Nature*, 325(1):31–36, January 1987.

[3] Anna Di Rienzo and Alan C. Wilson. Branching pattern in the evolutionary tree for human mitochondrial DNA. *Proceedings of the National Academy of Sciences, USA*, 88:1597–1601, 1991.

[4] David W. Frayer, Milford H. Wolpoff, Alan G. Thorne, Fred H. Smith, and Geoffrey G. Pope. Theories of modern human origins: The paleontological test. *American Anthropologist*, 95(1):14–50, 1993.

[5] Henry Harpending. Signature of ancient population growth in a low resolution mitochondrial DNA mismatch distribution. *Human Biology*, 66(4):591–600, 1994.

[6] Henry C. Harpending, Stephen T. Sherry, Alan R. Rogers, and Mark Stoneking. The genetic structure of ancient human populations. *Current Anthropology*, 34:483–496, 1993.

[7] Daniel L Hartl and Andrew G. Clark. *Principles of Population Genetics*. Sinauer, Sunderland, MA, 2nd edition, 1989.

[8] Masami Hasegawa, Anna Di Rienzo, and Alan C. Wilson. Toward a more accurate time scale for the human mitochondrial DNA tree. *Journal of Molecular Evolution*, 37:347–354, 1993.

[9] Richard R. Hudson. Gene genealogies and the coalescent process. In Douglas Futuyma and Janis Antonovics, editors, *Oxford Surveys in Evolutionary Biology*, volume 7, pages 1–44. Oxford University Press, Oxford, 1990.

[10] Albert Jacquard. *The Genetic Structure of Populations*. Springer-Verlag, New York, 1974.

[11] M. Kendall and A. Stuart. *The Advanced Theory of Statistics. II. Inference and Relationship*. MacMillan, New York, fourth edition, 1979.

[12] Motoo Kimura. Theoretical foundation of population genetics at the molecular level. *Theoretical Population Biology*, 2:174–208, 1971.

[13] Thomas Kocher and Allan Wilson. Sequence evolution of mitochondrial DNA in humans and chimpanzees: Control region and a protein-coding region. In S. Osawa and T. Honjo, editors, *Evolution of Life: Fossils, Molecules, and Culture*, pages 391–413. Springer-Verlag, New York, 1991.

[14] Wen-Hsiung Li. Distribution of nucleotide differences between two randomly chosen cistrons in a finite population. *Genetics*, 85:331–337, 1977.

[15] Paul Marjoram and Peter Donnelly. Pairwise comparisons of mitochondrial DNA sequences in subdivided populations and implications for early human evolution. *Genetics*, 136:673–683, February 1994.

[16] Masatoshi Nei and Nayouki Takahata. Effective population size, genetic diversity, and coalescence time in subdivided populations. *Journal of Molecular Evolution*, 37:240–244, 1993.

[17] Alan R. Rogers. Error introduced by the infinite sites model. *Molecular Biology and Evolution*, 9:1181–1184, 1992.

[18] Alan R. Rogers. Genetic evidence for a Pleistocene population explosion. *Evolution*, 1995. In press.

[19] Alan R. Rogers and Henry C. Harpending. Population growth makes waves in the distribution of pairwise genetic differences. *Molecular Biology and Evolution*, 9:552–569, 1992.

[20] Alan R. Rogers and Lynn B. Jorde. Genetic evidence on modern human origins. *Human Biology*, 67(1):1–36, Feb 1995.

[21] Stephen Sherry, Alan R. Rogers, Henry C. Harpending, Himla Soodyall, Trefor Jenkins, and Mark Stoneking. Mismatch distributions of mtDNA reveal recent

human population expansions. *Human Biology*, 66(5):761–775, Oct 1994.

[22] Montgomery Slatkin and Richard R. Hudson. Pairwise comparisons of mitochondrial DNA sequences in stable and exponentially growing populations. *Genetics*, 129:555–562, 1991.

[23] G. A. Watterson. On the number of segregating sites in genetical models without recombination. *Theoretical Population Biology*, 7:256–276, 1975.

[24] Milford H. Wolpoff. Multiregional evolution: The fossil alternative to Eden. In Paul Mellars and Chris Stringer, editors, *The Human Revolution: Behavioural and Biological Perspectives on the Origins of Modern Humans*, pages 62–108. Princeton University Press, Princeton, New Jersey, 1989.

DISTRIBUTION OF PAIRWISE DIFFERENCES IN GROWING POPULATIONS

GUNTER WEISS*, ANDREAS HENKING* , AND ARNDT VON HAESELER*

Abstract. A model is outlined that approximates the distribution of pairwise differences in growing populations. The statistical properties are analysed. An illustrative example is given.

Key words. coalescent, Finnish population, growing populations, distribution of pairwise differences, mitochondrial DNA.

1. Introduction. Following the recent development of the polymerase chain reaction and direct sequencing, it has become possible to amplify and sequence rapidly particular DNA segments starting from total genomic DNA preparations. The possibility of obtaining DNA sequences rapidly is a great advance for molecular evolutionary inference since it provides the highest possible resolution of variability. For these reasons, there are now several population studies of sequence variability of the control region of the mitochondrial DNA (mtDNA) and other parts of the molecule [5,19,20,21,22]. The control region, which in humans is about 1,100 base pairs long, contains promoters for transcription and the origin of replication of one of the DNA strands. Within species only base substitutions and very rarely length mutations occur. In general base substitutions accumulate at a much more rapid pace than in nuclear DNA [1]. In the control region, this rate is estimated to be 10–15 times higher than other mitochondrial regions [22]. Thus, the mtDNA control region is most suited to study intraspecies variation. Furthermore, mtDNA is almost exclusively maternally inherited. This mode of inheritance facilitates the statistical analysis, because the molecule represents maternal lineages that accumulate changes without the interference of recombination events.

One of the uses to which mitochondrial variability has been put to, is the study of early human origins. Cann et al. [2] used high resolution restriction mapping of mtDNA to show that all present–day mtDNA types stem from a single type that existed about 120,000 to 200,000 years ago in Africa. While the precise details of the time and the place of origin of this "mitochondrial Eve" (mt Eve) are the subject of much debate (cf. [17,18]), the general power of the methodology is well accepted.

In recent years, the study of DNA sequence variation within populations has been reduced to the analysis of the distribution of pairwise differences between individuals (e.g. [5,9,15,17,21,22]). Tacitly assuming that the observed distributions of pairwise differences or mismatch distributions [9] preserve enough information to infer population history. Theo-

---

* Institute for Zoology, University of Munich, P.O. Box 202136, 80021 Munich, Germany.

retical studies [9,12,15,16], based on the coalescent process [10], show that
at least some features of the past population dynamics can be retrieved
from pairwise difference distribution. Nevertheless, inference made from
pairwise differences has to be viewed with caution [12].

While the above mentioned studies use the coalescence process [10] to
analyse the dynamics of a population, we give another description of the development of an exponentially growing population. This approach is based
on the assumption of a randomly bifurcating tree. Section 2 outlines our
model, section 3 gives some results on the distribution of times back to the
most recent common ancestor of one pair of individuals. In the subsequent
section the goodness of the approximations made in section 3 is studied.
Finally, section 5 illustrates how to analyse pairwise difference distribution
with our model. As an example the population dynamics of the Finnish
population is studied. The data are the mtDNA difference distribution,
as computed from the hypervariable part one (HV-1) of the control region
(Sajantila et al., unpublished data). This section also illustrates the relation of our model and the coalescent based model of an exponentially
growing populations developed in [16].

2. The model. For a wide class of branching processes the existence
of a common ancestor for all contemporary individuals is a mathematical
necessity. Hence, we assume that a so–called most recent common ancestor
(mrca) exists for all individuals in a population under study. This can be
viewed as a formal extension of the coalescent idea, where one assumes
the existence of an mrca for a sample of individuals from a population.
Since we want to make statements about the distribution of pairwise differences between pairs of individuals in a population, we tailor the model
of a growing population starting with the mrca as follows:

- The mrca (mt Eve) of the population had exactly two daughters.
- Starting with the daughters of mt Eve, each female in the population begets one daughter and with probability λ a second daughter
  per generation.

This process ensures that the offspring of the mrca will not die out. After T generations the mrca will have $N_T = 2(1 + \lambda)^{T-1}$ descendants on
average[1]. Note, the definition of the mrca of the individuals of a population implies that the mrca must have at least two offspring. We ignore the
possibility that the mt Eve and all her female descendants had more than
two daughters, that transmit their genes to the population today. This
assumption, which is also made in the coalescence process, is certainly not
too restrictive, since λ in human population is usually small.

At this point it is also worth mentioning, that there may also exist
other individuals in the population that give rise to offspring, but these

---

[1] The parameter λ is somewhat different from the well known average number of
progeny per individual and generation [4] or the closely related Malthusian parameter r
[11]. As we will see later a simple relation exits between λ and r.

individuals by chance do not transmit their genes to the individuals living today. The development of mt Eve's offspring can be viewed as a randomly bifurcating tree.

In order to compute the distribution of pairwise differences between DNA sequences or the mismatch distribution [15] it is crucial to know the number of pairs of individuals in the contemporary population that had their most recent common ancestor t generations ago, where $t = 1, \ldots, T$ and T is the age of the population measured in generations. If the exact genealogy were known, this would be a straightforward approach. The true genealogy or tree however is not known, moreover the number of possible trees grows exponentially with the age of the population. For each of mt Eve's daughters exist

$$(1) \qquad\qquad S_T = S_{T-1}(S_{T-1} + 1)$$

possible rooted, non–commutative, unlabelled trees (cf. [8]), if the time back to mt Eve equals T and if $S_1 = 1$. Figure 1 shows the six possible trees for one of mt Eve's daughters if time back to mt Eve equals three generations. For $T = 6$ there are already 3,263,442 different trees. So far we have only used a purely combinatorial argument to count the number of possible trees. In our model that includes λ as "growth parameter", each tree has a well defined, but unfortunately not easily derived, probability of occurrence.

The trees $T_1, T_2, T_3, T_4, T_5,$ and T_6 in Figure 1 occur with probability

$$
\begin{aligned}
p_1 &:= Pr(T_1) &=& \quad \lambda^3 \\
p_2 &:= Pr(T_2) &=& \quad \lambda^2(1 - \lambda) \\
p_3 &:= Pr(T_3) &=& \quad \lambda^2(1 - \lambda) \\
p_4 &:= Pr(T_4) &=& \quad \lambda(1 - \lambda) \\
p_5 &:= Pr(T_5) &=& \quad \lambda(1 - \lambda)^2 \\
p_6 &:= Pr(T_6) &=& \quad (1 - \lambda)^2.
\end{aligned}
$$

T_1 is called the full tree, since each individual produces exactly two offspring. Each of the remaining five trees can be uniquely embedded in the full tree. In other words it corresponds to a subtree of T_1, where the correspondence is defined by the edgelist. For example the tree T_2 corresponds to the subtree characterized by edgelist $\{e_1, e_2, e_3, e_4, e_5, e_6\}$, whereas T_3 matches the subtree given by $\{e_1, e_2, e_3, e_4, e_6, e_7\}$. Embedding is unique because we consider the trees as non–commutative trees. If this assumption is dropped T_2 and T_3 are topologically identical, because we can view T_3 as the mirror image of T_2.

Given the probability of the trees we compute the probability $Pr(e_i)$ of observing edge e_i in the collection of all trees as the sum of the probabilities to observe each of the possible trees. Hence, the probability to observe edge

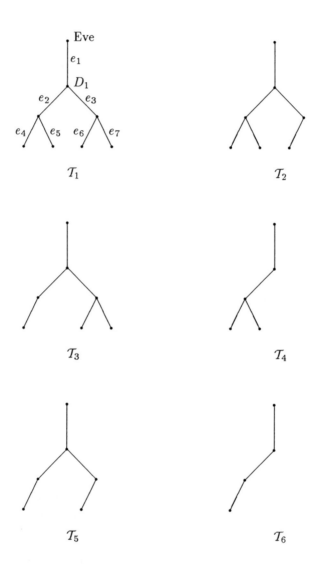

FIG. 1. *The six possible non–commutative trees for one of mt Eve's daughters, if time back to mt Eve equals three generations for the contemporay individuals.*

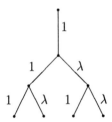

FIG. 2. *The "average tree" for one of mt Eve's daughters. Time back to mt Eve equals three generations.*

e_3 equals

$$Pr(e_3) = \sum_{i=1}^{6} p_i = \lambda^3 + \lambda^2(1-\lambda) + \lambda^2(1-\lambda) + \lambda(1-\lambda)^2 = \lambda.$$

Similarily, we compute $Pr(e_7) = \lambda^2$. If the conditional probabilities are computed for each edge of the full tree, the tree in Figure 2 together with the "edge weights" is obtained. This "average tree" serves as a model to compute an approximate formula for times back to the mrca for all pairs of individuals in the contemporary population.

3. Computing distribution of times back to the mrca. If $\lambda = 1$ the resulting tree is a strictly bifurcating tree and no stochastic element is involved in the generation of the genealogy. In this situation the number of recent pairs of individuals that had their mrca $t = 1, \ldots T$ generations ago is given by

$$(2) \qquad n_T(t) = 2\lambda(1+\lambda)^{T+t-3} = 2^{T+t-2}, \text{ where } t = 1, \ldots, T.$$

Hence, the probability to observe a pair of individuals among the $\binom{N_T}{2}$ possible pairs of individuals in the recent population that had their mrca $X_T = t$ generations ago is given by

$$(3) \qquad Pr(X_T = t) = \frac{n_T(t)}{\binom{N_T}{2}}.$$

If $\lambda < 1$ then $n_T(t)$ is a random variable, and we approximate $Pr(X_T = t)$ using the following equation

$$(4) \qquad Pr(X_T = t) = \frac{E[n_T(t)]}{\sum_{i=1}^{T} E[n_T(i)]}.$$

The exact value of $n_T(t)$ in equation (2) is now regarded as expectation $E[n_T(t)]$, where

$$(5) \qquad E[n_T(t)] = \begin{cases} 2\lambda(1+\lambda)^{T+t-3}, & \text{if } t = 1, \ldots, T-1 \\ (1+\lambda)^{2(T-1)}, & \text{if } t = T. \end{cases}$$

$$(6) \qquad \sum_{i=1}^{T} E[n_T(i)] = (1+\lambda)^{T-2}\left(2(1+\lambda)^{T-1} - 2 + (1+\lambda)^T\right)$$

Hence, we approximate the corresponding probabilities as follows

$$(7) \quad Pr(X_T = t) = \begin{cases} \frac{2\lambda(1+\lambda)^{t-1}}{2(1+\lambda)^{T-1}-2+(1+\lambda)^T}, & \text{if } t = 1, \ldots, T-1 \\ \frac{(1+\lambda)^T}{2(1+\lambda)^{T-1}-2+(1+\lambda)^T}, & \text{if } t = T. \end{cases}$$

As indicated above equation (7) is exact if $\lambda = 1$. Before we study the validity of the above approximations, we give the approximate moments of the distribution defined by equations (7). $E[X_T]$ is approximately given by

$$(8) \qquad E[X_T] \approx T - 2\frac{1+\lambda}{\lambda(3+\lambda)},$$

if $\frac{(1+\lambda)^{T-1}-1}{(1+\lambda)^{T-1}}$ is close to one. Under the same assumption the variance can be approximated as

$$(9) \qquad Var[X_T] \approx 2\frac{(1+\lambda)(4+3\lambda+\lambda^2)}{\lambda^2(3+\lambda)^2} + O\left(\frac{T}{(1+\lambda)^{T-1}}\right).$$

The variance is constant if T is large.

4. Testing the goodness of the approximation. This section describes the validity of the approximations made in the previous sections. Central to the formulae derived above is the assumption that the distribution of the times back to the mrca for each pair can be approximated by an average tree (e.g. Figure 2).

To test the goodness of formula (7) we performed a series of simulations for various probabilities λ of bifurcations and for different times T back to mt Eve. More explicitly, we have developed a computer program that generates an offspring population starting with the daughters of mt Eve. In each generation the number of offspring is determined according to the model of evolution described above. After T generations the process stops and the distribution of times back to the mrca is computed for the entire population at time T. This process is repeated independently. Table 1 shows our choice of parameters for five different values of λ.

Figures 3, 4, 5, 6 and 7 show the results of the simulations. It should be obvious that the approximation of the probability distribution predicted by equation (7) is reasonably good.

TABLE 1

Choice of parameters for the simulation results as depicted in Figures 3, 4, 5, 6 and 7.

Figure	growth rate λ	generations back to mt Eve	simulations
3	0.01	1,000	1,000
4	0.1	100	1,000
5	0.25	50	1,000
6	0.5	30	100
7	0.75	20	100

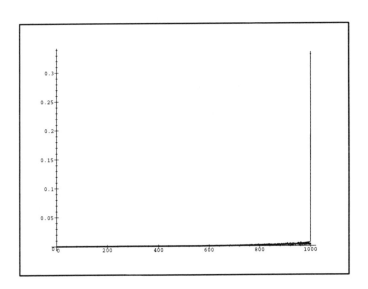

FIG. 3. $\lambda = 0.01$, *time since mt Eve equals 1000 generations. The dots indicate the distribution obtained as average of 1000 simulated population developments. The line represents the theoretical curve predicted from equation (7).*

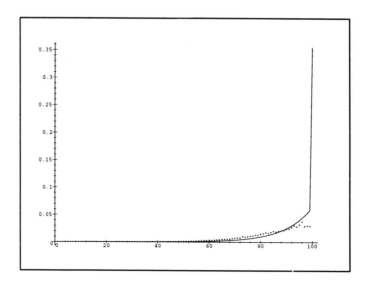

FIG. 4. $\lambda = 0.1$, *time since mt Eve equals 100 generations. See Figure 3 for details.*

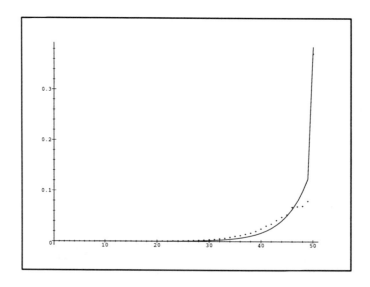

FIG. 5. $\lambda = 0.25$, *time since mt Eve equals 50 generations. See Figure 3 for details.*

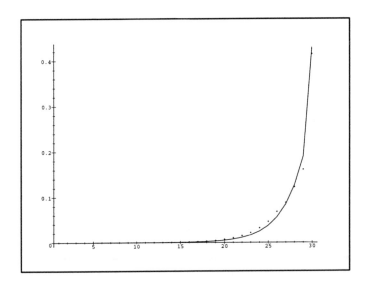

FIG. 6. $\lambda = 0.5$, *time since mt Eve equals 30 generations. The dots indicate the averaged distribution from 100 simulations. See Figure 3 for details.*

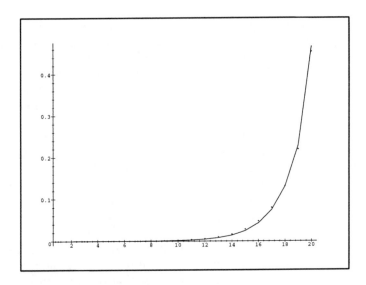

FIG. 7. $\lambda = 0.75$, *time since mt Eve equals 20 generations. The dots indicate the averaged distribution from 100 simulations. See Figure 3 for details.*

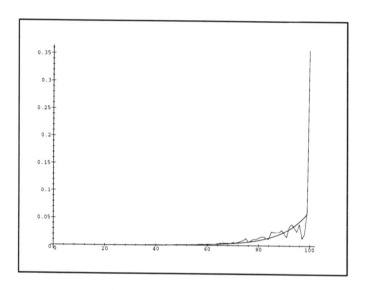

FIG. 8. *Averaged distribution of times back to the mrca for a sample of size n = 50 taken from a population that evolved according to the parameters used in Figure 4. The smooth curve displays the predicted probability of pairs of individuals with mrca t generations ago. The rugged curve represents the average of 100 simulations.*

The approximation gets better if λ is large. As λ approaches zero the fluctuations of the simulated curves increase. Independent of the choice of λ there seems to be the tendency that the theoretical curve underestimates the probabilities to observe two closely related individuals, whereas the probability to observe two individuals that had their mrca a long time ago is overestimated. The reasons for this particular deviation are as yet unclear.

It is remarkable that the probability to observe a pair of individuals in the population that had their mrca exactly T generations ago is in the range of 0.3 to 0.5. A fact that is more or less independent of the growth rate. This explains the observation that a reasonable fit of pairwise distances from a sample of n individuals to the distribution of distances for exactly two sequences from a growing population is observed [16,15].

For real data, however, there is a second approximation involved. It is impossible to sample the entire population. Instead only a small sample of size 20, 30 or 50 is taken, that is assumed to represent the population. Figure 8 shows for a representative simulation that the theoretical distribution is already in good agreement with a simulated distribution if a random sample of $n = 50$ individuals is taken from the entire population.

differences	0	1	2	3	4	5	6	7	8	9	≥ 10
observed	29	84	165	252	249	207	137	72	24	6	0
expected	25	96	186	241	237	187	123	70	35	16	9

TABLE 2

Distribution of pairwise differences for a sample of n = 50 individuals from the Finnish population (pairs observed). The last row displays the number of pairs one expects to see, if the pairwise difference distribution (10) is fitted to the data.

5. The Finnish population. So far we have only computed the distribution of times back to the mrca for each pair of individuals. To obtain a distribution of pairwise differences we have to assume a substitution process. For the remainder of the paper we suppose an infinite site model [23]. The process of nucleotide substitution is modelled as a Poisson process with parameter μ. Where μ is the substitution rate per sequence and per generation. More precisely, the probability $Pr(D = d|X_T = t)$ for d mutations depends on μ and t, where t is a complex function of λ in our model (see equation (7)).

The theorem of total probability and equation (7) allows us to compute the distribution of pairwise differences. The probability to observe $D = d$ differences is given by

$$(10) \quad Pr(D = d) = Pr(D = d|\lambda, \mu) = \sum_{t=1}^{T} Pr(D = d|X_T = t) \cdot Pr(X_T = t).$$

If we assume, that N_T, the number of contemporary individuals is known, equation (10) depends on λ and μ only. Hence given λ one can estimate μ from a sample distribution of pairwise differences or vice versa. λ and μ are compound parameters in the distribution (10), therefore it is not possible to estimate both parameters separately.

The number of expected pairwise differences is simply

$$(11) \qquad\qquad E[D] = 2\mu \times E[X_T]$$

Table 2 shows the distribution of pairwise differences from the hypervariable part number one (HV-1) of the control region of mtDNA for a sample of 50 individuals of the Finnish population (Sajantila *et al.*, unpublished data). Figure 9 shows the resulting frequency distribution. The sequences are 360 base–pairs long and we assume a substitution rate of $\mu = 0.0025$ [21]. We want to estimate the age of the Finnish mt Eve based on distribution (10). The distribution (Figure 9) is clearly unimodal, thus supporting the assumption that the population is growing. Moreover, as pointed out by Nevanlinna [13] the Finnish population shares some additional properties that makes it suitable for an analysis based on our model: There was apparently little mixing with other populations during historical times, the Finnish population has a uniform origin and the population size is increasing. Unfortunately the growth rate does not appear to be constant during

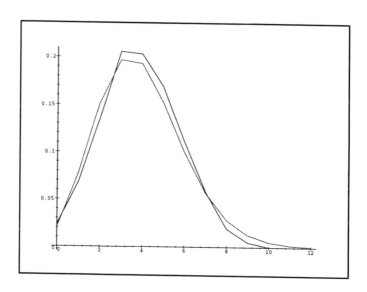

FIG. 9. *Fit of pairwise distance distribution to the observed pairwise distance distribution of a sample of size n = 50 taken from the Finnish population. For details see text.*

the history of the Finnish population [13], a fact that may limit the following conclusions. According to [13] we assume a population size of approximately five million individuals. Hence, the effective number of breeding females equals 2,500,000. Parameter estimation of λ can be obtained by minimizing the error function

$$\chi^2(\lambda, \mu) = \sum_{d \geq 0} \frac{(obs_d - kPr(D = d))^2}{kPr(D = d)},$$

where k is the sample size (i.e. the number of pairwise comparisons), obs_d is the number of observed pairs with difference d, and $Pr(D = d)$ equals the probability to observe d differences for the choice of parameters.

The third row of table 2 shows the result of the minimization procedure. The minimal χ^2 value is about 27.8, with an corresponding λ of 0.01714. Figure 9 illustrates the resulting frequency distribution. Both distributions almost superimpose each other. ¿From the relation $N_T = 2(1 + \lambda)^{T-1}$ we estimate that the mt Eve of the Finnish population lived about 20,675 (827 generations) years ago, assuming a generation time of 25 years. It is noteworthy to remember that the time back to the common ancestor of all Finns is not related to the time of the colonization of Finland. The colonization took place approximately 5,000 years ago [3]. At that time we estimate about 84,000 descendants from the mt Eve of the Finnish population. This estimate is most probably an overestimation, since our model supposes a constant λ. The demographic analysis indicates that the Finnish population has undergone a major expansion in the last two centuries [13]. However, the good fit of the observed distribution of pairwise differences with the expected one (Figure 9), gives us no reason to fit a more elaborate model.

Slatkin and Hudson [16] suggested a method to estimate the growth rate r of an exponentially growing population based on distribution of pairwise differences. If we equate their equation (9) with equation (11) the following relationship between λ (the probability to have two offspring) and the true growth rate r of the population is obtained

$$\frac{\ln(N_T/2)}{\ln(1+\lambda)} + 1 - 2\frac{1+\lambda}{\lambda(3+\lambda)} = \frac{1}{r}\ln(N_T r - \gamma),$$

where $\gamma \approx 0.577\ldots$ is the Euler–Mascheroni constant. This equation can be solved numerically. For the estimated value of $\lambda = 0.01714$, the growth rate of the population equals $r = 0.01321$. Based on the numerical value of r the number of females that lived 20,675 years ago is roughly equal to 45. By the time the ancestors of the todays Finnish population probably colonized Finland [3] the population consisted of approximately 92,000 females and 91 % were descendants of the mt Eve of all living Finns.

6. Summary and Discussion . We have outlined a method to compute the distribution of pairwise distances in an exponentially growing population. The parameters of the model are the substitution rate μ per DNA-segment and per generation and the probability λ that a female has exactly two offspring.

Based on this simplified model it is then possible to infer the actual value of λ from a set of observed sequence differences, if μ is known. The analysis of the distribution of pairwise distances of the Finnish population serves as an example.

In this particular example we have estimated that the most recent common ancestor of all living Finns existed about 20,000 years ago. Applying the theory of pairwise distances of exponentially growing populations as developed by Slatkin and Hudson [16], we estimated that at the time of the mrca of all Finns approximately 45 females had been living. Although the estimation does not contradict the history of the Finnish population, more work is certainly necessary to investigate the evolution of the Finnish population in more detail.

At present we are not able to put any error bars on these estimates. Figure 9 gives us the impression that the fit of the predicted and the observed distributions is reasonably good. This impression is supported by the extremely low χ^2 value. Since there are dependencies in the data, the χ^2 value cannot be used as a test statistics for the goodness of fit. One way to evaluate the goodness of fit (based on the minimizing criterion) is the simulation of an empirical χ^2 distribution for populations that evolve according to the parameters estimated from the data.

In this paper we have shown that an unimodal distribution can be explained by assuming an exponentially growing population. Unfortunately the reverse is not necessarily true, if we observe an unimodal distribution this may be due to a completely different evolutionary scenario (e.g. [15], [12]). In other words the distribution of pairwise differences is not sensitive enough to unveil the true history of the population. It is necessary to investigate the question of whether other statistics, for example based on comparisons of all possible triplets or quartets of sequences, provide more insights into the evolutionary history. In the context of phylogenetic tree reconstruction, such approaches have been helpful [6,7].

Acknowledgements: We would like to thank Antti Sajantila for fruitful discussions about the evolution of the Finnish population and for letting us use his unpublished sequence data. Finally, we appreciate Elizabeth Watson's help in improving the english manuscript. This research was supported by grant Ha1628/2-1 from the Deutsche Forschungsgemeinschaft.

REFERENCES

[1] BROWN,W. M., M. GEORGE, AND A. C. WILSON, *Rapid evolution of animal mitochondrial DNA*, Proc. Natl. Acad. Sci. USA 76:1971–1976, 1979.

[2] CANN, R. L., M. STONEKING, AND A. C. WILSON, *Mitochondrial DNA and human evolution*, Nature, 325: 31–36, 1987.

[3] CARPELAN, C., *Katsaus saamelaisten esihistoriaan*, In: Suomen väestön esihistorialliset juuret. Societas Scientiarum Fennica, 97 – 109, 1984.

[4] CROW, J. F. AND M. KIMURA, *An introduction to population genetics theory*, Alpha Editions, Edina MN, 1970.

[5] DI RIENZO A., AND A. C. WILSON, *The pattern of mitochondrial DNA variation is consistent with an early expansion of human population*, Proc. Natl. Acad. Sci. USA 88: 1597–1601, 1991.

[6] EIGEN, M., R. WINKLER–OSWATITSCH, AND A. DRESS, *Statistical geometry in sequence space: A method of quantitative comparative sequence analysis*, Proc. Natl. Acad. Sci. USA 85: 5913–5917, 1988.

[7] EIGEN, M., B. LINDEMANN, M. TIETZE, R. WINKLER-OSWATITSCH, A. DRESS, AND A. VON HAESELER, *How old is the genetic code? Statistical geometry provides an answer*, Science, 244: 673–679, 1989.

[8] HARDING, E. F., *The probability of rooted tree-shapes generated by random bifurcation*, Adv. Appl. Prob. 3: 44-77, 1971.

[9] HARPENDING, H. C., S. T. SHERRY, A. R. ROGERS, AND M. STONEKING, *The genetic structure of ancient human populations*, Current Anthropology, 34: 483–496, 1993.

[10] HUDSON, R. R., *Gene genealogies and the coalescent process*, Oxf. Surv. Evol. Biol. 7: 1–44, 1990.

[11] HUTCHINSON, G. E., *An introduction to population ecology*, New Haven and Londen, Yale Univ. Press 5th ed, 1980.

[12] MARJORAM P & P. DONNELLY, *Pairwise Comparisons of Mitochondrial DNA Sequences in Subdivided Populations and Implications for Early Human Evolu-*

tion, Genetics, 136: 673–683, 1987.

[13] NEVANLINNA, H. R., *The Finnish population structure – A genetic and genealogical study*, Hereditas 71: 195–236, 1972.

[14] ROGERS, A. R., *Error introduced by the infinite-sites model*, Mol. Biol. Evol. 9: 1181–1184, 1992.

[15] ROGERS, A. R., AND H. HARPENDING, *Population growth makes waves in the distribution of pairwise genetic differences*, Mol. Biol. Evol. 9:552–569, 1992.

[16] SLATKIN, M., AND D. HUDSON, *Pairwise comparisons of mitochondrial DNA sequences in stable and exponentially growing populations*, Genetics 129: 585–595, 1991.

[17] STONEKING, M., *Mitochondrial DNA and human evolution*, J. Bioenergetics and Biomembranes, 26:251–259,1994.

[18] TEMPLETON, A. R., *The "Eve" hypothesis: A genetic critique and reanalysis*, American Anthropologist, 95:51–72,1993.

[19] VIGILANT, L., R. PENNINGTON, H. HARPENDING, T. KOCHER, AND A. C. WILSON, *Mitochondrial DNA sequences in single hairs from South African population*, Proc. Natl. Acad. Sci. USA 86: 9350–9354, 1989.

[20] VIGILANT, L., M. STONEKING, H. HARPENDING, K. HAWKES, AND A. C. WILSON, *African population and the evolution of human mitochondrial DNA*, Science, 253:1503–1507, 1991.

[21] WARD, R. H., B. L. FRAZIER, K. DEW, AND S. PÄÄBO, *Extensive mitochondrial diversity within a single Amerindian tribe*, Proc. Natl. Acad. Sci. USA 88: 8720–8724, 1991.

[22] WARD, R. H., A. REDD, D. VALENCIA, B. L. FRAZIER, AND S. PÄÄBO, *Genetic and linguistic differentiation in the Americas*, Proc. Natl. Acad. Sci. USA 90: 10663–10667, 1993.

[23] WATTERSON, G. A., *On the number of segregating sites in genetical models without recombination*, Theor. Popul. Biol 7: 256–276, 1975.

BRANCHING AND INFERENCE IN POPULATION GENETICS

NEIL O'CONNELL*

Abstract. The probabilistic structure of the genealogy in branching processes is described; in particular, we present an analogue of Kingman's coalescent for near-critical branching processes. This result is applied to the problem of estimating the age of our most recent common ancestor using samples of mtDNA taken from contemporary humans. We also discuss more general issues concerning the use of models for making inference about the past of a population.

Key words. Branching, genealogy, evolution, Eve.

1. Introduction. One of the most exciting detective stories in modern times began with Darwin's theory of evolution. Scientists have since been investigating the mysteries of human origin with remarkable success using natural clues such as fossils, bones and, more recently, molecules. The molecule in question is DNA, which is difficult to interpret but potentially the most informative clue of all: we simply have to learn how to read it. Given that we understand the mechanisms by which DNA evolves, we can attempt to make inference about the past using DNA samples. In fact, geneticists have being doing this for some time.

In 1991, Vigilant *et al.* [24] claimed to have found molecular evidence for a recent African origin, using samples of mitochondrial DNA (mtDNA) from contemporary humans. Their estimate of the age of our most recent common mitochondrial ancestor, more affectionately referred to as 'Eve', is 200,000 years. It has since become a very controversial topic: there is a vast literature supporting variations of this hypothesis and an equally vast literature in opposition, finding fault in the methods used and quoting contradictory fossil evidence. Popular accounts can be found in [12,22,25]. This controversy is a symptom of the fact that here we have a difficult statistical problem.

One of the difficulties is that to make inference about the age and whereabouts of a most recent common ancestor it seems necessary to make assumptions about the genealogical dynamics of the population. This fact does not seem to be fully appreciated. For example, the approach of Vigilant *et al.* [24] assumes that the true phylogeny relating individuals in a sample is the one which requires fewest mutations in order to explain the variety of DNA types observed: this method of reconstructing phylogenies is known as *parsimony*, and has been subjected to criticism on various grounds (see, for example, [6]). One of the problems with this approach is

* Dublin Institute for Advanced Studies, 10 Burlington Road, Dublin 4, Ireland. Research supported by NSF grants MCS90-01710 and DMS91-58583, and by grants from EOLAS and Mentec Computer Systems Ltd.

that often there are several, equally 'parsimonious', possible phylogenies, each leading to a different conclusion; another is the fact that there is no theoretical basis for parsimony. There is also the added difficulty of rooting the inferred trees.

Some authors (see, for example, [8]) have attempted to use maximum likelihood methods, but these are difficult to formulate properly and become very complicated when a large sample is used.

In this paper we argue that it is not necessary to construct a tree in order to make inferential statements about the *age* of Eve. We will present an alternative approach, where the genealogy is modelled via a branching process. We do not attempt to consider the whereabouts of Eve as this would require a spatial component in the model, which we haven't included.

The idea of adopting a statistical model for genealogical dynamics in order to make inference about the past has also been applied by Lundstrom, Tavaré and Ward [13] and Griffiths and Tavaré [7], where traditional (Wright-Fisher/Moran type) population genetics models are used. The application of branching processes in evolution is a more recent development, and seems very promising; Jagers, Nerman and Taib [9,11,19,20,21] have done a considerable amount of work on this topic. For general background on biological applications of branching processes, see [10].

The relationship between branching and traditional models for evolution is well known (see, for example, [3,15,18]). Qualitatively, both models display similar behaviour unless there is some kind of spatial structure, in which case the behaviour can be radically different. We will discuss these issues in §5.

The outline of the paper is as follows. In §2 we present some informal arguments justifying the use of a statistical model for population dynamics. In §3 we describe the probabilistic structure of genealogical trees in branching processes; in particular we present an analogue of Kingman's coalescent as an approximation for the family tree in a near-critical branching process. In §4 we apply these results to the problem of estimating the age of Eve. Finally, in §5, we discuss the relationship between branching and traditional models for evolution.

2. Using a model to estimate the age of Eve. Suppose we have a model that describes the statistical nature of the past evolution of the human species, and suppose one of the parameters of the model is taken to be the time back, T, to Eve. Suppose further that there exists a constant γ such that the expected time back to the most recent common ancestor of two randomly chosen individuals alive today is γT. Then, under the usual assumptions of neutral evolution, the expected divergence between two randomly chosen individuals is $\delta \gamma T$, where δ denotes the mean rate of divergence along distinct lines of descent. (By 'divergence' we mean some measure of genetic distance. For example, in the infinite alleles model with mutation rate θ, $\delta = 2\theta$. In practice, the divergence between two DNA sequences is usually expressed as the proportion of sites that differ; the 'usual' assumptions are that substitutions occur at fairly steady rates along lineages, and that the occurrence of substitutions is independent of demography.) It follows that the expected mean pairwise divergence $\bar{d}$ in a random sample of individuals is also given by $\delta\gamma T$.

If δ and γ are known, this yields a straightforward moment estimate

$$(2.1) \qquad \hat{T} := \frac{\bar{d}}{\delta\gamma}.$$

In practice the parameters δ and γ are *not* known; however, given reliable estimates, the estimator (2.1) can be approximated.

In the next section we show that for branching processes there is such a constant γ that depends on the mean growth rate of the population. This fact is combined with (2.1) in §4 to provide a method for the simultaneous estimation of the mean population growth rate and T, the age of Eve.

3. The genealogy of branching processes. Let Z be a Markov branching process with mean lifetime 1, offspring distribution ν and let ξ be a realisation of ν. Fix $t > 0$, and for each $0 \leq s \leq t$, define $N_t(s)$ to be the number of individuals alive at time s with descendents alive at time t. The process N_t is called the *reduced branching process*, and can be thought of as the family tree relating the individuals alive at time t. It is also referred to as the *reduced family tree*. Note that N_t is also a Markov process.

When t is large, the only case where the genealogy is non-trivial is when the branching process is close to critical ($E\xi \approx 1$). To make this more precise, suppose $E\xi = 1+\alpha/t$, for some $\alpha \in \mathbb{R}$. In [14] it is shown that when t is large and the time units are taken as t generations, the reduced process can be approximated by a linear pure birth process $\{N(r),\ 0 \leq r < 1\}$ with jump rate $b(\alpha, r)N(r)$ at time r, where

$$(3.1) \qquad b(\alpha, r) = \begin{cases} \alpha (1 - e^{-\alpha})^{-1} (1 - r)^{-1} & \alpha \neq 0, \\ (1 - r)^{-1} & \alpha = 0. \end{cases}$$

This result provides a complete probabilistic description of the family tree relating those individuals alive at time t; it is thus the branching process analogue of Kingman's coalescent. Note that, just as the coalescent can be used to speed up simulations, so can this result. It can also be used to describe the genealogical relationship between two randomly chosen individuals. If S_t denotes the time back to the most recent common ancestor of two individuals chosen randomly from the population at time t ($S_t = t$ if they have no common ancestor), then [14, Theorem 2.3] the laws of S_t/t conditional on $\{N_t(0) = x\}$ converge weakly, as $t \to \infty$, to a limiting law

μ_x, say, on $(0, 1]$, defined by

$$(3.2) \qquad \mu_x(0, r) = \frac{2q_r^x}{(x-1)!} \left\{ -(1-q_r)^{-x} - F(x-1, 1-q_r) \right\},$$

for $0 \le r \le 1$, where

$$(3.3) \qquad q_r = \frac{e^{-(1-r)\alpha} - e^{-\alpha}}{1 - e^{-\alpha}},$$

and $F : \mathbb{Z}_+ \times (0, 1) \to \mathbb{R}$ is defined by

$$(3.4) \qquad F(n, y) = \frac{\partial^n}{\partial y^n} \left\{ \frac{\log(1-y)}{y^2} \right\}.$$

It follows (applying bounded convergence) that

$$(3.5) \qquad ES_t/t \to \gamma_x(\alpha),$$

where $\gamma_x(\alpha)$ denotes the mean of μ_x. In other words, if t is large,

$$(3.6) \qquad ES_t \simeq \gamma_x(\alpha)t.$$

We remark that γ is increasing and $\gamma(\alpha) \nearrow 1$ as $\alpha \to \infty$.

In the supercritical case, when the process is not 'close' to critical, individuals are typically distantly related. For example, it follows from results of Bühler [1], Zubkov [26] and Durrett [5], that if $E\xi > 1$, $(S_t/t|\ Z(t) > 0) \to 1$ in probability as $t \to \infty$. This fact can be extrapolated (in some sense) from (3.1) by letting $\alpha \to \infty$. In the subcritical case ($E\xi < 1$) individuals typically have very recent ancestors: in this case, $(S_t/t|\ Z(t) > 0) \to 0$ in probability [5,26]. Again this can be seen from (3.1) by letting $\alpha \to -\infty$. Essentially what we are doing here is describing the continuum of non-trivial possibilities in between, which arise when the process is close to critical.

4. Estimating the age of Eve. It is thought that mtDNA is inherited primarily from the mother. This assumption allows us to restrict our attention to single-sex populations, and so we are not forced to make questionable assumptions about the mating behaviour of people. We make the usual assumptions that mutations occur randomly along lines of descent at a constant rate, and that these mutations are selectively neutral. The divergence rate is very small, so over the time period we are considering here (the post-Eve period) we will assume that each substitution produces a new type, that is, reverse substitutions do not occur. Thus, if the most recent common ancestor of two individuals died s million years ago, the number of differences between their mtDNA types will be approximately Poisson with mean $2us$, where u is the substitution rate (in units of number of substitutions per million years). Now suppose two individuals are sampled randomly from the current population, and δ denotes the rate of divergence (in units of percentage divergence per million years). Note that if l denotes the sequence length, then $\delta = 2u/l$. If we have a model for the genealogical structure of the population, then the expected amount of divergence between the mtDNA sequences of the two individuals will be equal to the expected time back to the common ancestor of the two individuals (under our model, in units of millions of years), multiplied by the divergence rate, δ.

We will assume that the (effective) female population size follows a Markov branching process Z with mean offspring $1+\alpha/T$, where $T = T_a/\lambda$; T_a is the time to our most recent common ancestor, λ is the mean effective lifetime (or *generation time*) and $\alpha \in \mathbb{R}$ is our 'growth' parameter.

If we start time at the death of Eve then, in the notation of §3, $N_T(0) = 2$. (Eve, by definition, had at least 2 daughters with descendents alive today, and [14, Theorem 2.2] tells us that 3 such daughters is extremely unlikely: $N_T(0-) = 1$ and $N_T(0) \geq 2$ together imply that $N_T(0) = 2$ with high probability when T is large.) Note that $Z(T)$ is the current (effective) female population size.

Using our approximation results, we can simultaneously estimate α and T, based on the observations $Z(T)$ and the average pairwise divergence in a random sample of n contemporary individuals $\overline{d}_n$. We will assume for the moment that the divergence rate δ is known. Denote by λ the mean effective lifetime of an individual. By [14, Theorem 2.1] (an exponential limit law for near-critical branching processes),

$$(4.1) \qquad E(Z(T)\mid N_T(0) = 2) \simeq \frac{\sigma^2 T_a}{\lambda\alpha}(e^\alpha - 1).$$

We also have, by (3.6),

$$(4.2) \qquad E\left(\overline{d}_n \mid N_T(0) = 2\right) \simeq \delta T_a \gamma(\alpha),$$

where

$$(4.3) \quad \gamma(\alpha) = 1 - 2\alpha^{-1} \int_0^1 \frac{u}{(1-u)^3(u+\kappa(\alpha))}\left[1 - u^2 + 2u\log u\right] du,$$

and

$$(4.4) \qquad \kappa(\alpha) = \frac{e^{-\alpha}}{1 - e^{-\alpha}}.$$

Note that $\gamma(\alpha)$ is positive and increasing in α, $1/3 < \gamma(\alpha) < 1$, and $\gamma(\alpha) \nearrow 1$ as $\alpha \to \infty$.

For the simplest moment based estimates, assuming that δ, σ^2 and λ are known, just set

$$(4.5) \qquad Z(T) = \frac{\sigma^2 \hat{T}_a}{\lambda\hat{\alpha}}(e^{\hat{\alpha}} - 1),$$

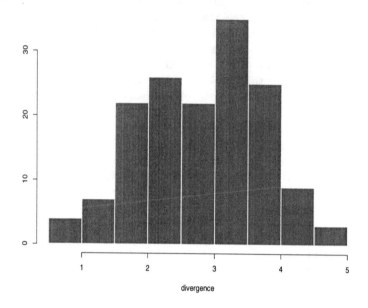

FIG. 4.1. *Pairwise divergences among sample of 19 individuals.*

$$(4.6) \qquad\qquad \hat{T}_a = \frac{\bar{d}_n}{\delta\gamma(\hat{\alpha})},$$

and solve for $(\hat{\alpha}, \hat{T})$. Although σ^2 is unknown, when α is sufficiently large the actual value (within reason) will not affect the estimates considerably. (This is due to the dominating exponential term in Equation 4.5.) The same is true for λ.

Note that in theory this approach assumes that α is small relative to T. However, since a large value of α corresponds to the (significantly) supercritical case, the estimate of T obtained from (4.6) will still make sense if the estimated value of α is large.

We would now like to apply our method to some data: but where does one find a *random* sample of individuals? Strictly speaking this is simply not available, as yet. However, we will do our best with what we have.

Of the 189 individuals considered by Vigilant *et al.* [23], we have hand-picked a somewhat representative sub-sample of 19, without being deliberately biased in any way. The larger the sub-sample, the less representative it becomes; the smaller it is, the less useful it becomes. Our sample consists of 6 Asians, 1 Native Australian, 1 Papa New Guinean, 6 Europeans and 5 Africans.

A histogram of the 171 pairwise divergences in this sample is shown in Figure 4.1. The average divergence was found to be 2.8%.

TABLE 4.1
Estimates for α and T_a.

$\lambda Z(T)/\sigma^2$	δ	$\hat{\alpha}$	$\hat{T}_a$
12.5 billion	1.8	11.33	1,706,103
	2.7	11.77	1,113,265
	4	12.21	762,437
5 billion	1.8	10.31	1,722,531
	2.7	10.76	1,143,245
	4	11.2	768,607
30 billion	1.8	12.29	1,693,286
	2.7	12.94	1,125,364
	4	13.17	757,516

In June 1992, according to the *Population Reference Bureau Estimates*, the human population size was approximately 5.412 billion. This gives us about 1 billion as a rough estimate for the current effective female population size, assuming that about half the population is female, and that the current female population represents approximately 2.7 generations. We will soon see that our estimates are quite insensitive to variations in this figure, so we needn't be very exact.

Note that the estimates $\hat{\alpha}$ and $\hat{T}_a$ are determined by $\lambda Z(T)/\sigma^2$ and δ; these are shown in Table 4.1, for various different values of $\lambda Z(T)/\sigma^2$ and δ. If $Z(T) = 1$ billion, $\sigma^2 = 2$ and $\lambda = 25$, then $\lambda Z(T)/\sigma^2 = 12.5$ billion. Although these choices seem somewhat arbitrary, we can see from Table 4.1 that any kind of realistic deviations from these values will have little or no effect on the estimates. The most important parameter is δ, the rate of divergence, and as yet there is no universally agreed 'best estimate' for δ. The values used in Table 4.1 are based on human-chimpanzee comparisons using a simple correction for multiple substitutions and possible dates of 9, 6 and 4 million years for human-chimpanzee divergence.

We conclude this section with some remarks on possible developments. To derive our estimates for the growth rate, α, and the age of Eve, T_a, we simply calculated the expected current population size and the expected average pairwise divergence in a sample of contemporary individuals, and assumed the other parameters were known. We are therefore not fully utilising the information contained in the sample. It might be helpful to know more about the joint distribution of the pairwise divergences (d_{ij}), or the joint distribution of the respective frequencies of distinct types, in a finite sample. The latter would be analogous to *Ewens' sampling formula* for the infinite-alleles Wright-Fisher model for neutral evolution. Ewens' sampling formula is not applicable to the Eve problem because it is based on the assumption that the population size is constant over time.

In particular, it may be possible to estimate α, T_a and δ simultaneously, without having to rely on human-chimpanzee comparisons, thus avoiding the assumption that the rate of divergence has been constant ever since the human and chimpanzee lines diverged.

Taib [20,21] has made some progress in this direction by obtaining an expression for the the asymptotic proportion of alleles (types) with exactly j representatives in the population, for a supercritical branching process with neutral mutations. Unfortunately, this result is not directly applicable here. The recent work of Pitman [16,17] on sampling distributions may be very useful here.

5. Branching versus traditional models. The essential difference between branching and traditional models is that the latter assumes the total population size to follow some deterministic function over time, in which case the genealogy is described by a corresponding time-change of Kingman's coalescent. The relationship between the two is given by the fact that if we condition a branching process on the evolution of its total population size, we have a traditional population genetics model (this has been known in the genetics literature for some time; for rigorous formulations in the context of Fleming-Viot processes, see [3,15,18]).

For this reason, the qualitative behaviour of the two models is similar if the mean population growth rates are comparable. However, if a spatial component is introduced—for example, if individuals are allowed to immigrate between geographically separated colonies—the two models can exhibit radically different behaviour. This is because in a branching process, if the dynamics of migration and reproduction are independent, the spatial component does not influence the genealogy at all; in the corresponding traditional model (the general stepping stone model) where the population size in each colony is restricted to follow a deterministic course, the genealogy is strongly influenced by the degree of separation between colonies [2,4].

A more realistic model would perhaps lie somewhere in between: population dynamics are certainly affected by geographical factors, but not to the extent which is assumed in traditional models. Although the details of such a model might be difficult to ascertain, a lot could be learned by studying the implied qualitative behaviour.

Acknowledgements. The author would like to thank all the participants of this workshop for many valuable discussions, and the organisers for making it happen. Thanks also to Peter Donnelly for helpful suggestions.

REFERENCES

[1] W. Bühler. The distribution of generations and other aspects of the family structure of branching processes. In *Proc. Sixth Berkeley Symp. Math. Statist. Prob., vol. III*, pages 463–480. University of California Press, Berkeley and Los Angeles, 1972.

[2] D. Dawson. Hierarchical and mean-field stepping stone models. Presented at this workshop.

[3] P. Donnelly and T.G. Kurtz. Particle representations for measure valued population models. Preprint.

[4] P. Donnelly and P. Marjoram. Population evolution and genealogy. Presented at this workshop.

[5] R. Durrett. The genealogy of critical branching processes. *Stoch. Proc. Appl.*, 8:101–116, 1978.

[6] J. Felsenstein. Cases in which parsimony or compatibility methods will be positively misleading. *Syst. Zool.*, 27:401–410, 1978.

[7] R. Griffiths and S. Tavaré. Ancestral inference in population genetics. *Statist. Sci.*, submitted.

[8] M. Hasegawa and S. Horai. Time of the deepest root for polymorphism in human mitochondrial DNA. *J. Mol. Evol.*, 32(1):37–42, 1990.

[9] P. Jagers. The growth and stabilisation of populations. *Statist. Sci.*, 6:269–283, 1991.

[10] P. Jagers. *Branching Processes with Biological Applications*. Wiley, Chichester, 1975.

[11] P. Jagers, O. Nerman, and Z. Taib. When did Joe's great ... grandfather live? Or on the timescale of evolution. In: I.V. Basawa and R.L. Taylor (eds.), Selected Proceedings of the Sheffield Symposium on Applied Probability. IMS Lecture Notes–Monograph Series, Vol. 18, 1991.

[12] L. Krishtalka. *Dinosaur Plots and Other Intrigues in Natural History*. Avon Books, New York, 1989.

[13] R. Lundstrom, S. Tavaré, and R.H. Ward. Estimating substitution rates from molecular data using the coalescent. *Proc. Natl. Acad. Sci. USA*, 89:5961–5965, 1992.

[14] Neil O'Connell. The genealogy of branching processes and the age of our most recent common ancestor. To appear in *Adv. Appl. Prob.*, July 1995.

[15] E.A. Perkins. Conditional Dawson-Watanabe processes and Fleming-Viot processes. Seminar in Stochastic Processes, 1991.

[16] J.W. Pitman. A two-parameter version of Ewens' sampling formula. Preprint.

[17] J.W. Pitman. Random discrete distributions invariant under size-biased permutation. *J. Appl. Prob.*, 1993. To appear.

[18] T. Shiga. A stochastic equation based on a Poisson system for a class of measure-valued diffusion processes. *Jour. Math. Kyoto. Univ.*, 30(2):245–279, 1990.

[19] Z. Taib. *Labelled Branching Processes with Applications to Neutral Evolution Theory*. PhD thesis, Chalmers University of Technology, Sweden, 1987.

[20] Z. Taib. The most frequent alleles in a branching process with neutral mutations. Preprint.

[21] Z. Taib. Branching processes and evolution. Presented at this workshop.

[22] A.G. Thorne and M.H. Wolpoff. The multiregional evolution of humans. *Scientific American*, April:76–83, 1992.

[23] L. Vigilant, R. Pennington, H. Harpending, T.D. Kocher, and A. Wilson. Mitochondrial DNA sequences in single hairs from a southern African population. *Proc. Natl. Acad. Sci. USA*, 86:9350–9354, 1989.

[24] L. Vigilant, L. Stoneking, H. Harpending, K. Hawkes, and A. Wilson. African populations and the evolution of human mitochondrial DNA. *Science*, 253:1503–1507, 1991.

[25] A.C. Wilson and R.L. Cann. Recent African genesis of humans. *Scientific Amer-*

ican, April:68–73, 1992.

[26] A. Zubkov. Limiting distributions of the distance to the closest common ancestor. *Theor. Prob. Appl.*, 20:602–612, 1975.

HUMAN DEMOGRAPHY AND THE TIME SINCE
MITOCHONDRIAL EVE

PAUL MARJORAM* AND PETER DONNELLY†

Abstract. A non-linear timescaling allows coalescent processes to be applied to populations of variable size. The dependence of the time to the most recent common ancestor on features of the population size is studied, and for a range of beliefs about human population sizes the distribution of the time since mitochondrial Eve is considered. Current beliefs about these sizes seem incompatible with recent estimates of the time since Eve. Population structure may also be incorporated into variable population sized models. Broadly speaking, increasing population structure increases the mean and the variability of the time since Eve.

1. Introduction. The relationship between the genealogical history of a population and its current genetic composition is now well established (e.g. Tavaré 1984, Hudson 1990). On occasions the genealogy itself is of interest, for example in studying times to a common ancestor. Otherwise it is central to understanding the structure and variability of many genetic data sets. The theory is now well developed in the case of the evolution of populations of constant size. Many real populations, in particular the human population, have evolved with variable (typically increasing) population sizes. Recent attention has been paid to the effects of such variability on some aspects of sample genealogy, in particular those which effect the distribution of pairwise genetic differences (Slatkin and Hudson 1991, Rogers and Harpending 1992, Marjoram and Donnelly 1994, Rogers 1995 and references therein). Our purpose here is to focus on the consequences of variation in population size and of geographical population structure, and on another aspect of genealogy, the time since the most recent common ancestor of the population.

The coalescent (Kingman 1982a,b,c) provides a robust description of the genealogy of samples from neutral haploid populations. In the coalescent approximation, it is necessary to measure time so that one unit of time in the coalescent accounts for approximately M generations in the real population, where M is the haploid population size. The coalescent still applies to (large) populations of variable size, the only difference being that the transformation from "coalescent time" to "real time" is more complicated, depending on the sequence of population sizes in a non-linear way. In section 3 we describe the particular time transformation which arises in connection with a large class of models.

The theoretical work here is motivated by a particular quantity of interest in human evolution: the time since Eve, the most recent common ancestor (MRCA) of human mitochondrial DNA. We argue that insight

---

* Department of Mathematics, Monash University, Clayton 3168, Australia.

† Departments of Statistics, and Ecology and Evolution, University of Chicago, 5734 University Avenue, Chicago, Il 60637 USA.

into this quantity is related both to analyses of mitochondrial DNA data from current human populations and to theoretical modelling in population genetics. From the second of these perspectives, we study the way in which the distribution of time since a common ancestor depends on properties of the sizes, through time, of the population under consideration. We consider the way in which aspects of the distribution depend on the qualitative behaviour of the sequence of population sizes in the case when the latter are initially small and approximately constant, before a possible bottleneck, followed by a period of exponential growth to current values. Simulation studies are used to provide information about the distribution of the time since Eve under various scenarios concerning population size and structure which might be thought plausible for human evolution.

Application of the theory to real populations is complicated by the fact that most populations (including humans) evolve in a geographically structured setting. The paper investigates the qualitative effects on some aspects of genealogy of various beliefs about the effects of population structure. It is argued that some general insights into human evolution are still possible.

Throughout we assume that the population under consideration is haploid and that its evolution is neutral. Of course standard arguments mean that some of the results apply in the (neutral) diploid case.

2. Statistical perspective.

Empirical studies of the time since human mitochondrial Eve use data on observed variation in the human mitochondrial genome in extant populations. See for example Cann *et al.* (1987), Vigilant *et al.* (1991), and Hasegawa and Horai (1991). Templeton (1993) provides a review and critique. In terms of statistical framework, most of these empirical studies (implicitly or explicitly) treat the value of the time in the current realization of evolution as a (fixed) parameter which is to be estimated, and inference takes the form of reports of point or interval estimates of its value.

In our view, such a framework is inappropriate. The quantity, T_{Eve}, in which we are interested, is not a parameter in the usual sense. Rather, it is a random variable, where the randomness relates to the fact that it takes different values in different realizations of evolution. Further, evolutionary models specify (in principle) the probability distribution of T_{Eve}. Such distributions represent our uncertainty about T_{Eve} (as a function of parameters of the model) before observing data, on the assumption that the model accurately describes evolution.

Now suppose we observe some data, D, which results from a (in our case, the single) version of evolution. Within the framework of classical statistics, it appears appropriate to report the conditional distribution of T_{Eve}, *given* the data D. A Bayesian statistician would reach the same conclusion, with different terminology. The (pre-data) distribution of T_{Eve} resulting from evolutionary modelling would lead to a prior distribution for this quantity. This would then be updated, given the data, to a posterior distribution. Note that for a given evolutionary model, these (classical and Bayesian) procedures would be identical in principle.

One consequence of this viewpoint is that interval estimates of T_{Eve} from genetic data must incorporate several sources of uncertainty. The first is model uncertainty: allowance for the underlying model not in fact (sufficiently accurately) describing the relevant evolutionary processes. The second concerns uncertainty over the values of parameters in the models, for example mutation rates and demographic features of the population. The third is the variability inherent in the conditional distribution of T_{Eve} given the data. (In most calculations, this third level of uncertainty implicitly assumes that the model, and values of parameters involved, are correct.) We note that it is far from clear that the interval estimates of T_{Eve} in many published studies do incorporate each of these levels of uncertainty. In the absence of such an approach, the intervals themselves are difficult to interpret. Furthermore, they may substantially under-represent the uncertainty associated with the estimates.

Implementation of the framework just described is far from straightforward. Although implicitly specified by an evolutionary model, the distributions of interest (notably the joint distribution of T_{Eve} and the data D, or the marginal distribution of D) is not known in explicit form. Griffiths and Tavaré (1994b) used Monte Carlo likelihood methods to investigate the conditional distribution of the time since the most recent common mtDNA ancestor of a sample from the Nuu-Chah-Nulth Amerindian population, on the basis of mitochondrial DNA sequence data. The evolutionary model was based on the coalescent with variable population size. For this data set, they had used these techniques elsewhere (Griffiths and Tavaré 1994a) to show that there was little evidence for a population expansion (see also Griffiths and Tavaré 1994b), and their inference for T_{Eve} used a constant population size model.

In connection with the ancestor of the entire population of human mitochondrial DNA, different assumptions about human demography would seem appropriate. It does not seem reasonable to assume that the human population size has been constant throughout its evolution. In addition, it may be appropriate to consider evolutionary models which allow for geographical substructure of the population.

At present, estimation of the conditional distribution of T_{Eve}, given mtDNA data, does not seem realistic for evolutionary models incorporating both variation in population size and geographical structure. In lieu of this, our concern here is with understanding the way in which the unconditional distribution of T_{Eve} depends on demographic structure of the population. This may be of inherent interest. If, in addition, reliable information were available about the value of T_{Eve} for our realization of evolution, it may shed light on the plausibility of competing demographic scenarios for the human population.

One traditional approach is to equate the population of varying size with one of constant size M_e, where M_e is an appropriate "effective" population size. While some aspects of the two populations will be the same, many will not and this approach will in general be misleading. In particular it is inappropriate to use this technique to investigate mitochondrial Eve.

3. The coalescent process in variable-sized populations. The behaviour of population genealogies in large constant-sized populations is well understood. If we look back through time and trace the ancestry of a particular group of individuals its behaviour may be accurately described via a stochastic process known as the coalescent. This keeps track of the ancestry of the group. It can be thought of as a model for the genealogical tree associated with the individuals. An event corresponds to the coalescence of two ancestral lines (i.e. an occasion when two ancestors of the sample themselves share an ancestor). We will describe the process in terms of a haploid population of individuals. Unless recombination is relevant, it also applies to the haploid genes in diploid populations. The coalescent is a continuous-time process but can also be used to model populations evolving through discrete generations. For further details see for example Tavaré (1984), or Hudson (1990).

Our interest here centres on using the coalescent to approximate the behaviour of real populations and in particular to investigate properties of the distribution of time to a MRCA. For convenience we will assume our population is evolving through discrete generations: we label the present generation as time 0 and denote the size of the population t generations ago by $M(t)$.

In a constant-sized population we have $M(t) = M(0) = M$ (say) for all t and the coalescent approximation proceeds via a rescaling of time. We consider the period of time between each generation as counting for $1/(M\sigma^2)$ units of time in the coalescent process, where σ^2 is the variance of the offspring number distribution. With this time-scaling we can "translate" events in the coalescent process into those in the real population. Henceforth, for convenience, we assume $\sigma^2 = 1$, as is the case in the well-known Wright-Fisher model. Thus an event happening at time s in the coalescent occurs in generation

$$\tilde{\tau}(s) = \min_t \left\{ t : \sum_{i=1}^{t} \frac{1}{M} = \frac{t}{M} \geq s \right\} = sM$$

in the corresponding approximation to the real population. In a constant-sized population, s units of time in the coalescent correspond to sM generations in real-time. Formal details of the argument may be found in (Kingman 1982c).

The position is more complicated for populations whose size changes with time. Kingman (1982c) proved that under certain assumptions about the mechanism governing the changes in population size, (and the fact that these sizes were large) the coalescent still gives a good approximate description of genealogy, provided a particular non-linear timescaling (described at (3.1) below) is used. Informally, the time between generations t and $t+1$ accounts for $1/M(t+1)$ units of coalescent time. This class of models includes those in which the population sizes follow a deterministic sequence (see also Griffiths and Tavaré 1994b and references therein).

Donnelly and Kurtz (1996) consider a very general class of (neutral) population processes. The class includes as special cases processes which arise from traditional genetics models (with possibly variable population sizes) and branching process models. They prove that for this general class, conditional on the realized values of the population size process, the genealogy of the population is (asymptotically) described by a time change of the coalescent. In this general setting, however, the functional form of the time change depends on aspects of the dynamics of the random process governing the size of the population. In particular, the time scaling which we describe below at (3.1) which has, to our knowledge, formed the basis for all work on the application of coalescent theory with variable population size to problems in human evolution, *is only appropriate under specific assumptions about the dynamics of the population.* Loosely speaking, the timescaling which is conventionally used applies to populations which develop as branching processes or as branching processes with fixed or independently varying population sizes. (Traditional genetics models can be thought of as branching processes conditioned to take particular values for the total population size, e.g. Karlin and McGregor 1962, 1965.) The conventional timescaling, and hence any results derived from it, *may not apply for large classes of models which might be thought plausible for human population dynamics.*

Thus any progress on the analysis of variable population size models requires specific assumptions about the dynamics of the process describing the total population size. In the interests of comparability with other authors, and definiteness, we will assume throughout the remainder of this paper that the demography of the populations we are considering is governed by one from the class of models for which the "usual" time change arguments are valid. We will explore elsewhere the consequences for the interpretation of human molecular genetic data of other demographic scenarios. Our perspective throughout this paper (motivated by applications to human evolution) will be to condition on the realized values of the population sizes.

Recall that the population size t generations ago is denoted by $M(t)$. Following the discussion above, it follows (e.g. Kingman 1982c) that an event at time s in the coalescent now occurs at time

$$(3.1) \qquad \tau(s) = \min_t \left\{ t : \sum_{i=1}^{t} \frac{1}{M(i)} \geq s \right\}$$

in our approximation to the real population. We can thus still describe the evolution of a real population via the coalescent process, then transform time via (3.1) from "coalescent time" back to numbers of generations.

It is on this basis that we simulate the behaviour of samples from populations whose size varies through time. We consider a sample of size n. The time taken for the coalescent process to decrease from K to $K-1$ is exponentially distributed with parameter $\binom{K}{2}$, so the total coalescent-time to reach the MRCA is given by

$$ s = \sum_{i=2}^{n} T(i) \qquad \text{where } T(i) \sim \text{ exponential} \left(\binom{i}{2} \right). $$

We then translate s into real-time (i.e. a number of generations) and record that for this realization $\tau(s)$ generations were required for a MRCA to be obtained. A large number of repetitions then provides us with an empirical estimate of the distribution of time to MRCA.

4. Time to the most recent common ancestor.

4.1. General conclusions. We now investigate the behaviour witnessed in simulations of the MRCA problem. With a view to drawing conclusions about human populations we typically simulate a population which has been of roughly constant size for a period of its history before undergoing exponential growth. For instance, for humans, one might assume growth to have begun 50Kyr. before the present (b.p.) (Weiss 1984). In later simulations we will vary this assumption. Since the time between generations $-t$ and $-(t+1)$ (i.e. between t and $t+1$ generations ago) is considered to take $1/M(t+1)$ units of coalescent-time, the values taken by $\{M(t)\}_{t=0,1,2,...}$ are of great importance. These exact values are subject to considerable debate. We begin by presenting a series of broad conclusions drawn from the simulations indicating the qualitative ways in which changes in the population sizes affect the distribution of time to MRCA. Later we investigate the consequences of particular population sizes. In the theory and simulations, time is measured in units of generations. When we come to discuss distributions of times based on specific scenarios for human demography, we will assume that each generation occupies 20 years.

We are able to draw the following qualitative conclusions about the dependence of the time to the MRCA on modelling assumptions.

1. **General form.** There are two general shapes for the density function of the time to the MRCA depending on whether the MRCA occurs during the period of approximately constant population size (as seems likely for humans) or during the period of rapid growth. In the former case the distribution is skewed to the right. Informally, this is because the length of the right-hand tail is heavily dependent on the time taken for the last few lines of descent to coalesce. Since these are time-transformations of exponential random variables with relatively low parameters they are very

variable and in particular are capable of taking large values. It should be noted that this density is of virtually the same shape as that of the generic coalescent process but 'stretched' (in the time axis) by a factor of M_c, where M_c is the constant size of the population before exponential growth began. Indeed if the population were of constant size for all time this would be exactly the case. As in the constant population size case, the distribution of time to the MRCA exhibits a substantial variance. Figure 1 in the next subsection gives the simulated distribution for a particular collection of population sizes, but the shape is typical.

If, however, the MRCA tends to occur during the period of growth then the right-hand tail shortens relative to the left-hand tail until (for exponential growth throughout the population's history) the density becomes symmetric. The explanation for this is again the inherent time-scaling. As before we are effectively time-transforming the density for the generic coalescent via the map $\tau(\cdot)$, but now note that, since the population size is increasing with time, the coalescent time taken up by each generation increases as we move back through time. Consequently the right-hand tail is contracted relative to the centre of the density whilst the left-hand tail is relatively stretched. Thus the density becomes more symmetric.

2. **Initial population size.** Increasing the initial (constant) size of the population before exponential growth (i.e. M_c) has two effects. Each results from the fact that in our formulation, an increase in M_c will increase the size of all generations. Firstly, the density is moved to the right. Informally, all generations are now using up less coalescent-time so that more generations are needed for coalescence. More formally, $\tau(s)$, where s is the time to MRCA for the coalescent, is increased. The intuition is that the probability of a coalescence t generations ago, which is proportional to $1/M(t)$, has decreased, for all t. The second effect of increasing M_c is that the density is "stretched". Again, this follows from the (non-linear) increase in $\tau(s)$.

A change from size M_c to size M_c' for the period of constant population size will lead to approximately an expected extra $\{2 - \sum_{i=1}^{t_G} 1/M(i)\}\{M_c' - M_c\}$ generations being needed to obtain a MRCA, where t_G is the number of generations into the past that the exponential growth of the population commenced. So note that if the MRCA is expected to occur shortly before the period of growth began, this change in population sizes is likely to make relatively little difference.

Decreases in the initial population size will have the opposite effect.

3. **Sample size.** For a constant-size population the expected time to a common ancestor of a sample of size n is close to that for a larger sample of size n', unless n is small. The same is true here. An increase in sample size from n to n' adds a random quantity of time

$$\sum_{i=n+1}^{n'} T(i) \qquad \text{where } T(i) \sim \exp\left(\binom{i}{2}\right)$$

to the coalescent time to the MRCA. This is equivalent to adding $\sum_{i=n+1}^{n'} M_c T(i)$ to the expected real number of generations taken (assuming the MRCA occurs before growth begins). The expected extra real time is thus $\sum_{i=n+1}^{n'} M_c 2/(i(i-1))$. This involves adding at most $2M_c/n$ generations (the value if $n' = \infty$) to the time taken. This is small relative to the expected time to the MRCA for our original sample, provided n is reasonably large. This latter time is bounded below by $2M_c(n-1)/n$, the bound being exact for a constant population but becoming much less tight if we allow for a period of growth as we do here.

Of course the time to the MRCA of a sample will be a lower bound for the time to the MRCA of the whole population. The preceeding paragraph bears on the expected difference between these two quantities. Note however, that for reasonable sample sizes, the MRCA of the sample will coincide with the MRCA of the whole population (and in particular the time since the sample MRCA will be the same as the time since the population MRCA) with high probability. In the coalescent approximation to the panmictic setting, with constant population size, it is known (Saunders, Tavaré and Watterson 1984, see also Watterson and Donnelly 1992) that the probability that the MRCA of a sample of size n coincides with the MRCA of the whole population is

$$(4.1) \qquad \frac{[M(0)+1][n-1]}{[M(0)-1][n+1]}.$$

In fact, this result also applies in the coalescent approximation to panmictic populations of variable size. To see this, note that the probability in question is the probability that when the jump chain of the coalescent for the whole population has exactly two equivalence classes, the labels of the individuals in the sample occur in both classes. The variation in population size changes the timescaling of the coalescent approximation. It does not, however, change the dynamics of its jump chain. Thus the probability in question (and any other probability which depends only on the jump chain) will be the same as in the panmictic case.

4. **Onset of growth.** If we move the time at which the population size begins to grow a further l generations into the past then the density function for time to the MRCA is translated to the right by a little less than l generations. We lose l generations of constant-sized population, which accounts for l/M_c units of coalescent-time. On its own this would account for a shift of exactly l generations but the coalescent-time used up by the period of growth $(\sum_{t=1}^{t_G+l} 1/M(t))$ increases slightly and thus the overall shift is of less than l generations (where t_G is the number of generations ago at which exponential growth begins).

5. **Current population.** The time taken to attain a MRCA is quite insensitive to changes in current population size. Since we are supposing exponential growth the coalescent-time accounted for by the growth period

$(\sum_{t=1}^{t_G} 1/M(t))$ is reduced by only small amounts even for increases in orders of magnitude in the current population size. For example, if we had considered a population of current size 1×10^7, which had grown from a size of a few hundred over the preceding 2500 generations, and then increased the current size to 1×10^9 the time to the MRCA would typically increase by only 10 generations. A further increase to 1×10^{12} would add a further 5 generations or so to the time to the MRCA.

There is a general point to be made here. Provided the time to the MRCA is expected to occur at a time before growth begins, the particular form of the growth function is of limited importance. The sole factor which influences the time to the MRCA is the amount of coalescent time the growth accounts for (i.e. $\sum_{t=1}^{t_G} 1/M(t)$).

4.2. Simulations for possible human population sizes. In this subsection we present a rather more detailed analysis of the results of our simulations for population sizes which might be hoped to be representative of the human population. Our strategy will be to investigate a range of assumptions.

Some information is available on historical human population sizes; see for example Weiss (1984), Smith *et al.* (1989), Stringer (1990). More recently, attempts have been made to infer effective population sizes from genetic data. See for example Rogers (1995) and references therein. We will assume a current population of the order of 5×10^9 individuals. We then suppose the figure decreases exponentially (although as noted earlier the specific form the decrease takes is relatively unimportant) to reach a figure of 1.3×10^6 individuals 50Kyr. b.p. Previous to this the population size is assumed to remain at this level. (Again we vary this constant population size in some later simulations). Note that these represent (rough) census population sizes. For various reasons (including an allowance for overlapping generations and maternal inheritance) the population sizes in the simulations should be smaller than this. As a first approximation we use a population size equal to 1/10th of the above figures. We go on to consider variations in this scenario which involve a reduction in the assumed population size for 50Kyr. b.p. and earlier. Alternatively we add a population bottleneck to reflect a possible calamitous event, or suppose that the period of growth began more recently than 50Kyr. b.p.

We initially assume our population is unstructured or such that the migration rates are sufficiently high to avoid structure effects. The implications of possible structure are explored later. The sample considered is of size 50. Thus the probability that the MRCA of the sample is identical to that of the whole population (i.e. mitochondrial Eve) is approximately 49/51.

Figure 1 gives the simulated distribution of the time to the MRCA for a population of current size 5×10^8 breeding females, which has grown exponentially from a fixed size of 130000 breeding females prior to 50Kyr.

b.p.

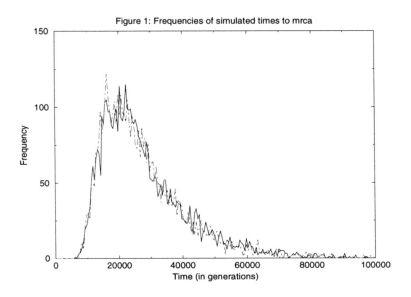

FIG. 1. *Simulated distribution of the time to the most recent common ancestor for a population of current size* 5×10^8 *breeding females. The population was assumed to be of constant size 130,000 breeding females until 50Kyr b.p. from which time it grew exponentially to the current value. The figure shows the results of two simulations, each based on 5,000 simulation runs.*

Table 1 summarizes the dependence of the distribution of time to the MRCA on the size of the population prior to its exponential growth. In all tables "population size" is defined as the number of breeding females present.

Table 1 illustrates the general point made earlier that unless the population size prior to the onset of growth is small, the time to the MRCA is dominated by the population size before growth. (For example with populations of constant size 130,000 and 13,000, the expected number of generations to the MRCA are 260,000 and 26,000 respectively.) The effect of the exponential growth and the details of that growth play a rather minor role. In these cases it is unrealistic to suppose a constant population size at all times prior to 50Kyr. b.p. While exponential growth also seems unrealistic over these time periods, there will have been some much slower population growth. The effect of this will be to reduce the times to the MRCA, perhaps by 50%, but not more markedly.

One might hope that the existence of a bottleneck might encourage a more recent MRCA. We investigate this by supposing once again a current population of 5×10^8 breeding females, but now we assume the occurrence of a sudden bottleneck 50Kyr. b.p. followed by a period of exponential growth

TABLE 1
Dependence of time to MRCA on population size at the onset of population growth. Growth begins 50Kyr. b.p. , previous to which the population has evolved at the indicated constant size.

Pop. size 50Kyr. b.p.	130000	13000	1300	130
Mean time to MRCA in Kyr.	5242	560	96	52
Lower 1% quantile	1600	160	60	47
Lower 5% quantile	2000	260	66	48
Upper 5% quantile	13400	1100	151	58
Upper 1% quantile	14800	1750	195	62

up to the present date. Alternatively, one could model the bottleneck by supposing that the population size remains at a constant small value for a fixed period of time before growing exponentially. Unless this fixed time period is large, this form of bottleneck will have much the same effect as the one considered here. As before we take the number of breeding females before 50Kyr. b.p. to be 130000 individuals. For further background on bottleneck effects, see for example Nei (1987).

Given these beliefs, Table 2 shows how the expected time to the MRCA depends on the size of the bottleneck.

TABLE 2
Dependence of time to MRCA on population size at the bottleneck (bottleneck occurs 50 Kyr. b.p., before this the population is of constant size 130000).

Pop. size at bottleneck	1000	100	50	20	10
Mean time to MRCA in Kyr.	4647	1554	401	46	44
Lower 1% quantile	1060	1020	44	42	41
Lower 5% quantile	1580	1480	45	43	42
Upper 5% quantile	10000	5900	2700	48	46
Upper 1% quantile	14300	11100	7100	48	48

There are effectively three possible types of behaviour which may be induced by a bottleneck, depending on the size of the population at the bottleneck, its size before the bottleneck, and the time of the bottleneck. For an extremely severe bottleneck (those of size 10 and 20 in Table 2) the MRCA occurs between the time of the bottleneck and the present, soon after the bottleneck. The major effect here is simply that the small population at and immediately after the bottleneck forces coalescences to occur quickly.

On the other hand, for most larger bottlenecks, the effect of the bottleneck is simply to reduce the number of ancestors of the population at the bottleneck. The time to the MRCA is then the time at which these ancestors themselves share a common ancestor, which depends on the population size prior to the bottleneck. For example the mean time (in generations) prior to the bottleneck for this to occur is between one and two times the size of the population prior to the bottleneck, so that the time to the MRCA may still be large. In Table 2 this behaviour occurs for bottlenecks of size 1000 or 100 (the latter hardly being "large"). The effect is rather like reducing the size of the sample in whose MRCA we are interested. A comparison with the first column of Table 1 shows a rather small reduction in time to MRCA for the bottleneck of size 1000. The bottleneck of size 100 gives a substantial reduction, but still results in large times to the MRCA.

For a small range of bottleneck sizes (the bottleneck of size 50 in Table 2) the process exhibits a mixture of each of the above behaviours. In some realizations the MRCA is reached between the bottleneck and the present. When this does not occur there are several ancestors at the time of the bottleneck and the additional time until the MRCA is large (of the order of the population size prior to the bottleneck, when measured in generations). The distribution of the time to the MRCA is then a mixture of the two distributions above. Note the great variability. The mean is an extremely misleading measure of the "location" of this distribution.

As the time of the bottleneck is moved further into the past, the size of bottleneck which results in the first or third type of behaviour grows. However even a bottleneck 100-150 Kyr. b.p. would need to be very small to ensure such behaviour. If the bottleneck is not small enough to ensure this, the time to the MRCA still depends largely on the size of the population before the bottleneck.

Note that in the absence of a bottleneck, the genealogy can be of either the first or second type above, depending on the size of the population prior to growth.

In the intermediate case one would expect some sample genealogies to be effectively star shaped and some quite similar to those predicted by the constant population size coalescent. If one could sample several independent genealogies it may be possible to recognize this phenomenon. This cannot be achieved by taking distinct samples from the same population at the same locus. The genealogies of the different samples are highly correlated, each being embedded in the population genealogy – for example if the samples are large, the MRCA of each is likely to be the *same* as the MRCA of the whole population so that (with reasonable probability) the different samples have *identical* times to their MRCA. In particular, there is in essence only one genealogy for all human mitochondrial DNA, and without re-running evolution it is impossible to obtain independent replications of this. Samples of unlinked nuclear DNA will, however, have effectively independent genealogies. Particularly if the recombination rate within each sampled region of the genome is small, comparisons between the patterns observed in each sample could be helpful in making inferences about population structure.

It may be plausible to suppose that the major period of growth within the human population has occurred more recently than 50Kyr. b.p. and that the growth has therefore been at a faster rate than simulated thus far. A consequence of an assumption of this nature would be that the human population becomes of relatively small size more quickly as we move back through time. This increases the probability of individuals sharing an ancestor in the previous generation and so leads to a more recent MRCA. To see whether this kind of situation gives a substantial reduction in time to MRCA we model a case where the human population evolved at a constant size until 5Kyr. b.p. whence it undergoes rapid exponential expansion until reaching the current level of 5×10^8 breeding females.

Table 3 illustrates the dependence of the distribution of time to the MRCA upon the size of the population prior to this more recent exponential growth.

TABLE 3

Dependence of time to MRCA on population size at the onset of population growth (growth begins 5Kyr. b.p. previous to which the population has evoloved at the indicated constant size).

Pop. size 5Kyr. b.p.	130000	13000	1300	130
Mean time to MRCA in Kyr.	5083	515	56	10
Lower 1% quantile	1409	136	19	6
Lower 5% quantile	2199	236	25	7
Upper 5% quantile	8800	871	110	15
Upper 1% quantile	14767	1504	151	19

By comparing Tables 3 and 1 we can see that a more recent onset of growth makes some difference to the time taken to attain a MRCA. The proportional reduction is greatest when the constant population size is small. Indeed, when the constant population is of size 1300 or 130 the reduction is quite substantial. This results from the fact that we are forcing a large part of the recent evolution of our population to have occurred at a small total size. However for more realistic values of population size for the period prior to growth (130000 or 13000 breeding females) the reduction is insignificant. Even prior to the onset of growth, the population is a large one and so coalescence is a rare event and the attainment of the MRCA takes a long time.

Finally we combine the two effects just explored to try to further encourage a more recent MRCA. Thus we suppose that the human population was a constant size of 130000 breeding females per generation until 5000 years ago whence a sudden bottleneck occurred followed by growth to the present value of 5×10^8.

Table 4 shows the dependence of time to the MRCA on the size of the bottleneck for this scenario of more recent growth.

TABLE 4

Dependence of time to MRCA on population size at the bottleneck (bottleneck occurs 5 Kyr b.p., previous to which the population evolves at a constant size of 130000).

Pop. size at bottleneck	1000	100	50	20	10
Mean time to MRCA in Kyr.	5054	4643	4269	3148	1679
Lower 1% quantile	1422	1055	207	4.9	4.7
Lower 5% quantile	1890	1530	1424	22	4.8
Upper 5% quantile	10455	10168	9785	8675	7750
Upper 1% quantile	14800	14050	13696	12740	11320

Once again we see that unless the bottleneck is very severe it has relatively little impact on the time to the MRCA. Even with a bottleneck as small as 50 breeding females we see only a reduction of around 15% in the expected time to MRCA. This is because in none of the 10000 realizations of evolution per simulation do we reach a MRCA before we pass back through the bottleneck. Thus there are always at least 2 lines of descent immediately prior to the bottleneck and, since the population then returns to 130000 breeding females per generation, the coalescence of these last two or more lines takes a considerable period of time. We notice more effect with bottlenecks of size 20 or 10. Here there is a comparatively large reduction in expected time to MRCA although these times are still of the order of several millions of years. However the figures for the expected time are not an entirely adequate summary of the behaviour here. With a bottleneck of size 10 we find that approximately 38% of the time the MRCA is found within the last 5000 years (typically between 4800 and 5000 years ago), whereas the other 62% of the time there are still several lines of descent after passing back through the bottleneck and in these instances the expected time to MRCA is around 2800Kyr. With a bottleneck of size 20 we observe similar results although now we find a MRCA more recently than 5000 years ago less than 5% of the time. Again it is advisable to recall that there has in actuality been only a single realization of human mitochondrial evolution. If there has indeed been a recent bottleneck in human evolution it may well be that either a very short *or* a very long time to MRCA resulted depending on whether there is more than one line of descent present once we have passed back through the bottleneck. This is a random event. Given an identical set of circumstances the random nature of aspects of human evolution leads to a MRCA more recently than the bottleneck with some probability (cf. the 38% mentioned above), but the other 62% (say) of the time we see a *much* longer time to the MRCA. To estimate the size of the bottleneck in the absence of other evidence requires knowledge of the probability of finding a MRCA more recently than this bottleneck. This knowledge is impossible to attain from the single realization of evolution at one locus.

5. Population structure with variation in population sizes.

5.1. Theory. Thus far we have considered only panmictic populations. Many real populations are structured. We still wish to study genealogy and so once again a coalescent approach is appropriate. To simplify the discussion we consider a 2-dimensional lattice model. That is we suppose a so-called "stepping stone" model in which there are C^2 equally sized colonies arranged in a $C \times C$ grid. We refer to the colony in the ith row and the jth column as colony (i, j). Each colony is of equal size and migration is assumed to occur between a colony and its four adjacent neighbours, each adjacent colony being equally likely to be chosen as the destination of a migrating individual. To avoid edge effects we assume that individuals migrating off one edge of the lattice reappear on the opposite edge. Individuals leave their current colony at rate m_t in generation t. To keep track of genealogy it is necessary to count the number of ancestors of the sampled genes at each time t in the past, in each of the colonies. We thus describe the state of this ancestral process t generations ago by $\mathbf{A}_t = (A_{(1,1),t}, A_{(1,2),t}, \dots, A_{(C,C-1),t}, A_{(C,C),t})$ where $A_{(i,j),t}$ is the number of ancestors of our original sample in colony (i, j) at time t in the past.

Recent progress on the coalescent in structured populations has been made in the case of constant population size, see for example Takahata (1988, 1991), Notohara (1990), Herbots (1995), Hudson (1990), and Nath and Griffiths (1993). If the (haploid) population size is M, all colonies are of equal size, and the migration probability m_t is constant over time and of order M^{-1} (say $m_t = m/M$), then the dynamics of the 'structured' coalescent are as follows: at rate 1 any pair of ancestors in a particular colony coalesce, and at rate m any particular ancestor migrates, choosing a new colony uniformly at random from among the four neighbouring colonies. That is, the process $\mathbf{A}_t$ is Markov, with transitions from

$$(A_{(1,1),t}, \dots, A_{(i,j),t}, \dots, A_{(C,C),t}) \text{ to } (A_{(1,1),t}, \dots, A_{(i,j),t}-1, \dots, A_{(C,C),t}),$$

at rate $M^{-1}C^2\binom{A_{(i,j),t}}{2}$, $i,j = 1, 2, \ldots, C$, and from

$$(A_{(1,1),t}, \ldots, A_{(i,j),t}, \ldots, A_{(C,C),t}) \quad \text{to} \quad (A_{(1,1),t}, \ldots, A_{(i,j),t} - 1, \ldots, A_{(i',j'),t}$$
$$+1, \ldots, A_{(C,C),t}),$$

at rate $mA_{i,t}/(4M)$, $i,j = 1, 2, \ldots, C$, for all i', j' such that colony (i', j') is adjacent to colony (i, j).

The evolution of a real (constant population size) structured population is then approximated by transforming time, exactly as in the non-structured constant population size case, via the map $\tau(s) = Ms$, so that events at time s in the structured coalescent occur at time Ms in the real population.

The situation becomes more complicated with variation in population size. For simplicity both in describing the theory and in subsequent simulations we suppose the colonies are of equal size $C^{-2}M(t)$ (or strictly speaking the integer part thereof). The extension to more general structures is technically straightforward and the behaviour is qualitatively similar. We also assume that m_t is constant for all t. A consequence of this is that the relative probabilities of coalescences and migrations change with time. Given that we are currently considering the situation t generations ago and the state of the process is $\mathbf{A}_t = (A_{(1,1),t}, \ldots, A_{(C,C),t})$, the probability ($\pi_t$ say) of a migration or coalescence in the previous generation is given, up to terms of order $O(M(t+1)^{-2})$, by

$$\pi_t = \sum_{i=1}^{C} \sum_{j=1}^{C} \left[A_{(i,j),t+1} m_{t+1} + \frac{\binom{A_{(i,j),t+1}}{2}}{C^{-2}M(t+1)} \right],$$

(assuming, as before, that migration probabilities are of order $M(t)^{-1}$). Hence we derive the probability that the first event of interest (denoted by E) occurs a further s generations ago to be $\left[\prod_{l=1}^{s-1}(1 - \pi_{t+l}) \right] \pi_{t+s}$. Given that an event first occurs a further s generations ago we have $\mathbf{A}_{t+l} = \mathbf{A}_t$ for all $l = 1, 2, \ldots, s-1$ and the conditional probability that the event is either a migration or a coalescence is as follows:

$$P(\text{a particular ancestor migrates} \mid E) = m_{t+s}/\pi_{t+s}$$

$$P\left(\begin{array}{c} \text{coalescence} \\ \text{in colony } (i,j) \end{array} \mid E \right) = \binom{A_{(i,j),t+s}}{2} \bigg/ (C^{-2}M(t+s)\pi_{t+s}),$$

$i,j = 1, \ldots, C$. By iterating this procedure it is possible to simulate the behaviour of the population. The analysis described above applies exactly to the case in which the size of the population changes in a deterministic way. It should also apply, conditional on the realized population sizes, for any of the branching-type models we have been considering in this paper. Note that if the population is growing in real-time (i.e. $M(t)$ is decreasing as t increases) and $m_t = m$ for all t (i.e. migration probabilities are constant), then the conditional probability of a coalescence increases as we look further into the past. This is because coalescence probabilities are proportional to $1/[C^{-2}M(t+s)]$. Equivalently this may be viewed as an apparent reduction in the migration rates as we look further into the past of the underlying coalescent process. The apparent reduction in migration rates t generations ago compared to time 0 (i.e. the present) is by a factor of $M(0)/M(t)$. Populations whose size decreases forward in time behave in a complementary fashion.

In a constant-sized structured population it is known that for models with enough symmetry in the migration and colony structure (and migration rates that are constant through time) the expected time until the MRCA of two individuals from the same colony is independent of the migration rate (e.g. Slatkin 1991). This somewhat surprising result is also rather misleading. While the *expected* time to the MRCA of two individuals from the same colony does not depend on migration rate, the *distribution* of this time does change with migration rate and hence so do many quantities of genetic interest. Loosely speaking, as the migration rate decreases, the distribution exhibits more variability. A migration rate of ∞ corresponds to the panmictic case.

In the variable population size case with a constant number of migrations per colony per generation ($m_t = m/M(t)$), the underlying structured coalescent process is the same (except for timescaling) as in the constant population size case. However if T denotes the (coalescent) time until two individuals from the same colony share a MRCA, although $E(T)$ does not depend on the migration parameter m, $E(\tau(T))$ the expected number of *generations* until until two individuals from the same colony share a MRCA *does* depend on the migration rate. In the case of a constant population which then grows exponentially, the function $\tau(\cdot)$ is in fact concave. It is likely that the distribution of T becomes more variable as m decreases. If this were true, $E(\tau(T))$ would then decrease as m decreases. We stress again that most quantities of interest depend on the entire distribution of $\tau(T)$, not just on its mean.

For the increasing population we are considering, a constant migration rate in real time corresponds to a decreasing migration rate as time increases in the structured coalescent. It is no longer true that even the mean (coalescent) time to the MRCA of a pair of individuals from the same colony is independent of migration rates if these are changing with (coalescent) time. If they are decreasing, as here, this mean (coalescent) time until the MRCA should be increased. It seems difficult to be precise about the effect on the number of generations until the MRCA. The effect of structure is to increase its variability, and for increasing populations this time may be much longer than in the equivalent non-structured population. Provided migration rates are not extremely low, it is clear that even for a sample all drawn from one colony we can expect some migration to have occurred before a MRCA is attained. Thus the last two, or possibly more, lines of descent will be in different colonies. If the population size is increasing in real-time then the apparent migration rates in the coalescent are decreasing as we move back through time. But before a MRCA can occur we need a migration of one of the two individuals in the line of descent. This event is of probability $2m'_t$ where m'_t is the migration rate in the coalescent. So as m'_t gets very small we may wait a great deal of time for the necessary final migration, and hence for the MRCA, to occur.

5.2. Simulation of structured populations. We return to the results of our computer simulations, but henceforth all simulations include population structure. The common format is the $C \times C$ lattice model. Other forms of population structure were also considered but the results from these were not sufficiently distinct to merit inclusion. Again our sample is of size 50, drawn from the same colony, and we assume a current population of 5×10^8 breeding females per generation. Initially we consider a situation where the human population was of a constant (effective) size of 10000 breeding females up until 5Kyr. b.p. whence it began a period of exponential growth to reach the current value. We note that the exact values we choose to assume for the changing population size are not of the greatest significance. Our primary interest is to determine the qualitative changes in the distribution of time to the MRCA resulting from the introduction of structure.

We begin by exploring the importance of the degree to which the population is structured. Specifically we vary the size of the lattice upon which the population evolves. For the present we arbitrarily assume a migration probability of 1×10^{-5} per individual per generation. We examine below the effect of varying this parameter. Table 5 shows how the distribution of time to the MRCA depends on the size of the lattice.

TABLE 5

Dependence of time to MRCA on the lattice size. Migration rate is 1×10^{-5}, population evolves at a constant size of 10000 breeding individuals until commencing growth 5Kyr. b.p. to reach a present size of 5×10^8.

Lattice size	1×1	2×2	3×3	4×4	5×5	10×10
Mean time to MRCA (Kyr.)	399	1025	1587	2463	3214	10166
Lower 1% quantile	118.4	33.8	16.8	11.4	8.5	4.5
Lower 5% quantile	157.2	45.0	21.8	14.1	10.4	6.0
Upper 5% quantile	874	5934	10840	18960	22340	24760
Upper 1% quantile	1132	11400	25500	48140	70200	295900

Table 5 demonstrates that the degree of structure is of significant importance. We see a tendency for the time to MRCA to increase as the number of colonies into which the population is subdivided increases. There is a complicated interaction between two opposing effects when the number of colonies is increased. The first is that because colonies are smaller coalescences within colonies happen more rapidly. This effect tends to reduce the number of ancestors of the sample. The second effect is that when migrations do occur, there are more colonies amongst which the migrants will eventually be spread, so that outside the original colony, coalescences will happen more slowly as descendents of our sample will occasionally migrate and thus disperse throughout the whole population lattice. When there are only a few ancestors left, it will also take longer for them to be in the same colony, and hence longer, with more colonies, for the last few coalescences. The second of these effects is clearly the dominant one for the range of parameters we have considered.

We now investigate the effect of the migration rate in this setting. We consider the time to the MRCA in a sample with the same population size parameters as before and assume the population to be structured as a 3×3 lattice. Table 6 illustrates how the results depend on the migration rate.

TABLE 6

Dependence of time to MRCA on the migration rate. The population is structured as a 3×3 lattice and evolves at a constant size of 10000 breeding individuals until commencing growth 5Kyr. b.p. to reach a present size of 5×10^8.

Migration rate	1×10^{-7}	1×10^{-6}	1×10^{-5}	1×10^{-4}	1×10^{-3}	1×10^{-2}
Mean time to MRCA (Kyr.)	1827	1612	1587	1183	520	403
Lower 1% quantile	16.9	16.7	16.8	22.5	134.3	119.8
Lower 5% quantile	20.9	20.9	21.8	34.6	191.4	158.8
Upper 5% quantile	98.0	106.4	10840	3460	1096	858.0
Upper 1% quantile	138.0	47340	22500	5180	1524	1120

Table 6 shows that the time to MRCA increases as the migration rate decreases. The behaviour is more subtle than such a summary suggests however. For small migration rates few ancestors are likely to have been migrants. Thus coalescences are initially more likely since most ancestors are in the colony from which the sample was drawn. However it is likely that for any of these migration rates a small number of ancestors will be migrants. As a consequence the loss of the last few ancestors will take longer than otherwise expected since we have to wait for appropriate migration events to place coalescing ancestors in the same colony. Hence, for very small migration rates (e.g. 1×10^{-7}), we often find that we attain a MRCA before a single migration event has occurred. This leads to a high proportion of simulations with a recent MRCA. But in a small proportion of the simulations at least one ancestor of the sample migrates before the MRCA is reached and then we have to wait for further migration events to bring ancestors into the same colonies. This time is of the order of m_t^{-1} and so be may large. Our focus is simply on the depth of the genealogical tree associated with the sample. Of course, changes in the parameters we are investigating can have substantive effects also on the shape of the tree, which might be reflected in the patterns of observed molecular genetic data.

All of the above results on times to the MRCA in structured populations relate to the setting in which the sample with whose ancestry we are concerned are all drawn from a single colony. Were we sampling randomly from the entire population the first effect of reduced migration rate described above (i.e. early coalescences within the same colony) is much reduced and so the expected time to MRCA increases more markedly as the migration rate decreases. Table 7 describes simulation results when the sample is drawn from the entire population rather than from a single colony.

TABLE 7

Dependence of time to MRCA on the migration rate. The population is structured as a 3 × 3 lattice and evolves at a constant size of 10000 breeding individuals until commencing growth 5Kyr. b.p. to reach a present size of 5×10^8. The sample is drawn at random from all subpopulations.

Migration rate	1×10^{-6}	1×10^{-5}	1×10^{-4}	1×10^{-3}	1×10^{-2}
Mean time to MRCA (Kyr.)	162942	16338	1993	549	405
Lower 1% quantile	3371	3520	491	163	119
Lower 5% quantile	5284	5300	694	217	156
Upper 5% quantile	341000	35420	4400	1095	839
Upper 1% quantile	476000	50920	6140	1550	1170

6. Discussion. Variation in population sizes may be incorporated into the analysis of a large class of genetics models by using a non-linear timescale in the coalescent approximation to the genealogy of the sample or population. Effectively, a generation of (haploid) size $M(t)$ uses up $1/M(t)$ units of coalescent time. (Recall that if the population were of constant size M, one unit of coalescent time would correspond to M generations of real time.) For most questions of interest it can be extremely misleading to approximate the evolution of a variable sized population by that of a constant sized population whose size is given by an appropriate effective population size.

For populations which are approximately constant before growing exponentially, there are two general forms for the distribution of time to the

MRCA. If the initial size is small and/or the period of growth is extended, the MRCA occurs during the period of growth. In this case (Slatkin and Hudson 1991) the genealogy will be star-shaped and the distribution of time to the MRCA reasonably sharply peaked. The other type of behaviour occurs when the sample still has several ancestors at the time of onset of growth, in which case the time to the MRCA is comparable with that for a sample from the population (of constant size) before growth. In particular it has a large mean and a large variance. Unless the size of the population before growth is extremely small, there is thus a substantial difference between an assumption of exponential growth throughout the evolution of the population (as in Slatkin and Hudson 1991), and the assumption of a population of approximately constant size prior to its exponential growth to current size. In the light of current beliefs the second assumption seems more appropriate for the human population. The current population size and the size of the sample have relatively little effect on times to their MRCA, in either case.

If it is assumed that the population undergoes a bottleneck and decreases sharply in size before the period of growth, then the two effects above still occur, depending on the size and timing of the bottleneck. In this case there is also an intermediate possibility (for only a small range of parameter values) which is effectively a mixture of the two effects. There will be some realizations in which the MRCA occurs soon after the bottleneck (in particular relatively recently), otherwise if there are still several ancestors at the time of the bottleneck the MRCA will occur much further into the past. The distribution of times to the MRCA is then effectively bimodal and extremely variable, and its mean is an inappropriate measure of location.

The incorporation of population structure in models with variation in population sizes means that the appropriate coalescent process is time inhomogeneous, and more difficult to analyze. Broadly speaking when considering the ancestry of a sample from a single colony, decreasing the migration rate speeds up the early coalescences, but may substantially slow later coalescences, and the distribution of time to a MRCA becomes more variable. (Exactly the same effect occurs when population structure is introduced into populations of fixed size.) If the sample under consideration is not all taken from the same colony, then population structure substantially changes its underlying genealogy. With variation in population size it is no longer true that the mean time to the MRCA of two individuals from the same colony does not depend on migration rates.

There has been considerable recent interest in, and controversy surrounding, "Eve", the most recent common ancestor of human mitochondrial DNA. This centres in particular on her date of birth and her place of residence. Apart from intrinsic interest, knowledge of both of these would, for example, have a considerable bearing on the current dispute concerning the evolution of modern humans. See for example Takahata (1991). Placing

Eve in Africa in the period 100 - 300 Kyr. b.p. would be strong evidence in favour of the "Out of Africa" theory, that modern humans evolved in Africa and then migrated to other parts of the world around 100K years ago. It is argued (e.g. by Vigilant *et al.* 1991) that even in the absence of direct evidence about her location, her existence in this range of times is inconsistent with the competing "Multiregional" theory.

We have argued that in "estimating" the time, T_{Eve}, since Eve it is inappropriate to regard this time as a parameter in the usual statistical sense. Instead, it is a random variable and inference should take the form of reporting its conditional distribution given the data (with the usual caveats about investigating the sensitivity of the conditional distribution to modelling assumptions). This conditional distribution is the normalization of the product of the unconditional distribution of T_{Eve} multiplied by the conditional distribution of the data given T_{Eve}. Such an analysis is difficult in practice for at least two reasons. Firstly, the distributions depend on unknown parameters such as mutation rates, mutation mechanisms, population sizes, and population structure. While these could also be estimated from the data, it would be better if reliable estimates of these were available from other sources. The second problem is that even if these were assumed known, the dimensionality of the underlying genealogies means that a full analysis for realistic evolutionary models is computationally infeasible. Current developments (Griffiths and Tavaré 1994a,b, Kuhner *et al.* 1995) may make such a method accessible in the foreseeable future.

In lieu of a full analysis, the results of this paper give insight into the unconditional distribution of the time to Eve. Recent point estimates of the time since Eve lie around 200Kyr., and interval estimates vary in width from about 80Kyr. to 200Kyr. (e.g. Hasagawa and Horai 1991, Vigilant *et al.* 1991). Current beliefs about human population sizes lead to (pre-data) distributions which are concentrated on much larger values and exhibit substantial variability (for example the first column of Table 1). For these population sizes a bottleneck has little effect on the distribution of the time since Eve. In order to move the mean time into the range of current estimates, a bottleneck would need to be both severe and last for a long period, at around 150 - 200 Kyr. b.p. While it is plausible under the Out-of-Africa theory that the ancestral population of modern non-African groups underwent a severe bottleneck during its exodus from Africa, it seems much more difficult to reconcile a severe bottleneck for the *entire* human population with current beliefs about early human populations.

It is even more difficult to find patterns for human demography which are consistent with recent estimates of the time since Eve when population subdivision is considered. Population structure increases the mean time since Eve and substantially increases variability. All of the simulations reported earlier concern the effect of subdivision on the time to the MRCA of a sample from the same subpopulation. The time to the MRCA of the whole population may in general be substantially larger. Of course if there are relatively high levels of gene flow, a subdivided population will behave like a panmictic one. For smaller migration rates, the time until Eve may be large even in small populations. It depends crucially on the time for the last few ancestors to be traced to the same subpopulation; see for example Takahata (1991).

In order to reconcile the unconditional distribution of the time since Eve with recent estimates based on other methods, one must argue for an extremely small population prior to the exponential growth and possibly a late onset of growth (or a severe population bottleneck), in addition to high levels of gene flow between subpopulations, particularly around the time of the MRCA. Each seems unrealistic unless the entire human population occupied a small physical area 200,000 years ago.

An alternative explanation of the discrepancy between the existing estimates of the time since Eve and the unconditional distributions obtained here, is the possibility of a selective sweep (around 200,000 years ago) through the human mitochondrial population. Such an explanation was posited in Excoffier (1990) and Di Rienzo and Wilson (1991). See also Marjoram and Donnelly (1994) and Marjoram (1996). In view of the growing evidence for non-neutrality of mitochondrial evolution in other species (Ballard and Kreitman 1994, 1995; Rand, Dorfsman, and Kann 1994) it would appear that this explanation warrants serious consideration in this context.

Note that throughout we have assumed that the variance of offspring numbers is one, as given for example by the Wright-Fisher model. This variance effects all the times discussed, and for example doubling the variance of offspring numbers would halve both the mean and the standard deviation of the distribution of times since Eve. Because it is female offspring numbers that are relevant to the discussion of human mitochondrial Eve, a variance approximately equal to the value used here may not be unreasonable.

If the primary interest is in using genetic data to make inferences about the size of the human population at various times, then data from nuclear DNA would appear more promising than that from mitochondrial DNA. The latter is a consequence of a single realization of genealogy for human mitochondria, so that any inferences based on this will be affected by the considerable variability inherent in this genealogy. Samples of unlinked nuclear DNA effectively result from independent replicates of the genealogy, so that their use should decrease the variability due to genealogy. Particularly for neutral sequences within which recombination is unlikely (so that the genealogical theory is more transparent) such data may give insight into population size and structure.

It must be remembered that the analysis of this paper (and those by other authors) assumes that the underlying human demographic process belongs to the branching-type class described in section 3, so that the relevant coalescent approximation applies. Further, many of our (and others') conclusions are extremely sensitive to this assumption. It is far from clear *a priori* that this assumption is justified for human populations. The difficulty of reconciling predictions from such models (for plausible scenarios for population sizes and structure) with apparent conclusions from molecular data adds weight to this view.

Acknowledgements. It is a pleasure to thank Simon Tavaré for helpful comments on an early draft of the manuscript. The first author was supported by SERC grant GR/F 32561 and by Australian Research Grant A19131517. The second author was supported in part by SERC grants GR/F 32561, GR/G 11101, GR/H 22880, and B/AF/ 1255, and by a Block Grant from the University of Chicago.

REFERENCES

Ballard, J.W.O. and Kreitman, M. (1994). Unravelling selection in the mitochondrial genome of Drosophila. *Genetics* **138** 757–772.

Ballard, J.W.O. and Kreitman, M. (1995). The bell tolls for the neutral evolution of mitochondrial DNA. *Trends in Ecology and Evolution.* In press.

Cann, R.L., Stoneking, M. and Wilson, A.C. (1987). Mitochondrial DNA and human evolution. *Nature* **325** 31–36.

Di Rienzo, A. and Wilson, A.C. (1991). The pattern of mitochondrial DNA variation is consistent with an early expansion of the human population. *Proc. Natl. Acad. Sci. USA* **88** 1597–1601.

Donnelly, P.J. and Kurtz, T. (1996). Particle representations for measure-valued population models. In preparation.

Excoffier, L. (1990) Evolution of human mitochondrial DNA: evidence for departure from a pure neutral model of populations at equilibrium. J. Mol. Evol. 30 125–139.

Felsenstein, J. (1992). Estimating effective population size from samples of sequences: Inefficiency of pairwise and segregating sites as compared to phylogenetic analysis. *Genet. Res.* **59** 139–147.

Griffiths, R. and Tavaré, S. (1994a). Sampling theory for neutral alleles in a varying environment. *Phil. Trans. R. Soc. Lond.* B **344** 403–410.

Griffiths, R. and Tavaré, S. (1994b). Ancestral inference in population genetics. *Stat. Sci.* **9** 307–319.

Hasagawa, M. and Horai, S. (1991). Time of the deepest root for polymorphism in human mitochondrial DNA. *J. Mol. Evol.* **32** 37–42.

Herbots, H. (1995). The structured coalescent. In this volume.

Hudson, R.R. (1990). Gene genealogies and the coalescent process. *Oxf. Surv. Evol. Biol.* **7** 203–217.

Karlin, S. and McGregor, J. (1962) Direct product branching processes and related Markov chains. *P.N.A.S.* **51** 598–602

Karlin, S. and McGregor, J. (1965) Direct product branching processes and related induced Markov chains. I. Calculation of rates of approach to homozygosity. *Bernoulli, Bayes Laplace Anniversary Volume*, Springer-Verlag, Berlin pp11-145.

Kingman, J.F.C. (1982a). On the genealogy of large populations. *J. Appl. Prob.* **19A** 27–43

Kingman, J.F.C. (1982b). The coalescent. *Stoch. Proc. Appl.* **13** 235–248.

Kingman, J.F.C. (1982c). Exchangeability and the evolution of large populations. In *"Exchangeability in Probability and Statistics"*, (Koch, G. and Spizzichino, F. ed), pp 96–112, North Holland, Amsterdam.

Kuhner, M.K., Yamato, J. and Felsenstein J. (1995) Applications of Metropolis-Hastings genealogy sampling. This volume.

Marjoram, P. (1996). The effect of selective sweeps on pairwise difference distributions. In preparation.

Marjoram, P. and Donnelly P. (1994). Pairwise comparisons of mitochondrial DNA sequences in subdivided populations and implications for early human evolution. *Genetics* **136** 673–683.

Nath, M. and Griffiths, R.C. (1993) The coalescent in two colonies with symmetric migration. *J. Math. Biol.* **31** 841–852.

Nei, M. (1987) *Molecular Evolutionary Genetics* Columbia University Press, New York.

Notohara, M. (1990). The coalescent and the genealogical process in geographically structured population. *J. Math. Biol.* **29** 59–75.

Rand, D.M., Dorfsman, M. and Kann, L.M. (1994). Neutral and non-neutral evolution of Drosophila mitochondrial DNA. *Genetics* **138** 741–756.

Rogers. A.R. (1995). Population structure and modern human origins. In this volume.

Rogers, A.R. and Harpending, H. (1992). Population growth makes waves in the distribution of pairwise differences. *Mol. Biol. Evol.* **9** 552–562.

Saunders, I.W., Tavaré, S. and Watterson, G.A. (1984). On the genealogy of nested subsamples from a haploid population. *Adv. Appl. Prob.* **16** 471–491.

Slatkin, M. (1991). Inbreeding coefficients and coalescence times. *Genet. Res. Camb.* **58** 167–175.

Slatkin, M. and Hudson, R.R. (1991). Pairwise comparisons of mitochondrial DNA sequences in stable and exponentially growing populations. *Genetics* **129** 555–562.

Smith, F.H., Falseth, A.B. and Donnelly, S.M. (1989). Modern human origins. *Yearbook of Physical Anthropology* **32** 35–68.

Stringer, C.B. (1990). The emergence of modern humans. *Scient. Am. (Dec.)* 98–104.

Takahata, N. (1988). The coalescent in two partially isolated diffusion populations. *Genet. Res.* **52** 213–222.

Takahata, N. (1991). Genealogy of neutral genes and spreading of selected mutations in a geographically structured population. *Genetics* **129** 585–595.

Tavaré, S. (1984). Line-of-descent and genealogical processes and their applications in population genetics. *Theor. Pop. Biol.* **25**, 119–164.

Templeton, A.R. (1993). The "Eve" hypothesis: a genetic critique and reanalysis, *Am. Anthrop.* **95** 51–72.

Vigilant, L., Pennington R., Harpending H., Kocher T. D. and Wilson A. C. (1991). African populations and the evolution of human mitochondrial DNA. *Science* **253** 1503–1507.

Watterson, G.A. and Donnelly, P.(1992). Do Eve's alleles live on? *Genet. Res.* **60** 221–234.

Weiss, K.M. (1984). On the number of members of the genus Homo who have ever lived, and some evolutionary implications. *Hum. Biol.* **56:4**, 637–649.

MOLECULAR POPULATION GENETICS OF A PHENOTYPICALLY MONOMORPHIC PROTEIN IN DROSOPHILA*

STEPHEN W. SCHAEFFER[†]

Abstract. Advances in molecular biology and population genetics theory have created a powerful methodology to understand the relative contribution of the major evolutionary forces that modulate genetic variation. It will be shown how simple measures of nucleotide diversity may be used to estimate important population genetic parameters. This paper summarizes how nucleotide sequences in the alcohol dehydrogenase region have been used to determine how much mutation, migration, recombination, and selection have occurred in the past evolutionary history of 12 populations of *Drosophila pseudoobscura*.

Key words. Population genetics, neutral theory, mutation, migration, recombination, selection, linkage disequilibrium.

1. Introduction. The modern study of genetics consists of two subdisciplines, molecular and population genetics. Molecular genetics considers how genes are expressed in a temporal and tissue-specific manner. Aspects of a gene's expression pattern are important if any attempt is to be made to cure a patient of a genetic disease. Knowing the expression pattern for a single gene is insufficient to explain the cause of all genetic diseases because individuals within populations may differ in the genes that they carry and a cure for one individual may not be applicable for a second patient. Population genetics is the study of why genes are variable within populations and what forces create and maintain observed diversity.

Two mechanisms in population genetics have been used to explain the levels of genetic variation in natural populations. The neutral mutation hypothesis states that most allelic alternatives are selectively equivalent and reflect a transient phase of molecular evolution where: (a) the rate of fixation or loss of an allele is largely determined by random genetic drift; (b) nucleotide substitutions or amino acid replacements will accumulate in the nuclear genome at a rate proportional to the mutation rate; (c) rates of nucleotide substitution or amino acid replacement will be greater in regions of lowered functional constraint; and (d) regions of the genome that diverge rapidly between species should also be highly polymorphic within species [39]. The alternative to the neutral model is a selection model where the alternative alleles cause individuals in populations to vary in their survival or fecundity. Some researchers have suggested that genetic heterogeneity within populations is so important that it is actively maintained by natural selection [21, 47]. The studies of protein variation in the 1960s and

---

* This research was supported by the National Institutes of Health (GM-42472).

† Department of Biology, The Pennsylvania State University, 208 Mueller Labs, University Park, PA 16802-5301.

70s made little progress in discriminating the roles of neutral and selective forces because both hypotheses were found to be consistent with the observed variation data.

Two major advances in experimental and theoretical population genetics provide powerful tools that can discriminate between neutral and selective hypotheses. Experimentalists may generate population samples of nucleotide sequences because of advances in rapid sequencing methods due to the polymerase chain reaction [29, 59, 60]. Nucleotide sequence data may discriminate between hypotheses because the redundancy of the genetic code superimposes two histories on a nucleotide sequence. The first is a selective history written in the amino acid sequences of proteins and the second is a neutral history described by synonymous and noncoding sites [48]. In addition, low levels of recombination between nucleotides [15] correlate the inheritance patterns of linked selected and nonselected nucleotides. Thus, when balancing or directional selection acts on amino acid diversity, the number of variants in linked neutral sites may be increased or decreased [5, 6, 43, 45].

The statistical methods and theoretical models necessary to analyze and understand nucleotide sequence data have kept pace with or surpassed experimental studies of DNA in populations. Critical population parameters and test statistics have been designed that are capable of rejecting expectations of the neutral theory [35, 73]. Coalescent methods allow neutral gene genealogies to be simulated under various population genetic models without an extensive need for computer time or space [32, 34, 38, 41, 42, 69, 72, 74]. Simulation results allow strong inferences to made about the past evolutionary history of populations. Thus, experimental and theoretical tools exist to determine how important natural selection has been in modulating genetic diversity in the genomes of several model organisms.

The alcohol dehydrogenase (*Adh*) gene of *Drosophila* has been a model system for evolutionary studies at the molecular level [44, 46, 49]. *D. pseudoobscura* encodes a single form of the ADH enzyme in natural populations [57]. The molecular evolution of nucleotide diversity of a gene that encodes a single allelic form should be consistent with expectations of a neutral model because selection acts predominantly on amino acid polymorphisms and ADH lacks protein variants. Thus, the study of a phenotypically simple gene should provide valuable insights about the process of neutral evolution. The evolutionary histories of phenotypically more complicated genes may then be compared with the *Adh* data used as a neutral control locus. This paper will summarize how simple estimates of nucleotide diversity have been used to infer the strength of mutation, recombination, migration, and selection in the evolutionary history of a phenotypically simple gene.

2. The fine structure of the *Adh* **region.** The alcohol dehydrogenase region in *D. pseudoobscura* is composed of two genes, *Adh* and *Adh-Dup*, that resulted from an ancient duplication of an ancestral gene (Fig. 2.1) [63]. The *Adh* gene is transcribed from two promoters during development, but the sequence that encodes the enzyme in the two transcripts is identical [9]. The 5′ or distal promoter is used primarily in adults, while the 3′ or proximal promoter is used primarily in larvae [61, 62]. In addition, two tissues, fat body and Malpighian tubules, show the same developmental utilization pattern of the two promoters [62]. Transcription factors responsible for the regulation of the distal and proximal promoters are currently being isolated [8, 28, 50]. The *Adh* gene encodes an enzyme that is 254 or 256 amino acids long depending on the *Drosophila* species that is studied [71]. Natural populations of *D. pseudoobscura* encode a single protein form of *Adh* that is 254 amino acids long [57, 63].

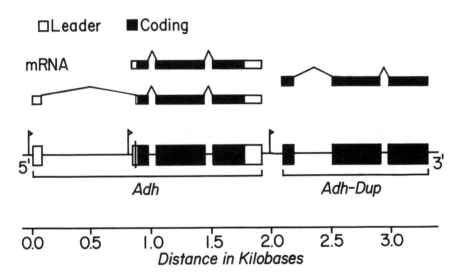

FIG. 2.1. *Fine structure of the Adh region of D. pseudoobscura. The region is subdivided into 17 sequence domains from 5′ to 3′ : 5′ flanking, Adh adult leader, Adh adult intron, Adh larval leader, Adh exon 1, Adh intron 1, Adh exon 2, Adh intron 2, Adh exon 3, Adh 3′ leader, intergenic, Adh-Dup exon 1, Adh-Dup intron 1, Adh-Dup exon 2, Adh-Dup intron 2, Adh-Dup exon 3, and 3′ flanking.*

The *Adh-Dup* gene, also referred to as *Adh-related* [36], was discovered when the sequences ∼ 300 base pairs downstream of *Adh* were found to be highly conserved between *D. pseudoobscura* and *D. mauritiana* [63].

The exact size of the mRNA transcript produced by *Adh-Dup* has not been determined. The ratio of synonymous to nonsynonymous substitutions, the level of codon bias, and the detection of cDNA clones all indicate that *Adh-Dup* produces a gene product, but the function of the protein is unknown. The conserved open reading frames between *D. pseudoobscura* and *D. mauritiana* suggest that *Adh-Dup* encodes a protein that is 278 amino acids long. *Adh-Dup* has been reported in numerous members of the subgenus *Sophophora* [71], and has recently been detected in the *Scapto-drosophila* and *Drosophila* subgenera [2]. *Adh* and *Adh-Dup* were derived from a common ancestral gene because the amino acid sequences of the two proteins are 38% similar and the positions of the two small introns are conserved between the two genes.

3. Nucleotide sequence data and measures of nucleotide diversity. The complete nucleotide sequence (3.5 kilobases) of the *Adh* region has been determined for 109 strains of *D. pseudoobscura* and its close relatives [63, 64, 66, 67]. These strains were drawn from the North American range of *D. pseudoobscura*, which extends from British Columbia, Canada to Guatemala and from the Rocky Mountains to the Pacific Ocean. Ninety nine strains of *D. pseudoobscura* were collected from 12 populations in the North American distribution and the sample sizes varied from one to 26 strains (Fig. 3.1). The samples of *Adh* sequences from *D. pseudoobscura* may be considered random samples because the strains were not be chosen because of electrophoretic mobility or other classification systems for strains. The nucleotide sequences have been aligned by minimizing the numbers of gaps and mismatches assumed in the sequences.

Nucleotide site heterozygosity was estimated either with segregating sites (S) or pairwise differences (k_{ij}). A segregating site is defined as any aligned base in the sample that has two or three nucleotides present. A pairwise difference is defined as any nucleotide change that is observed between two sequences, i and j (Fig. 3.2). S is a biased estimator of heterozygosity because it is sensitive to sample size, while k_{ij} is an unbiased estimator of heterozygosity because it is sensitive to the differences in allele frequency between segregating sites. These two estimates of nucleotide variation are used to quantify the mutation, migration, and recombination parameters ($4N\mu$, Nm, and $4Nc$, where N is the effective population size, μ is the neutral mutation rate per generation, m is the neutral migration rate per generation, and c is the neutral recombination rate per generation). These population parameters may also be used for statistical tests of the neutral mutation hypothesis. Nucleotide variation within a genetic locus that leads to the statistical rejection of a neutral model may indicate the past action of natural selection in the history of the population.

FIG. 3.1. *Geographic range and collection localities of Drosophila pseudoobscura with sample sizes greater than three. The geographic range is designated by (- - - -). The arrow shows the southern boundary of the range which is coincident with the southern border of Guatemala. The names of populations are: SW, Stemwinder Provincial Park, British Columbia, Canada; AH, Apple Hill, California; GB, Gundlach-Bundschu Winery, Sonoma Valley, California; BR, Bryce Canyon National Park, Utah; MV, Mesa Verde National Park, Colorado; KB, Kaibab National Forest, Arizona; SB, San Bernadino Mountains, California; MX, Mexico; and BG, Bogota, Colombia.*

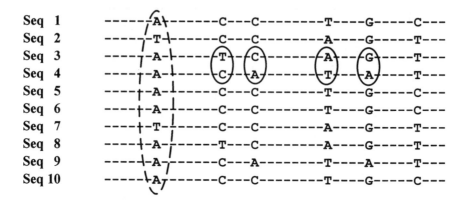

FIG. 3.2. *A hypothetical alignment of* 10 *nucleotide sequences (SEQ). The dashes indicate monomorphic nucleotides that do not vary among the 10 sequences. The aligned nucleotides that are variable are represented in vertical columns. There are a total of six segregating sites (S) in this alignment, one such site is indicated by the dashed circle. The four pairwise differences between sequences 3 and 4 are shown with solid circles. The average number of pairwise differences for all possible pairs of the ten sequences is 2.51.*

4. The past action of selection in the *Adh* region. *Adh* and *Adh-Dup* have been observed to differ in their levels of amino acid polymorphism found in *D. pseudoobscura* populations. Populations of *D. pseudoobscura* encode one major form of the ADH enzyme [14, 56] and nine forms of the ADH-DUP enzyme (S. W. Schaeffer, unpublished data). What explains the differences in amino acid diversity in the two proteins? ADH could be monomorphic either because of a low neutral mutation rate at amino acid positions or because of a recent directional selection event that fixed a new beneficial allele. ADH-DUP could be polymorphic either because of a high neutral mutation rate at amino acid positions or because of some type of balancing selection maintaining alternative forms of the enzyme. The neutral and selective hypotheses may be discriminated by an examination of linked neutral variation [35, 73]. Levels of variation in linked neutral sites such as synonymous or intron sites may be decreased or increased by directional or balancing selection acting on amino acid variants [1, 7, 11, 34, 46], however, the evolution of linked synonymous sites will be independent of nonsynonymous sites if amino acid polymorphisms are selectively neutral [12].

Two statistical tests use estimates of the neutral mutation parameter to determine if variation in synonymous and intron sites departs from expectations of an equilibrium neutral model [35, 73]. The test statistic, D, of Tajima [73] is the difference in two estimates, $M(k)$ and $M(S)$, of the neutral mutation parameter, $4N\mu$. $M(S)$ is a function of the number of segregating sites S and the sample size n. $M(k)$ is the mean of pairwise differences k_{ij} estimated from all comparisons of two sequences. The Tajima test is performed on the two estimates of $4N\mu$ determined from samples of nucleotide sequences collected from within species. $M(S)$ and $M(k)$ should be equivalent under the null hypothesis of an equilibrium neutral model such that D will be zero. The action of balancing selection will tend to increase the number of intermediate frequency segregating sites such that D will be significantly larger than zero because $M(k)$ will outweigh $M(S)$. On the other hand, the rapid fixation of a beneficial mutation by directional selection will tend to leave no variable sites or many rare frequency variants and D will be significantly smaller than zero because $M(k)$ will less than $M(S)$.

The Hudson, Kreitman, and Aguade (HKA) [35] method estimates a goodness-of-fit test statistic, X^2, that determines if the neutral mutation parameter is constant among loci sampled from within the genome. Different genetic loci in the genome may vary in the neutral mutation parameter because various loci are not expected to have similar selective constraints on nucleotide sequence. The HKA test controls for differences in selective constraint between tested loci by comparing the neutral mutation parameter in each genetic locus with an estimate of between species divergence, $4N\mu$ scaled by the coalescence time. The neutral theory predicts that within species polymorphism will be correlated with between species divergence across different genetic loci [39].

Neutral polymorphisms in *Adh* and *Adh-Dup* fail to reject an equilibrium neutral model as the cause of the observed amino acid variation [64, 67]. These results hold for each of the nine populations that were sampled from *D. pseudoobscura* (Fig. 3.1) or for the 99 sequences of *Adh* and *Adh-Dup* drawn from the North American range of the species. The Tajima test shows that the frequency distribution of segregating sites is consistent with neutral expectations. The HKA test shows that highly polymorphic regions observed within species also tend to diverge rapidly between species. Thus, the ADH enzyme is monomorphic because of a low neutral mutation rate at amino acid positions while the ADH-DUP protein is highly polymorphic because of a higher neutral mutation rate at amino acid positions.

The approaches of Tajima and Hudson, Kreitman, and Aguade have provided the first generation of tests used to find departures from an equilibrium neutral model, however, each method has associated problems. The D statistic of Tajima is a useful metric for describing the frequency distribution of nucleotide data. Estimates of D less than zero indicate overrepresentation of low frequency variants compared to intermediate frequency sites, while estimates of D greater than zero suggest an excess of intermediate frequency polymorphisms compared to rare frequency segregating sites. The Tajima test suffers from low power to detect departures from a neutral model when the number of segregating sites is small [27, 33]. The HKA test is an effective test for detecting selection in a genomic region if one locus used in the test is known to evolve by neutral forces and if the genetic loci examined are of an appropriate size to reflect the hitchhiking effects associated with positive Darwinian selection [33, 46]. These limitations of the HKA test are problematic for the biologist who must make careful choices when the test is used.

5. Population structure of *D. pseudoobscura.* The observation of gene frequency differences between populations may be explained by neutral or selective hypotheses. Random genetic drift in small populations isolated by low levels of gene flow may explain why populations become genetically different. On the other hand, species that exchange large number of migrants between populations each generation, will differentiate only if natural selection alters the genotypic composition of the population.

A classic case of genetic differentiation is observed in *D. pseudoobscura* when one observes chromosome morphology between populations. Dobzhansky and his colleagues found that one chromosome, III, in *D. pseudoobscura* differed in the arrangement of its genes among individuals collected from natural populations [19, 23, 24]. Segments of the chromosome were found to be inverted between individuals and the inversions differed in frequency between populations within a season [23, 24] and within a population between seasons [20]. Estimates of gene flow that measured the direct movement of flies have yielded small [18, 22, 25, 55] and large [16, 17, 37] values for the migration rate. Thus, previous studies have failed to resolve whether neutral or selective forces explain the geographic pattern of inversion frequencies.

A neutral genetic region, such as *Adh*, is an excellent candidate to indirectly estimate levels of gene flow between populations. Nei's [52] measure of interpopulation nucleotide diversity, γ_{st} may be used to estimate the neutral migration parameter, Nm, among all populations in the North American range of *D. pseudoobscura* [69]. γ_{st} is analogous to Wrights F_{st}, which is the ratio of the average intrapopulation heterozygosity to the total heterozygosity of the entire population [77]. In this case, γ_{st} is a function of the number of pairwise differences, k_{ij} averaged across populations. All geographic populations will tend to have uniform gene frequencies if Nm is greater than one because gene flow is extensive. On the other hand, geographic populations will tend to become differentiated by random genetic drift if Nm is less than one because gene flow is limited.

The overall value of Nm for *D. pseudoobscura* was 2.38 migrants exchanged among populations per generation [67], which is sufficient to homogenize gene frequencies among all populations of *D. pseudoobscura* [76]. In fact, evidence for extensive gene flow is dramatically demonstrated by the near identity of two sequences that were collected from British Columbia, Canada and Tulincingo, Mexico. The uniformity of allozyme frequencies is also consistent with extensive gene flow among populations [47]. Thus, any examples of genetic differentiation observed in *D. pseudoobscura* given such high levels of gene flow should be considered as strong evidence for the past action of natural selection [10, 26, 67, 68].

The paracentric inversions of the third chromosome do seem to provide the most striking example of genetic differentiation in *D. pseudoobscura*. The selective forces imposed by different habitats and variable seasons tend to overcome the extensive migration that acts to homogenize the frequencies of the inversions among populations. For example, Arizona populations are nearly fixed for the Arrowhead inversion while populations collected from Washington, Oregon, or California have at least four inversions found at moderate frequency [4, 19]. Immigration of Pacific coast flies should maintain inversion diversity in Arizona populations, however, selection seems to remove individuals that carry all inversion types except for the Arrowhead chromosome.

Pacific coast populations of *D. pseudoobscura* are highly polymorphic for third chromosomal inversions, however, the frequencies of the inversions vary widely among populations. The Treeline (TL) inversion tends to be in high frequency in northern latitudes and in low frequency in southern populations, while the reverse is true of the Chiricahua (CH) chromosome. These data suggest that selection in heterogeneous environments is acting in coastal populations as it probably does in Arizona. What prevents inversions from becoming fixed in any of the populations in California, Oregon, or Washington? Dobzhansky showed that inversion frequencies in populations in the San Jacinto mountains cycle with the seasons [20]. This may suggest that the climate of the Arizona desert is relatively constant through time, while the environments of Pacific coast populations undergo regular climatic changes with each season. Powell and Anderson have reviewed field and laboratory experiments that suggest that inversion homo- and heterozygotes differ in their fitness [3, 54], which may explain the geographic and seasonal cycling.

The methods to estimate levels of gene flow have not changed much since the time of Sewall Wright. The major change has been in the data used to infer levels of gene flow. Proteins were not the best molecules to use to estimate Nm because comparisons of two proteins collected from two populations could not distinguish whether the sequences were identical-by-descent. Nucleotide sequences have removed the ambiguity of protein studies because identity-by-descent can readily be determined.

6. Estimates of linkage disequilibrium and the recombination parameter. Linkage disequilibrium is a statistical measure of nonrandom associations among genetic markers within a population [75]. The expected value of linkage disequilibrium for a pair of selectively neutral sites in a finite population is zero, although the variance of the distribution is quite large [30, 53]. The nucleotide diversity in the *Adh* region provided an excellent opportunity to determine how much linkage disequilibrium exists in a genetic region that seemed to evolve predominantly by mutation and random genetic drift.

The *Adh* region has 359 segregating sites in the sample of 99 sequences that may be tested for statistical associations [65]. Schaeffer and Miller [65] used Fisher's exact test to determine which pairs of nucleotide sites were nonrandomly associated [70]. One problem with this large number of segregating sites is that many comparisons of two sites are possible. The number of independent tests for statistical association between two segregating sites i and j is equal to $(n_i - 1)(n_j - 1)$, where n_i is the number of segregating nucleotides at site i and n_j is the number of segregating nucleotides at site j [75]. The nucleotide sites found by Schaeffer and Miller fell into two classes, sites with two or three nucleotides. The majority of sites, 332, had two nucleotides segregating, while 27 sites had three nucleotides segregating. Thus, 74,278 independent comparisons of the 359 sites are possible for the *Adh* data. The multiple comparison problem was overcome for this data set by applying a sequential Bonferroni method [58].

Only 127 of the 74,278 pairwise comparisons of the 359 segregating sites were found to be nonrandomly associated with Fishers exact test [65] These data support the theoretical expectation that linkage disequilibrium will be zero [30,53] because the majority of nucleotide sites are randomly associated in the *Adh* region. The 46 sites that are in linkage disequilibrium are distributed across the 3.5 kilobases of the *Adh* region, although the sites found to be associated were usually adjacent to each other. Two sets of sites in linkage disequilibrium are unusual because they are clustered in two introns of the *Adh* gene. The first cluster is in the adult intron and is composed of eight segregating sites. The second cluster is in the second intron and is composed of 14 sites.

Low recombination rates, random genetic drift, population subdivision, epistatic selection, and positive Darwinian selection each may create the nonrandom associations that are observed among the sites in the two *Adh* introns [40, 47, 51]. Previous sections of this paper have reviewed how pairwise difference measures were used to rule out random genetic drift, positive Darwinian selection, and population subdivision as explanations for the clusters of linkage disequilibrium. Mutation is the process that introduces nonrandom associations between sites along a chromosome, but recombination is the force that breaks up nonrandom associations between sites. Thus, regions of the genome with low recombination rates will tend to have high levels of linkage disequilibrium among nucleotide sites.

The nucleotide sequence data in the *Adh* region has been used to estimate the recombination parameter to determine if the nonrandom associations in the intron sites are due to low recombination rates. Brown *et al.* [13] have shown that the variance of the number of pairwise differences, k_{ij}, may be written as a function of the pairwise linkage disequilibrium among all pairs of segregating sites and of the variance in segregating site heterozygosity, h_i. The variance of pairwise differences and of segregating sites will be equal if nucleotide sites are associated at random or are in linkage equilibrium. The variance of pairwise differences will be greater than the variance of segregating sites if nucleotides are nonrandomly associated. Hudson [31] used the Brown *et al.* equation for the variance of the pairwise difference distribution and developed a method to estimate the recombination parameter, $4Nc$.

Schaeffer and Miller [65] found that recombination rates are high in the *Adh* region compared to mutation rates. Seven to 17 recombination events occur in this region for each mutation that is introduced in the DNA. These data indicated to Schaeffer and Miller that linkage disequilibrium is short-lived in the *Adh* region because as new associations are introduced by mutation, recombination breaks up the associations. While it is possible that recombination could vary in this 3.5 kilobase region, it is unlikely to have occurred. Thus, recombination may not be used to explain the two clusters of linkage disequilibrium in the introns of *Adh*.

Lewontin [47] has suggested that similar patterns of significant linkage disequilibrium observed between populations is a sensitivE indicator of natural selection. The nonrandom associations observed in the introns are found across all populations of *D. pseudoobscura* consistent with Lewontin's argument. The most likely explanation for the two clusters of linkage disequilibrium in the introns of *Adh* is strong epistatic selection. Epistatic selection implies that certain combinations of the nucleotides in the introns are favored by selection and that recombination events between the major types are slightly deleterious. What is the nature of the selection that maintains these associations in the introns? Wolfgang Stephan and David Kirby (University of Maryland, personal communication) have suggested that the intron clusters of linkage disequilibrium form stem-loop structures in messenger RNA. The nonrandom associations among sites are due to compensatory mutations that maintain secondary structure in mRNA. Thus, populations of *D. pseudoobscura* are segregating for two or three major forms of stem-loop structures. Recombination events between the two alternate forms of the stem loop structure do not appear to be favored by natural selection.

Analyses of linkage disequilibrium should be viewed with caution. The number of significance tests that were performed for the *Adh* data was enormous. With this many tests, one is likely to get rejection of Fisher's exact test just by chance. In addition, some linkage disequilibrium values will tend to be correlated because many tests are performed where one segregating site is compared to all other sites. This may elevate the fraction of tests that show statistical significance. The most troublesome problem is that the power to detect significant linkage disequilibrium between a pair of sites changes with gene frequency. Comparisons of rare frequency sites have low power to detect significant linkage disequilibrium, while comparisons of intermediate frequency sites will have much greater power to find statistical associations. One of the appeals of the Hudson [31] estimator of the recombination parameter is that it summarizes the multilocus structure for a genomic region. Hudson's recombination parameter may not be usd for statistical tests at this time because the variance for the $4Nc$ is not available and is likely to be quite large.

7. Conclusions. Simple estimates of nucleotide diversity may be used to understand the relative contribution of mutation, migration, recombination, and selection in a genomic region. The *Adh* region of *D. pseudoobscura* encodes two proteins that differ in the level of constraint of their amino acid sequences. ADH has a low neutral mutation rate while ADH-DUP has a large neutral mutation rate. Both loci are consistent with a neutral model of molecular evolution where mutation and random genetic drift dictate levels of nucleotide polymorphism. *D. pseudoobscura* is a species capable of extensive gene flow among populations in North America, which should lead to homogeneous gene frequencies among populations. Recombination is a powerful force in the *Adh* region. It is estimated that 7 to 17 recombination events occur for each mutation event that is introduced. This high level of recombination should prevent nonrandom associations among nucleotide sites from accumulating in this region. Coupling the tools of molecular and statistical biology have provided a powerful method to understand the role that evolutionary forces play in modulating genetic diversity.

Acknowledgments. The author would like to thank Ms. Ellen L. Miller for her excellent technical assistance.

REFERENCES

[1] M. AGUADE, N. MIYASHITA, AND C. H. LANGLEY, Reduced variation in the *yellow-achaete-scute* region in natural populations of *Drosophila melanogaster*, Genetics, 122 (1989), pp. 607-615.

[2] R. ALBALAT AND R. GONZALEZ-DUARTE, *Adh* and *Adh-dup* sequences of *Drosophila lebanonensis* and *D. immigrans*: interspecies comparisons, Gene, 126 (1993), pp. 171-178.

[3] W. W. ANDERSON, Selection in natural and experimental populations of *Drosophila pseudoobscura*, Genome, 31 (1989), pp. 239-245.

[4] W. W. ANDERSON, et al., Four decades of inversion polymorphism in *Drosophila pseudoobscura*, Proc. Natl. Acad. Sci. USA, 88 (1991), pp. 10367-10371.

[5] C. F. AQUADRO, Why is the genome variable? Insights from *Drosophila*, Trends in Genet., 8 (1992), pp. 355-362.

[6] C. F. AQUADRO, Molecular population genetics of *Drosophila*, in "Molecular Approaches to Fundamental and Applied Entomology," J. Oakeshott and M. J. Whitten, J. Oakeshott and M. J. Whitten, Springer-Verlag, 1993.

[7] D. J. BEGUN AND C. F. AQUADRO, Molecular population genetics of the distal portion of the X chromosome in Drosophila: Evidence for genetic hitchhiking of the *yellow-achaete* region, Genetics, 129 (1991), pp. 1147-1158.

[8] C. BENYAJATI, *et al.*, Characterization and purification of *Adh* distal promoter factor 2, Adf-2, a cell-specific and promoter-specific repressor in *Drosophila*, Nucleic Acids Res., 20 (1992), pp. 4481-4489.

[9] C. BENYAJATI, *et al.*, The messenger RNA for alcohol dehydrogenase in Drosophila melanogaster differs in its $5'$ end in different developmental stages, Cell, 33 (1983), pp. 125-133.

[10] A. BERRY AND M. KREITMAN, Molecular analysis of an allozyme cline: alcohol dehydrogenase in *Drosophila melanogaster* on the east coast of North America, Genetics, 134 (1993), pp. 869-893.

[11] A. J. BERRY, J. W. AJIOKA, AND M. KREITMAN, Lack of polymorphism on the Drosophila fourth chromosome resulting from selection, Genetics, 129 (1991), pp. 1111-1117.

[12] C. W. BIRKY AND J. B. WALSH, Effects of linkage on rates of molecular evolution, Proc. Natl. Acad. Sci. USA, 85 (1988), pp. 6414-6418.

[13] A. H. D. BROWN, M. W. FELDMAN, AND E. NEVO, Multilocus structure of natural populations of HORDEUM SPONTANEUM, Genetics, 96 (1980), pp. 523-536.

[14] G. K. CHAMBERS, *et al.*, Alcohol-oxidizing enzymes in 13 *Drosophila* species, Biochem. Genet., 16 (1978), pp. 757-767.

[15] A. CHOVNICK, W. GELBART, AND M. MCCARRON, Organization of the Rosy locus in *Drosophila melanogaster*, Cell, 11 (1977), pp. 1-10.

[16] J. A. COYNE, *et al.*, Long-distance migration of *Drosophila*, Am. Nat., 119 (1982), pp. 589- 595.

[17] J. A. COYNE, S. H. BRYANT, AND M. TURELLI, Long-distance migration of *Drosophila*. 2. Presence in desolate sites and dispersal near a desert oasis, Am. Nat., 129 (1987), pp. 847-861.

[18] D.. W. CRUMPACKER AND J. S. WILLIAMS, Density, dispersion, and population structure in *Drosophila pseudoobscura*, Ecol. Monogr., 43 (1973), pp. 499-538.

[19] T. DOBZHANSKY, Chromosomal races in Drosophila pseudoobscura and Drosophila persimilis, Carnegie Inst. Washington Publ., 554 (1944), pp. 47-144.

[20] T. DOBZHANSKY, Genetics of natural populations. IX. Temporal changes in the composition of populations of *Drosophila pseudoobscura*, Genetics, 28 (1943), pp. 162-186.

[21] T. DOBZHANSKY, A review of some fundamental concepts and problems of population genetics, Cold Spring Harbor Symp. Quant. Biol., 20 (1955), pp. 1-15.

[22] T. DOBZHANSKY AND J. R. POWELL, Rates of dispersal of *Drosophila pseudoobscura* and its relatives, Proc. R. Soc. Lond. B., 187 (1974), pp. 281-298.

[23] T. DOBZHANSKY AND M. L. QUEAL, Genetics of natural populations. I. Chromosome variation in populations of *Drosophila pseudoobscura* inhabiting isolated mountain ranges, Genetics, 23 (1938), pp. 239-251.

[24] T.. DOBZHANSKY AND A. H. STURTEVANT, Inversions in the chromosomes of *Drosophila pseudoobscura*, Genetics, 23 (1938), pp. 28-64.

[25] T. DOBZHANSKY AND S. WRIGHT, Genetics of natural populations. X. Dispersion rates in *Drosophila pseudoobscura*, Genetics, 28 (1943), pp. 304-340.

[26] J. A. ENDLER, Gene flow and population differentiation, Science, 179 (1973), pp. 243-250.

[27] Y.-X. FU AND W.-H. LI, Statistical tests of neutrality of mutations, Genetics, 133 (1993), pp. 693-709.

[28] U. HEBERLEIN, B. ENGLAND, AND R. TJIAN, Characterization of Drosophila transcription factors that activate the tandem promoters of the alcohol dehydrogenase gene, Cell, 41 (1985), pp. 965-977.

[29] R. G. Higuchi and H. Ochman, Production of single-stranded DNA templates by exonuclease digestion following the polymerase chain reaction, Nucleic Acids Res., 17 (1989), 5865.

[30] W. G. Hill and A. Robertson, Linkage disequilibrium in finite populations, Theor. Appl. Genet., 38 (1968), pp. 226-231.

[31] R. R. Hudson, Estimating the recombination parameter of a finite population model without selection, Genet. Res., Camb., 50 (1987), pp. 245-250.

[32] R. R. Hudson, Gene genealogies and the coalescent process, Oxf. Surv. Evol. Biol., 7 (1990), pp. 1-44.

[33] R. R. Hudson, et al., Evidence for positive selection in the superoxide dismutase (Sod) region of Drosophila melanogaster, Genetics, 136 (1994), pp. 1329-1340.

[34] R. R. Hudson and N. L. Kaplan, The coalescent process in models with selection and recombination, Genetics, 120 (1988), pp. 831-840.

[35] R. R. Hudson, M. Kreitman, and M. Aguade, A test of neutral molecular evolution based on nucleotide data, Genetics, 116 (1987), pp. 153-159.

[36] P. S. Jeffs, E. C. Holmes, and M. Ashburner, The molecular evolution of the alcohol dehydrogenase and alcohol dehydrogenase-related genes in the Drosophila melanogaster species subgroup, Mol. Biol. Evol., 11 (1994), pp. 287-304.

[37] J. S. Jones, et al., Gene flow and the geographic distribution of a molecular polymorphism in Drosophila pseudoobscura, Genetics, 98 (1981), pp. 157-178.

[38] N. L. Kaplan, T. Darden, and R. R. Hudson, The coalescent process in models with selection, Genetics, 120 (1988), pp. 819-829.

[39] M. Kimura, The Neutral Theory of Molecular Evolution, Cambridge University Press, New York, 1983.

[40] M. Kimura and T. Ohta, Theoretical Aspects of Population Genetics, Princeton University Press, Princeton, NJ, 1971.

[41] J. F. C. Kingman, On the genealogy of large populations, J. Appl. Prob., 19A (1982), pp. 27-43.

[42] J. F. C. Kingman, The coalescent, Stochast. Proc. Appl., 13 (1982), pp. 235-248.

[43] M. Kreitman, Molecular population genetics, Oxf. Surv. Evol. Biol., 4 (1987), pp. 38-60.

[44] M. Kreitman, Nucleotide polymorphism at the alcohol dehydrogenase locus of Drosophila melanogaster, Nature, 304 (1983), pp. 412-417.

[45] M. Kreitman, Detecting selection at the level of DNA, in Evolution at the Molecular Level, R. K. Selander, A. G. Clark, and T. S. Whittam, R. K. Selander, A. G. Clark, and T. S. Whittam, Sinauer, 1991.

[46] M. Kreitman and R. R. Hudson, Inferring the evolutionary histories of the Adh and Adh-Dup loci in Drosophila melanogaster from patterns of polymorphism and divergence, Genetics, 127 (1991), pp. 565-582.

[47] R. C. Lewontin, The Genetic Basis of Evolutionary Change, Columbia University Press, New York, 1974.

[48] R. C. Lewontin, Population genetics, Ann. Rev. Genet., 19 (1985), pp. 81-102.

[49] J. H. McDonald and M. Kreitman, Adaptive protein evolution at the Adh locus in Drosophila, Nature, 351 (1991), pp. 652-654.

[50] K. Moses, U. Heberlein, and M. Ashburner, The Adh gene promoters of Drosophila melanogaster and Drosophila orena are functionally conserved and share features of sequence structure and nuclease-protected sites, Mol. Cell. Biol., 10 (1990), pp. 539-548.

[51] M. Nei, Molecular Evolutionary Genetics, Columbia University Press, New York, 1987.

[52] M. Nei, Evolution of human races at the gene level, in Human Genetics. Part A: The Unfolding Genome, B. Bonne-Tamir, T. Cohen, and R. M. Goodman, B. Bonne-Tamir, T. Cohen, and R. M. Goodman, Alan R. Liss, Inc., 1982.

[53] T. Ohta and M. Kimura, Linkage disequilibrium at steady state determined by random genetic drift and recurrent mutation, Genetics, 63 (1969), pp. 229-238.

[54] J. R. POWELL, Inversion polymorphisms in *Drosophila pseudoobscura* and *Drosophila persimilis*, in Drosophila Inversion Polymorphism, C. B. Krimbas and J. R. Powell, C. B. Krimbas and J. R. Powell, CRC Press, 1992.

[55] J. R. POWELL, *et al.*, Genetics of natural populations. XLIII. Further studies on rates of dispersal of *Drosophila pseudoobscura* and its relatives, Genetics, 82 (1976), pp. 493-506.

[56] S. PRAKASH, Genetic divergence in closely related sibling species *Drosophila pseudoobscura*, *Drosophila persimilis*, and *Drosophila miranda*, Evolution, 31 (1977), pp. 14-23.

[57] S. PRAKASH, Further studies on gene polymorphism in the mainbody and geographically isolated populations of *Drosophila pseudoobscura*, Genetics, 85 (1977), pp. 713-719.

[58] W. R. RICE, Analyzing tables of statistical tests, Evolution, 43 (1989), pp. 223-225.

[59] R. K. SAIKI, *et al.*, Primer-directed enzymatic amplification of DNA with a thermostable DNA polymerase, Science, 239 (1988), pp. 487-491.

[60] F. SANGER, S. NICKEN, AND A. R. COULSON, DNA sequencing with chain-terminating inhibitors, Proc. Natl. Acad. Sci. USA, 74 (1977), pp. 5463-5467.

[61] C. SAVAKIS AND M. ASHBURNER, A simple gene with a complex pattern of transcription: the alcohol dehydrogenase gene of *Drosophila melanogaster*, Cold Spring Harbor Symp. Quant. Biol., 50 (1985), pp. 505-514.

[62] C. SAVAKIS, M. ASHBURNER, AND J.H. WILLIS, The expression of the gene coding alcohol dehydrogenase during the development of *Drosophila melanogaster*, Dev. Biol., 114 (1986), pp. 194-207.

[63] S. W. SCHAEFFER AND C. F. AQUADRO, Nucleotide sequence of the Adh gene region of *Drosophila pseudoobscura*: Evolutionary change and evidence for an ancient gene duplication, Genetics, 117 (1987), pp. 61-73.

[64] S. W. SCHAEFFER AND E. L. MILLER, Molecular population genetics of an electrophoretically monomorphic protein in the alcohol dehydrogenase region of *Drosophila pseudoobscura*, Genetics, 132 (1992), pp. 163-178.

[65] S. W. SCHAEFFER AND E. L. MILLER, Estimates of linkage disequilibrium and the recombination parameter determined from segregating nucleotide sites in the alcohol dehydrogenase region of *Drosophila pseudoobscura*, Genetics, 135 (1993), pp. 541-552.

[66] S. W. SCHAEFFER AND E. L. MILLER, Nucleotide sequence analysis of *Adh* genes estimates the time of geographic isolation of the Bogota population of *Drosophila pseudoobscura*, Proc. Natl. Acad. Sci. USA, 88 (1991), pp. 6097-6101.

[67] S. W. SCHAEFFER AND E. L. MILLER, Estimates of gene flow in *Drosophila pseudoobscura* determined from nucleotide sequence analysis of the alcohol dehydrogenase region, Genetics, 132 (1992), pp. 471-480.

[68] G. M. SIMMONS, *et al.*, Molecular analysis of the alleles of alcohol dehydrogenase along a cline in *Drosophila melanogaster*. I. Maine, North Carolina, and Florida, Evolution, 43 (1989), pp. 393-409.

[69] M. SLATKIN, Inbreeding coefficients and coalescence times, Genet. Res., Camb., 58 (1991), pp. 167-175.

[70] R. R. SOKAL AND F. J. ROHLF, Biometry, W. H. Freeman and Co., New York, 1981.

[71] D. T. SULLIVAN, P. W. ATKINSON, AND W. T. STARMER, Molecular evolution of the alcohol dehydrogenase genes in the genus *Drosophila*, Evol. Biol., 24 (1990), pp. 107-147.

[72] F. TAJIMA, Evolutionary relationship of DNA sequences in finite populations, Genetics, 105 (1983), pp. 437-460.

[73] F. TAJIMA, Statistical method for testing the neutral mutation hypothesis by DNA polymorphism, Genetics, 123 (1989), pp. 585-595.

[74] S. TAVARE, Line-of-descent and genealogical processes, and their applications in population genetic models, Theor. Pop. Biol., 26 (1984), pp. 119-164.

[75] B. S. WEIR, Genetic Data Analysis, Sinauer Associates, Inc., Sunderland, MA,

1990.

[76] S. WRIGHT, Evolution in mendelian populations, Genetics, 16 (1931), pp. 97-159.
[77] S. WRIGHT, The genetical structure of populations, Annals of Eugenics, 15 (1951), pp. 323-354.

ESTIMATION OF THE AMOUNT OF DNA POLYMORPHISM AND STATISTICAL TESTS OF THE NEUTRAL MUTATION HYPOTHESIS BASED ON DNA POLYMORPHISM

FUMIO TAJIMA*

1. Introduction. Whether the amount of genetic variation in a population is maintained by natural selection or by random genetic drift of neutral mutants is one of the most important issues in population genetics.

In the past, the genetic variation was studied mainly by electrophoretic mobility of protein and serological typing of blood group, and Harris [8] and Lewontin and Hubby (Hubby and Lewontin [9]; Lewontin and Hubby [17]) showed that a large amount of genetic variation is maintained at the protein level in human and *Drosophila pseudoobscura* populations, respectively. Since then, a large amount of genetic variation has been observed in populations of various organisms. The amount of polymorphism is determined by many factors, such as mutation, natural selection, population size, population structure, migration, random genetic drift, and so on (Crow and Kimura [2]; Nei [19]). The neutral mutation-random genetic drift hypothesis (or the neutral theory) holds that at the molecular level most evolutionary changes are not caused by Darwinian selection but random drift of mutant alleles that are selectively neutral or nearly neutral (Kimura [11,13], King and Jukes [15], Kimura and Ohta [14]). This means that the amount of polymorphism is mainly determined by the balance between mutational input and random extinction. Since the proposal of the neutral theory, many researchers have tested the neutral theory by examining the amount and pattern of polymorphism revealed by electrophoretic mobility of protein or by serological typing of blood group (e.g., Yamazaki and Maruyama [38,39]; Nei, Fuerst and Chakraborty [20]). These studies have indicated that the general pattern of protein polymorphism can be explained by the neutral theory. Since the statistical power of these tests is not very high, however, we cannot exclude the possibility that a substantial proportion of polymorphic alleles are under some kind of natural selection, such as overdominance.

We can now study the genetic variation at the molecular level by using restriction enzyme technique or DNA sequencing (e.g., Shah and Langley [27], Brown [1], Kreitman [16]). Because of a high resolving power of these techniques, we can study the amount and pattern of polymorphism in detail. Recently, Tajima [31] and Fu and Li [6] have proposed the statistical methods for testing the neutrality of mutations, which can be applied to data obtained by these techniques. In this note, I will briefly

* Department of Population Genetics, National Institute of Genetics, Mishima, Shizuoka-ken 411, Japan.

summarize how to estimate the amount of DNA polymorphism from DNA sequence data and will examine statistical properties of the tests of neutrality. Throughout this note, we use the infinite site model (Kimura [12]), since this model is known to be a good approximation (see Rogers [25]).

2. Estimation of the amount of DNA polymorphism. Two different quantities are often used for measuring the amount of DNA polymorphism, i.e., the number of segregating (or polymorphic) sites and the average number of nucleotide differences among a sample of DNA sequences. For the measurement of DNA polymorphism detectable by restriction endonucleases, see Nei and Tajima [23] and Nei et al.[24]. In order to show how to measure the amount of DNA polymorphism clearly, consider the following DNA sequences with a length of 20 nucleotides:

Sequence 1   ACTGGCTAAGCGCATACTAG

Sequence 2   ACTGGCGAAGCCCATGCTAG

Sequence 3   ACCGGTGAAGTCCATGCTTG

Sequence 4   ACCGGCGAAGCCCATGCTAG

2.1. Number of segregating (or polymorphic) sites. The number of segregating sites (S) is the number of sites which are occupied by at least two different nucleotides. In the above example, there are seven such sites as shown by bold letters, so that $S = 7$. It is easy to see that the number of segregating sites depends on the number of DNA sequences used. As the number of sequences increases, the expectation of the number of segregating sites also increases. There is no simple way to convert the number of segregating sites into a measure which is independent of the number of sequences. Only when mutations are selectively neutral and when the population is panmictic and at equilibrium, can we easily convert the number of segregating sites into a measure which does not depend on the number of sequences.

Watterson [37] showed that when mutations are selectively neutral and when the population is panmictic and at equilibrium, the expectation of the number of segregating sites among a sample of n DNA sequences is given by

(1) $$E(S) = a_1 M$$

where

(2) $$a_1 = \sum_{i=1}^{n-1} \frac{1}{i},$$

$M = 4Nv$, N is the effective population size, and v is the mutation rate per DNA sequence per generation. Since the expectation of S/a_1 is independent of n, it can be used as an estimate of M, namely,

$$(3) \qquad \hat{M} = \frac{S}{a_1}.$$

The variance of $\hat{M}$ depends on the recombination rate. When there is no recombination, the variance of $\hat{M}$ can be estimated by

$$(4) \qquad V(\hat{M}) = \frac{a_1{}^2 S + a_2 S^2}{a_1{}^2 (a_1{}^2 + a_2)}$$

where

$$(5) \qquad a_2 = \sum_{i=1}^{n-1} \frac{1}{i^2}$$

(Tajima [33,34]), which can be obtained by using $V(S) = a_1 M + a_2 M^2$ (Watterson [37]). When there is recombination, a lower bound of $V(\hat{M})$ is given by

$$(6) \qquad V_{\min}(\hat{M}) = \frac{S}{a_1{}^2}$$

(Tajima [33,34]).

Incidentally, values of a_1 and a_2 can be computed by

$$(2a) \qquad a_1 \approx \gamma + \log_e(n - 0.5) + \frac{1}{24n(n-1)}$$

$$(2b) \qquad \approx \gamma + \log_e(n - 0.5)$$

and

$$(5a) \qquad a_2 \approx \frac{\pi^2}{6} - \frac{1}{n - 0.5}$$

(Tajima [33,34]), where $\gamma = 0.5772...$ (Euler's constant) and $\pi^2/6 = 1.6449....$ These formulae give good approximations to (2) and (5).

In the above example, we have $S = 7$ and $n = 4$ so that we obtain $a_1 = 1.8333...$ from (2), $a_1 \approx 1.833$ from (2a), $a_1 \approx 1.830$ from (2b), $a_2 = 1.3611...$ from (5), and $a_2 \approx 1.359$ from (5a). Then, $\hat{M} = 3.818$, $V(\hat{M}) = 5.684$, and $V_{\min}(\hat{M}) = 2.083$ are obtained from (3), (4), and (6), respectively.

It should be noted that these formulae can be used only when mutations are selectively neutral and when the population is panmictic and at equilibrium, since $\hat{M}$ depends on n when these conditions are not satisfied (Tajima [30,32]). In other words, we should not use the number of segregating sites to estimate the amount of DNA polymorphism when we do not know whether these conditions are satisfied or not.

2.2. Average number of nucleotide differences. The amount of DNA polymorphism can be measured by the average number of pairwise nucleotide differences among a sample of DNA sequences, which is defined by

$$(7) \qquad k = \frac{2\sum\limits_{i=1}^{n-1} \sum\limits_{j=i+1}^{n} k_{ij}}{n(n-1)}$$

where k_{ij} is the number of nucleotide differences between sequences i and j and n is the number of DNA sequences sampled from a population as before. In the above example, we have $k_{12} = 3, k_{13} = 7, k_{14} = 4, k_{23} = 4, k_{24} = 1$, and $k_{34} = 3$ so that we obtain $k = 2 \times (3+7+4+4+1+3)/(4 \times 3) = 3.667$. k can be also computed by

$$(8) \qquad k = \sum_{i=1}^{S} h_i$$

(Tajima [31], Nei and Miller [21]), where h_i is the estimate of the expected heterozygosity at the i-th segregating site, which is given by

$$(9) \qquad h_i = n(1 - \sum_{j=1}^{4} x_{ij}{}^2)/(n-1)$$

(Nei and Roychoudhury [22]), in which x_{ij} is the observed frequency of nucleotide j at the i-th segregating site ($j = 1, 2, 3$, and 4 correspond to A, T, G, and C, respectively). The two definitions of k, (7) and (8), always give the same value. In the previous example, the first segregating site is site 3, at which we have $x_{11} = 0, x_{12} = 0.5, x_{13} = 0$, and $x_{14} = 0.5$ so that we obtain $h_1 = 4 \times (1 - 0^2 - 0.5^2 - 0^2 - 0.5^2)/3 = 0.667$. In the same way, we have $h^2 = h^3 = h^4 = h^5 = h^6 = h^6 = 0.5$. Then, we have $k = 0.667 + 0.5 + 0.5 + 0.5 + 0.5 + 0.5 + 0.5 = 3.667$, which is the same as that obtained previously. Which method is easier to calculate depends on the number of segregating sites (S) and the number of DNA sequences (n).

It should be noted that, unlike the number of segregating sites, the average number of nucleotide differences not only has a clear biological meaning but also can be directly used as a measure of the amount of DNA polymorphism, since it does not depend on the sample size.

When mutations are selectively neutral, when the population is panmictic and at equilibrium, and when there is no recombination, Tajima [29] showed that expectation and variance of the average number of nucleotide differences among a sample of n DNA sequences are given by

$$(10) \qquad E(k) = M \text{ and } V(k) = b_1 M + b_2 M^2,$$

where b_1 and b_2 are given by

$$(11) \qquad b_1 = \frac{n+1}{3(n-1)} \quad \text{and} \quad b_2 = \frac{2(n^2+n+3)}{9n(n-1)}.$$

The variance can be divided into two components, i.e., the stochastic and sampling variances, which are given by

$$(12) \qquad V_{st}(k) = \frac{1}{3}M + \frac{2}{9}M^2$$

and

$$(13) \qquad V_{s}(k) = \frac{2}{3(n-1)}M + \frac{2(2_n+3)}{9n(n-1)}M^2,$$

respectively (Tajima [29]). The sampling variance approaches zero as n increases, whereas the stochastic variance is constant.

The total, stochastic, and sampling variances of k can be estimated by

$$(14) \qquad \hat{V}(k) = \frac{3n(n+1)k + 2(n^2+n+3)k^2}{11n^2 - 7n + 6},$$

$$(15) \qquad \hat{V}_{st}(k) = \frac{(3n^2 - 3n + 2)k + 2n(n-1)k^2}{11n^2 - 7n + 6},$$

and

$$(16) \qquad \hat{V}_{s}(k) = \frac{2(3n-1)k + 2(2n+3)k^2}{11n^2 - 7n + 6},$$

respectively (Tajima [33,34]). In the previous example, $k = 3.667$ and $n = 7$, so that we have $\hat{V}(k) = 5.444$, $\hat{V}_{st}(k) = 3.000$, and $\hat{V}_{s}(k) = 2.444$. Formulae (14), (15), and (16) give overestimates if there is recombination. The minimum estimates can be given by

$$(17) \qquad \hat{V}_{min}(k) = \frac{n+1}{3(n-1)}k,$$

$$(18) \qquad \hat{V}_{st,min}(k) = \frac{1}{3}k,$$

and

$$(19) \qquad \hat{V}_{s,min}(k) = \frac{2}{3(n-1)}k.$$

(Tajima [33,34]). In the above example, we have $\hat{V}_{\min}(k) = 2.037$, $\hat{V}_{\text{st},\min}(k) = 1.222$, and $\hat{V}_{\text{s},\min}(k) = 0.815$, which are substantially smaller than those for the case of no recombination.

k depends on the length of DNA sequence (m), so that the average number of nucleotide differences per site can be used instead, which is obtained by dividing k by m. In the above example, the average number of nucleotide differences per site is $k/m = 3.667/20 = 0.183$, which can be called the average heterozygosity per site.

3. Test of neutrality. The neutral mutation hypothesis (Kimura [11], [13]) can be tested by using the pattern of DNA polymorphism in a population. Here we consider a statistical method developed by Tajima [31] and two of four methods developed by Fu and Li [6]. The other two methods developed by Fu and Li [6] can be used only when an outgroup is available, so that they are not considered here.

Incidentally, Hudson et al. [10] and McDonald and Kreitman [18] also developed methods for testing neutrality. These methods, however, utilize not only the within-population data but also the between-population data, so that these methods also are not considered.

3.1. Tajima's test. Tajima [31] developed a statistical method for testing the neutral mutation hypothesis. This method utilizes the difference between the average number of nucleotide differences (k) and $\hat{M}(= S/a_1)$ estimated from the number of segregating sites. Since the expectations of k and $\hat{M}$ are both $M(= 4Nv)$ under the assumptions that mutations are selectively neutral and that the population is panmictic and at equilibrium, we expect $k \approx \hat{M}$ if these assumptions are correct. Namely, the following test statistic is used:

$$(20) \qquad D = \frac{k - S/a_1}{\sqrt{e_1 S + e_2 S(S - 1)}},$$

where

$$e_1 = \frac{b_1 - 1/a_1}{a_1},$$

$$e_2 = \frac{b_2 - (n + 2)/(a_1 n) + a_2/a_1{}^2}{a_1{}^2 + a_2}.$$

Then, D follows the beta distribution with mean 0 and variance 1 approximately. Thus, when the population is known to be panmictic and at equilibrium, if D is significantly different from zero, then the neutral mutation hypothesis is rejected. Furthermore, if D is significantly smaller than zero, then it suggests that purifying selection is operating against deleterious mutation. In contrast, if D is significantly larger than zero, it suggests that balancing selection such as overdominant selection is operating. The confidence limit of D for a given n can be obtained from Table 2 of Tajima [31]. It should be noted that this test is conservative since it depends on the assumption that there is no recombination between sites. In some cases we can detect intragenic recombination (Stephens [28], Sawyer [26], Takahata [35]). For the effects of population subdivision and change in population size on D, see Tajima [30,32,33,34].

3.2. Fu and Li's test. Let s_i be the number of sites at which the nucleotide frequencies are i/n and $1 - i/n$ where $i \leq n/2$. In the previous example, we have $s_1 = 6$ and $s_2 = 1$. Then, the following statistic was proposed by Fu and Li [6]:

$$(21) \qquad D^* = \frac{\frac{n}{n-1}S - a_1 s_1}{\sqrt{V(\frac{n}{n-1}S - a_1 s_1)}},$$

where $V[nS/(n - 1) - a_1 s_1]$ is the variance of $nS/(n - 1) - a_1 s_1$, for the estimation of which, see Fu and Li [6]. Fu and Li [6] also proposed the following test statistic:

$$(22) \qquad F^* = \frac{k - \frac{n-1}{n}s_1}{\sqrt{V(k - \frac{n-1}{n}s_1)}}.$$

For the estimation of $V[k - (n-1)s_1/n]$, see Fu and Li [6]. These statistics are based on the fact that the expectation of k is M, the expectation of S is $a_1 M$, and the expectation of s_1 is $nM/(n - 1)$ (Tajima [31], Fu and Li [6]). Under the same assumptions as those of D, if D^* and F^* are significantly different from zero, the neutral mutation hypothesis is rejected. Furthermore, if D^* and F^* are significantly smaller than zero, then purifying selection is suggested. In contrast, if D^* and F^* are significantly larger than zero, balancing selection such as overdominant selection is suggested. The confidence limits of D^* and F^* are given in Tables 2 and 4 of Fu and Li [6], respectively.

3.3. Difference among the three tests. Fu and Li [6] claimed that their tests are more powerful than Tajima's [31] test, since the correlation between k and S is much stronger than that between S and s_1 or that between k and s_1. In order to know which method is better, I have examined statistical properties of these tests.

All the test statistics can be written as

$$(23) \qquad \sum_{i=1}^{[n/2]} w_i s_i,$$

Table 1. Values of w_i in $\sum_{i=1}^{[n/2]} w_i s_i$ for three methods in the case of S = 10

	Fu and Li		Tajima		Fu and Li		Tajima
i	D*	F*	D	i	D*	F*	D
(a) n = 5				(d) n = 40			
1	−0.119	−0.131	−0.119	1	−0.441	−0.441	−0.235
2	0.179	0.196	0.179	2	0.140	0.046	−0.175
				3	0.140	0.068	−0.118
(b) n = 10				4	0.140	0.088	−0.064
1	−0.225	−0.246	−0.192	5	0.140	0.107	−0.014
2	0.146	0.125	0.003	6	0.140	0.125	0.034
3	0.146	0.164	0.142	7	0.140	0.141	0.077
4	0.146	0.188	0.226	8	0.140	0.157	0.118
5	0.146	0.195	0.253	9	0.140	0.171	0.156
				10	0.140	0.184	0.190
(c) n = 20				11	0.140	0.195	0.221
1	−0.334	−0.352	−0.226	12	0.140	0.206	0.248
2	0.141	0.078	−0.115	13	0.140	0.215	0.273
3	0.141	0.111	−0.017	14	0.140	0.223	0.294
4	0.141	0.139	0.068	15	0.140	0.229	0.312
5	0.141	0.163	0.140	16	0.140	0.235	0.326
6	0.141	0.183	0.199	17	0.140	0.239	0.338
7	0.141	0.198	0.245	18	0.140	0.242	0.346
8	0.141	0.209	0.277	19	0.140	0.244	0.351
9	0.141	0.216	0.297	20	0.140	0.245	0.352
10	0.141	0.218	0.303				

Table 2. Analysis of restriction site polymorphism in $D.$ $melanogaster$

Locus	Pop.	Estimate of M			D^*	F^*	D
		k	S	s_1			
Adh	South	1.47±0.91	0.92±0.51	0	1.02	1.32	1.39
	North	1.17±0.77	0.69±0.43	0	0.91	1.25	1.46
	Pooled	1.42±0.88	0.80±0.43	0	0.95	1.37	1.61
Amy	South	1.65±0.99	1.16±0.59	0.98±0.98	0.17	0.52	1.06
	North	1.79±1.05	1.62±0.74	0.98±0.98	0.50	0.50	0.29
	Pooled	1.74±1.02	1.39±0.62	0.99±0.99	0.35	0.51	0.59
Pu	South	1.45±0.90	1.39±0.67	2.93±1.95	−1.32	−1.02	0.13
	North	1.02±0.70	0.92±0.51	1.95±1.49	−1.13	−0.84	0.23
	Pooled	1.25±0.80	1.19±0.56	1.98±1.47	−0.74	−0.55	0.11
$Gpdh$	South	4.41±2.22	3.00±1.17	1.95±1.49	0.53	0.97	1.44
	North	3.97±2.03	3.70±1.37	1.95±1.49	0.75	0.68	0.23
	Pooled	4.31±2.15	3.58±1.20	2.97±1.88	0.29	0.48	0.59
Sum	South	8.97±2.74	6.47±1.56	5.86±2.64	0.23	0.93	1.98
	North	7.94±2.51	6.93±1.70	4.88±2.32	0.70	0.83	0.72
	Pooled	8.72±2.66	6.96±1.52	5.93±2.58	0.37	0.77	1.16

Data from Takano et al. (1991).   South: Ogasawara population.   North: Aomori population.

where $[n/2]$ is the largest integer which is not larger than $n/2$. It should be noted that, unlike an ordinary weighting factor, w_i can take a negative value. In the case of D, w_i is given by

$$(24) \qquad w_i = \frac{\frac{2i(n-i)}{n(n-1)} - \frac{1}{a_1}}{\sqrt{e_1 S + e_2 S(S-1)}}.$$

In the case of D^*, w_i is given by

$$(25) \qquad w_1 = \frac{\frac{n}{n-1} - a_1}{\sqrt{A}} \quad \text{and} \quad w_i = \frac{n}{(n-1)\sqrt{A}} \quad \text{for } 2 \le i \le [n/2],$$

where $A = V[nS/(n-1) - a_1 s_1]$. On the other hand, w_i for F^* is given by

$$(26) \qquad w_1 = \frac{3-n}{n\sqrt{B}} \quad \text{and} \quad w_i = \frac{2i(n-i)}{n(n-1)\sqrt{B}} \quad \text{for } 2 \le i \le [n/2],$$

where $B = V[k - (n-1)s_1/n]$. It might be clear from (25) and (26) that in the cases of D^* and F^* w_i is always positive for $i \ge 2$ whereas w_1 is always negative. [Note that the tests cannot be applied when $n \le 3$.] Furthermore, w_i for D^* is constant for $i \ge 2$. On the other hand, in the case of D, w_i can be negative even for $i \ge 2$. Table 1 gives an example, where $S = 10$ is assumed. It is clear from this table that w_i is always positive for $i \ge 2$ in D^* and F^*, whereas it can be negative in D even for $i \ge 2$. This difference is quite important. For example, when we observed a large number of sites where nucleotide frequencies are $2/n$ and $1 - 2/n$, D^* and F^* are both positive, and these values, especially D^*, may be significantly larger than zero. On the other hand, D is negative in the case of $n > 10$, and it may be significantly smaller than zero. Thus, the conclusion obtained from D^* and F^* may be completely different from that of D.

The above situation can be seen in the restriction site and insertion/deletion polymorphisms in *Drosophila melanogaster* obtained by Takano et al. [36]. They studied the regions of *Adh*, *Amy*, *Pu* (Punch) and *Gpdh* by using eight 6-cutter restriction endonucleases. Forty-three DNA sequences ($n = 43$) were collected each from two Japanese populations, Ogasawara and Aomori populations. The results of analyses are given in Tables 2 and 3, in which three test statistics are shown together with values of M estimated from three measurements. In the case of restriction site polymorphism, all three test statistics are not significantly different from zero. Thus, we can concluded that the restriction site polymorphism is consistent with the neutral mutation hypothesis. We can see from Table 2, however, that D is slightly different from D^* and F^*. Namely, in *Pu* locus D^* and F^* are negative (e.g., $D^* = -1.32$ in the Ogasawara population) whereas D is positive, although they are not significantly different from zero. This is because $S = 6$ and $s_1 = 3$ in the Ogasawara population and $S = 4$ and $s_1 = 2$ in the Aomori population. The distributions of s_i for

Table 3. Analysis of insertion/deletion polymorphism in *D. melanogaster*

Locus	Pop.	Estimate of M			D^*	F^*	D
		k	S	s_1			
Adh	South	0.36±0.35	0.92±0.51	1.95±1.49	−1.13	−1.43	−1.44
	North	1.24±0.80	1.85±0.82	0.98±0.98	0.62	0.15	−0.92
	Pooled	0.82±0.59	1.99±0.78	1.98±1.47	0.01	−0.62	−1.54
Amy	South	0.93±0.65	2.08±0.89	5.86±3.23	−2.47	−2.57[a]	−1.60
	North	0.14±0.21	0.46±0.34	0.98±0.98	−0.85	−1.14	−1.30
	Pooled	0.59±0.48	1.99±0.78	4.94±2.63	−2.05	−2.34[a]	−1.84[a]
Pu	South	0.05±0.12	0.23±0.23	0.98±0.98	−1.80	−1.85	−1.12
	North	0.09±0.17	0.23±0.23	0	0.56	0.18	−0.85
	Pooled	0.07±0.14	0.40±0.29	0.99±0.99	−1.03	−1.31	−1.31
Gpdh	South	1.32±0.84	2.54±1.03	6.84±3.64	−2.43	−2.49	−1.44
	North	0.87±0.62	1.62±0.74	0.98±0.98	0.50	−0.07	−1.27
	Pooled	1.10±0.73	2.98±1.05	7.91±3.67	−2.61[a]	−2.76[a]	−1.79[a]
Sum	South	2.65±1.13	5.78±1.47	15.63±5.18	−3.90[b]	−4.08[b]	−2.62[c]
	North	2.34±1.05	4.16±1.18	2.93±1.69	0.60	−0.23	−1.95[a]
	Pooled	2.58±1.06	7.36±1.55	15.81±4.85	−2.98[b]	−3.60[b]	−3.13[c]

Data from Takano et al. (1991).   South: Ogasawara population.   North: Aomori population.

a, b and c: Significant at 5%, 2% and 0.1% levels, respectively.

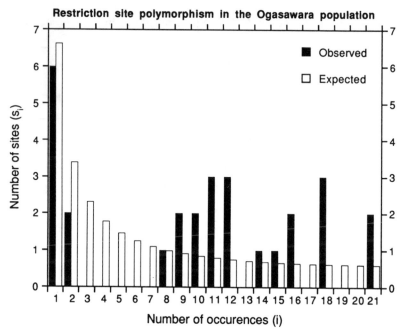

FIG. 1. Distribution of si for the restriction site polymorphism in the Ogasawara population of Drosophila melanogaster. Data from Takano et al. [36].

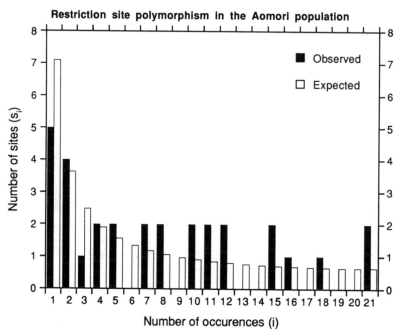

FIG. 2. Distribution of si for the restriction site polymorphism in the Aomori population of Drosophila melanogaster. Data from Takano et al. [36].

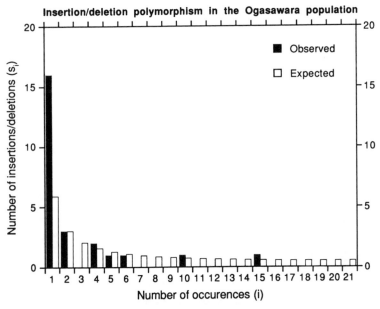

FIG. 3. Distribution of si for the insertion/deletion polymorphism in the Ogasawara population of Drosophila melanogaster. Data from Takano et al. [36].

pooled data of restriction site polymorphism are shown in Figures 1 and 2, together with the expectations of s_i for a given S, which are given by

$$(27a) \qquad E(s_i|S) \quad = \quad \frac{nS}{i(n-i)a_1} \quad \text{for } i < n/2$$

$$(27b) \qquad\qquad\qquad = \quad \frac{nS}{2i(n-i)a_1} \quad \text{for } i = n/2$$

(Tajima [31]). Figure 1 indicates that the observed distribution of s_i seems to be different from the expected one in the Ogasawara population. Namely, s_i tends to be smaller than the expectation for $i \leq 7$ and larger for $i \geq 9$. In fact, we obtained $D = 1.98$, which is almost significantly different from zero at the 5% level. On the other hand, we obtained $D^* = 0.23$ and $F^* = 0.93$ since $s_1 = 6$ is close to the expectation.

Table 3 and Figures 3 and 4 give the results of analyses for the insertion/deletion polymorphism. It is clear from Table 3 that all the test statistics are significantly smaller than zero in the Ogasawara population, so that we reject the neutrality. In the Aomori population, D is significantly smaller than zero, whereas D^* and F^* are close to zero. This difference can be seen in Figure 4. Namely, $s_1 = 3$ is smaller than the expectation and $s_2 = 9$ is substantially larger than the expectation. If we consider the fact that s_i is smaller than the expectation for $i \geq 6$ with the exception of

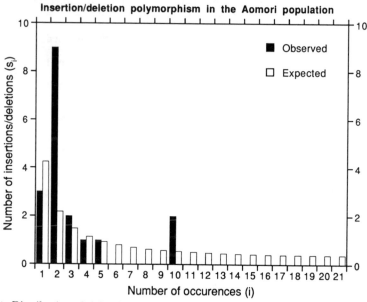

FIG. 4. Distribution of si for the insertion/deletion polymorphism in the Aomori population of Drosophila melanogaster. Data from Takano et al. [36].

s_{10}, the conclusion obtained from D might be more reasonable in this case. In any case, it may be dangerous to rely on s_1 too much.

4. Conclusion. In this note I have examined statistical properties of two measures of the amount of DNA polymorphism, i.e., the number of segregating sites (S) and the average number of nucleotide differences (k) among a sample of DNA sequences. The number of segregating sites can be used only when the following conditions are satisfied: (1) mutations are selectively neutral and (2) the population is panmictic and at equilibrium. On the other hand, the average number of nucleotide differences can be used even when the above conditions are not satisfied. Recently Felsenstein [3,4], Fu and Li [7], and Fu [5] developed different methods for estimating the amount of DNA polymorphism. However, these methods also can be used only when the above conditions are satisfied.

I have also examined the three statistical methods for testing the neutral mutation hypothesis, i.e., Tajima's [31] and Fu and Li's [6] tests. Fu and Li [6] claimed that their tests are more powerful than is Tajima's [31] test since the correlation between k and S is much stronger than that between S and s_1 or that between k and s_1. However, this is not the case, because there is no reason to believe that a strong correlation weakens a statistical power as long as the covariance is taken into account. Fu and Li's [6] tests, especially one based on D^*, strongly rely on the difference between the observed and expected values of s_1, whereas Tajima's [31]) test does not. If we consider the distribution of s_i, Tajima's [31] test seems to be better than Fu and Li's [6] tests. More studies, however, might be necessary before we draw a rigid conclusion.

Acknowledgment. I would like to thank an anonymous reviewer for her/his useful suggestions and comments.

REFERENCES

[1] Brown, W. M., 1980, Polymorphism in mitochondrial DNA of humans as revealed by restriction endonuclease analysis. Proc. Natl. Acad. Sci. USA **77**: 3605-3609.

[2] Crow, J. F. and M. Kimura, 1970, *An Introduction to Population Genetics Theory.* Harper and Row, New York.

[3] Felsenstein, J., 1992, Estimating effective population size from samples of sequences: inefficiency of pairwise and segregating sites as compared to phylogenetic estimates. Genet. Res. **59**: 139-147.

[4] Felsenstein, J., 1992, Estimating effective population size from samples of sequences: a bootstrap Monte Carlo integration method. Genet. Res. **60**: 209-220.

[5] Fu, Y.-X., 1994, A phylogenetic estimator of effective population size or mutation rate. Genetics **136**: 685-692.

[6] Fu, Y.-X. and W.-H. Li, 1993, Statistical tests of neutrality of mutations. Genetics **133**: 693-709.

[7] Fu, Y.-X. and W.-H. Li, 1993, Maximum likelihood estimation of population parameters. Genetics **134**: 1261-1270.

[8] Harris, H., 1966, Enzyme polymorphisms in man. Proc. R. Soc. Lond. B **164**: 298-310.

[9] Hubby, J. L. and R. C. Lewontin, 1966, A molecular approach to the study of genic heterozygosity in natural populations. I. The number of alleles at different loci in *Drosophila pseudoobscura*. Genetics **54**: 577-594.

[10] Hudson, R. R., M. Kreitman and M. Aguadé, 1987, A test of neutral molecular evolution based on nucleotide data. Genetics **116**: 153-159.

[11] Kimura, M., 1968, Evolutionary rate at the molecular level. Nature **217**: 624-626.

[12] Kimura, M., 1969, The number of heterozygous nucleotide sites maintained in a finite population due to steady flux of mutations. Genetics **61**: 893-903.

[13] Kimura, M., 1983, *The Neutral Theory of Molecular Evolution.* Cambridge Univ. Press, Cambridge.

[14] Kimura, M. and T. Ohta, 1971, Protein polymorphism as a phase of molecular evolution. Nature **229**: 467-469.

[15] King, J. L. and T. H. Jukes, 1969, Non-Darwinian evolution. Science **164**: 788-798.

[16] Kreitman, M., 1983, Nucleotide polymorphism at the alcohol dehydrogenase locus of *Drosophila melanogaster*. Nature **304**: 412-417.

[17] Lewontin, R. C. and J. L. Hubby, 1966, A molecular approach to the study of genic heterozygosity in natural populations. II. Amount of variation and degree of heterozygosity in natural populations of *Drosophila pseudoobscura*. Genetics **54**: 595-609.

[18] McDonald, J. H. and M. Kreitman, 1991, Adaptive protein evolution at the *Adh* locus in *Drosophila*. Nature **351**: 652-654.

[19] Nei, M., 1975, *Molecular Population Genetics and Evolution.* North-Holland, Amsterdam.

[20] Nei, M., P. A. Fuerst and R. Chakraborty, 1976, Testing the neutral mutation hypothesis by distribution of single locus heterozygosity. Nature **262**: 491-

493.

[21] Nei, M. and J. C. Miller, 1990, A simple method for estimating average number of nucleotide substitutions within and between populations from restriction data. Genetics 125: 873-879.

[22] Nei, M. and A. K. Roychoudhury, 1974, Sampling variances of heterozygosity and genetic distance. Genetics 76: 379-390.

[23] Nei, M. and F. Tajima, 1981, DNA polymorphism detectable by restriction endonucleases. Genetics 97: 145-163.

[24] Nei, M., F. Tajima and T. Gojobori, 1984, Classification and measurement of DNA polymorphism. pp. 307-330 in *Human Population Genetics: The Pittsburgh Symposium*, edited by A. Chakravarti. Van Nostrand Reinhold, New York.

[25] Rogers, A., 1992, Error introduced by the infinite-sites model. Mol. Biol. Evol. 9: 1181-1184.

[26] Sawyer, S., 1989, Statistical tests for detecting gene conversion. Mol. Biol. Evol. 6: 526-538.

[27] Shah, D. M. and C. H. Langley, 1979, Inter- and intraspecific variation in restriction maps of *Drosophila* mitochondrial DNAs. Nature 281: 696-699.

[28] Stephens, J. C., 1985, Statistical methods of DNA sequence analysis: detection of intragenic recombination or gene conversion. Mol. Biol. Evol. 2: 539-556.

[29] Tajima, F., 1983, Evolutionary relationship of DNA sequences in finite populations. Genetics 105: 437-460.

[30] Tajima, F., 1989, DNA polymorphism in a subdivided population:the expected number of segregating sites in the two subpopulation model. Genetics 123: 229-240.

[31] Tajima, F., 1989, Statistical method for testing the neutral mutation hypothesis by DNA polymorphism. Genetics 123: 585-595.

[32] Tajima, F., 1989, The effect of change in population size on DNA polymorphism. Genetics 123: 597-601.

[33] Tajima, F., 1993, Measurement of DNA polymorphism. pp. 37-59 in *Mechanisms of Molecular Evolution,* edited by N. Takahata and A. G. Clark. Japan Sci. Soc. Press, Tokyo/Sinauer Associates, Sunderland, MA.

[34] Tajima, F., 1993, Statistical analysis of DNA polymorphism. Jpn. J. Genet. 68: 567-595.

[35] Takahata, N., 1994, Comments on the detection of reciprocal recombination or gene conversion. Immunogenetics 39: 146-149.

[36] Takano, T. S., S. Kusakabe and T. Mukai, 1991, The genetic structure of natural populations of *Drosophila melanogaster.* XXII. Comparative study of DNA polymorphisms in northern and southern natural populations. Genetics 129: 753-761.

[37] Watterson, G. A., 1975, On the number of segregating sites in genetic models without recombination. Theor. Popul. Biol. 7: 256-276.

[38] Yamazaki, T. and T. Maruyama, 1972, Evidence for the neutral hypothesis of protein polymorphism. Science 178: 56-58.

[39] Yamazaki, T. and T. Maruyama, 1974, Evidence that enzyme polymorphisms are selectively neutral, but blood group polymorphisms are not. Science 183: 1091-1092.

COMPUTATIONAL METHODS FOR THE COALESCENT*

ROBERT C. GRIFFITHS[†] AND SIMON TAVARÉ[‡]

Abstract. This paper describes recent work on computational methods for the coalescent. We show how integro-recurrence relations for sampling distributions and related quantities may be solved by a simple Markov chain Monte Carlo method. We describe the method in the context of the coalescent process for a population that is evolving according to a deterministic population size function. The usual constant population size models appear as a special case of this approach. One of the appealing features of the approach is its generic nature: many apparently different problems may be attacked with this one approach. A wide variety of examples are discussed, among them maximum likelihood estimation of parameters.

Key words. Coalescent process, Markov chain Monte Carlo, Population genetics, Sampling distributions, Variable population size.

AMS(MOS) subject classifications. 60G35, 92A05, 92A10.

1. Introduction. Kingman's introduction of the coalescent in 1982 [15] [16] has had the effect of focusing attention on the role played by genealogy in the evolution of populations. Coalescent arguments are now standard in the population genetics literature, and many novel applications and extensions continue to be found. For recent reviews of a range of applications, see [13], [14] and [26] for example. One important feature of this genealogical approach is that it provides a simple way to simulate the behavior of samples taken from populations undergoing very complicated mutation mechanisms, usually without having to simulate the structure of the entire population. In general terms, the idea is to generate the genealogy of the sample back to the common ancestor, and then simulate the effects of mutation down the ancestral tree from this most recent common ancestor to the individuals in the sample. This approach is very useful for studying the statistical behavior of allelic configurations in a sample.

In contrast, we discuss methods that intrinsically use the mutation process in the other direction, from the individuals in the sample back to the common ancestor. This approach has proven useful for computing the probability that a sample of genes has a *particular* allelic configuration. Such sampling probabilities, epitomized by the Ewens Sampling Formula [4], play an important role in the statistical analysis of genomic variability, as they form the basis of a likelihood approach to inference in the coalescent.

Our approach to sampling probabilities and related quantities is via the

---

* Supported in part by Australian Research Grant A19131517, and NSF grant DMS 90-05833.

† Mathematics Department, Monash University, Clayton 3168, Australia.

‡ Departments of Mathematics and Biological Sciences, University of Southern California, Los Angeles, CA 90089-1113, USA.

derivation of certain integro-recurrence equations that they satisfy. These recursions are determined by what happens as we look back up the ancestral tree towards the root. Either we see a coalescence event or a mutation (or perhaps a recombination) event, and these different possibilities lead to a recursive formula for the sampling probability. These recursions are usually difficult to solve, either explicitly or by conventional numerical analysis techniques. We have developed a Markov chain Monte Carlo technique by which the solutions to such recursions can be approximated. The idea, typical of Monte Carlo methods, is to represent the quantity of interest as the mean of a functional of a non-homogeneous Markov chain, and to estimate this mean by repeated simulations of the chain.

There are many variants on this theme, among them a surface simulation technique that uses a single Markov chain to generate solutions of recurrences with different parameters (this being particularly useful to compute Monte Carlo approximants to likelihood surfaces), and a version that can be used to solve non-homogeneous recursions. We discuss these issues, and give a variety of examples, later in the paper.

The setting for the subsequent development is the coalescent evolving with a deterministically varying population size [16] [25]. We present the structure of this process as a deterministic time change of the usual coalescent, and use this representation to derive the appropriate sampling distributions. The results extend those in [8]. In particular, we show how the method can be used to study models for DNA sequence data in which the stationary distribution of a sequence is arbitrary, for example a high-order Markov measure. This extends earlier work in [6].

2. The coalescent. The ancestry of the genes in a random sample from a population is often modeled by a continuous-time stochastic process known as the *coalescent*. This process was introduced by Kingman [15] [16] as an approximation, valid in the limit of large population size, to the ancestral structure of a wide variety of selectively neutral reproduction models, including the Wright-Fisher model. Specifically, assume that a random sample of n genes is taken from a large random mating population with non- overlapping generations that have been of constant size N genes for a long time. Label the sampling generation as generation 0, and let $A_n^N(r)$ be the number of distinct ancestors the sample has r generations into the past. For the Wright-Fisher model, which can be thought of as the case where genes choose their parent genes uniformly and at random, Kingman [15] showed that as $N \to \infty$ the process $\{A_n^N(\lfloor Nt \rfloor), t \geq 0\}$ converges in distribution to a Markov pure death process $\{A_n(t), t \geq 0\}$ on the integers $n, n - 1, \ldots, 1$. $A_n(\cdot)$ moves from state k to $k - 1$ at rate $k(k - 1)/2$.

The same limiting process applies to many other neutral models of reproduction, as long as the genealogy does not collapse or expand too fast [15]. Let σ^2 being the (limiting) variance of the number of offspring of a typical gene, and suppose $0 < \sigma^2 < \infty$. When time is scaled in units of $\sigma^{-2}N$ generations, then the process $\{A_n^N(\lfloor \sigma^{-2}Nt \rfloor), t \geq 0\}$ converges in distribution to $\{A_n(t), t \geq 0\}$.

The coalescent may be thought of as a random tree, with leaves representing the n sample genes, and vertices where ancestral lines join. In this continuous-time approximation the tree is binary, and the topology of the tree is obtained by randomly merging pairs of individuals. The root of the tree is the most recent common ancestor (MRCA) of the sample of genes. Let T_j be the amount of time the sample has j distinct ancestors, for $j = n, n - 1, \ldots, 2$. The T_j are independent exponential random variables, with means $\mathbb{E}T_j = 2/(j(j-1))$, from which it follows that the time $T_{MRCA} = T_n + \cdots + T_2$ back to the MRCA has mean

$$(2.1) \qquad \mathbb{E}T_{MRCA} = 2\left(1 - \frac{1}{n}\right).$$

The coalescent may be modified to account for the effects of stochastic or deterministic variation in the population size [16]. In the determinsitic case, suppose that the population is of size $M(0) \equiv N$ genes at the time of sampling, and is of size $M(r)$ in the rth generation before sampling. Let $\sigma^2(r)$ be the variance of the number of offspring born to an individual in generation r. We concentrate here on fluctuations in the population size that are of order N. We assume that there is a relative size function v, and a variance function τ^2 such that for all $x \geq 0$

$$v(x) \equiv \lim_{N \to \infty} \frac{M(\lfloor Nx \rfloor)}{M(0)} > 0,$$

and

$$\tau^2(x) = \lim_{N \to \infty} \sigma^2(\lfloor Nx \rfloor).$$

For the Wright-Fisher model, $\tau^2(x) \equiv 1$. Define the population size intensity function Λ and its density λ by

$$(2.2) \qquad \Lambda(t) = \int_0^t \frac{\tau^2(x)}{v(x)} dx, \quad \lambda(t) = \frac{\tau^2(t)}{v(t)}, \quad t > 0.$$

With time measured in units of N generations, we may once more approximate the distribution of the ancestral process by a continuous time non-homogeneous death process $\{A_n^v(t), t \geq 0\}$, whose structure is most simply defined as a deterministic time change of the process $A_n(\cdot)$:

$$(2.3) \qquad A_n^v(t) = A_n(\Lambda(t)), \quad t \geq 0.$$

All of the properties of the variable-size process can be calculated using the representation in (2.3). For example, this shows immediately that if

$\Lambda(\infty) = \infty$ the sample may be traced back to a common ancestor with probability one. It also defines the joint distribution of the times $T_j, j = n, n-1, \ldots, 2$ (which are no longer independent), and so allows us to study properties of T_{MRCA}. These distributions are somewhat unmanageable, but they are very simple to simulate. Let $S_n < S_{n-1} < \cdots < S_2$ be the times at which $A_n^v(\cdot)$ moves to $n-1, n-2, \ldots, 1$, and set $S_{n+1} \equiv 0$. The sequence $S_n, S_{n-1}, \ldots, S_2$ is Markovian, and the coupling in (2.3) shows that

$$\Lambda(S_j) - \Lambda(S_{j+1}) \overset{d}{=} E_j, j = n, n-1, \ldots, 2$$

where the E_j are independent exponential random variables with parameter $j(j-1)/2$. To simulate the sequence of jump times, we need only
  (i) Set $s = 0, j = n$
  (ii) Generate E_j exponential $j(j-1)/2$
  (iii) Solve for t the equation $\Lambda(t) - \Lambda(s) = E_j$
  (iv) Set $S_j = t, s = t, j = j - 1$
  (v) If j is greater than 1, go to step (ii).
The topology of the ancestral tree in the variable population size case is formed just as before: merge a random pair of individuals at each of the times $S_n, \ldots, S_2$.

3. The effects of mutation. The effects of mutation are superimposed on the ancestral tree of the sample. These mutations occur according to Poisson processes of rate $\theta/2$ along each edge of the tree, the processes in different edges being conditionally independent given the length of the edges. As a limit from the Wright-Fisher model, if u is the probability of a mutation per gene per generation, then $\theta = \lim_{N \to \infty} 2Nu$, where N is the size of the population of genes from which the sample was taken.

The effects of each mutation may be modeled in many different ways. For example, to describe the evolution of a sample of DNA sequences in which the sites in the sequence are completely linked, it is convenient to consider a general mutation scheme in which there are d possible types of gene, labeled $1, 2, \ldots, d$. When a mutation arises in a lineage, a transition is made from type i to j according to the entry p_{ij} in a transition matrix P. It is convenient to allow entries on the diagonal of P to be non-zero, thereby providing for different overall mutation rates for the different types. The mutation rates in the model are uniquely determined by the generator matrix

(3.1)
$$R = (r_{ij}) \equiv \frac{\theta}{2}(P - I).$$

The configuration of types in the sample is determined by the mutations in the tree from the root to the leaves. It is usually assumed that P is a regular matrix with a stationary distribution π. If the common ancestor is chosen from a stationary population, then her type has the distribution π.

3.1. Mutation models for DNA sequences. If the sequences are of length s, and the alphabet of bases at each site has a elements (typically $a = 2$ or 4), then $d = a^s$. Types are denoted by sequences $\boldsymbol{i} = (i_1, \ldots, i_s)$ with entries in $[a] \equiv \{1, 2, \ldots, a\}$. We assume that there is no recombination between sites.

The standard model assumes that mutations cause just a single base change, site l being chosen with probability $h_l > 0$, $l = 1, 2, \ldots, s$. The lth site has transition matrix M_l, with stationary distribution $\boldsymbol{\pi}^l$. The mutation matrix P is then given by

$$(3.2) \qquad P = \sum_{l=1}^{s} h_l I \otimes I \otimes \cdots \otimes M_l \otimes \cdots \otimes I,$$

where $\otimes$ denotes direct product, I is the identity matrix, and

$$\boldsymbol{\pi} = \boldsymbol{\pi}^1 \otimes \cdots \otimes \boldsymbol{\pi}^s.$$

This model allows for variable rates at different sites (the overall rate at site l being $\theta h_l / 2$), and for arbitrary substitution probabilities at each site. Nonetheless, the stationary distribution of this mutation mechanism corresponds to independent trials, so that a single gene sampled at random from a stationary population will appear to have independent sites. In its simplest form, the model takes $h_l \equiv 1/s$ and $M_l \equiv M$, so that there are no mutational hotspots, and a gene sampled at stationarity appears to have i.i.d. sites.

For many gene regions, this independence of sites feature is clearly violated (cf. [1], [28]), and a more complex model of the substitution process is required to fit observed data. These authors note that many DNA sequences exhibit a Markovian structure. With this in mind, we describe a simple model, explored in more detail in Tavaré [27], that may be arranged to have an arbitrary stationary measure. To illustrate, suppose that we require the stationary measure $\boldsymbol{\pi}$ to be a Markov measure, that is for each $\boldsymbol{i} = (i_1, \ldots, i_s)$

$$(3.3) \qquad \pi(\boldsymbol{i}) = \mu(i_1) \prod_{j=2}^{s} t(i_{j-1}, i_j)$$

where $T = (t(l, m), 1 \leq l, m \leq a)$ is a strictly positive stochastic matrix, and $(\mu(l), 1 \leq l \leq a)$ is a probability distribution.

Suppose that the sequence is currently of type $\boldsymbol{i}$. A potential mutation changes a sequence of type $\boldsymbol{i}$ to a sequence of type $\boldsymbol{j}$ with probability $p(\boldsymbol{i}, \boldsymbol{j})$ determined by (3.2). Thus the potential mutant sequence differs from its parent in at most a single coordinate. The net effect of this potential mutation depends, however, on the bases at neighboring sites. For two sequences $\boldsymbol{i}$ and $\boldsymbol{j}$ that differ at a single coordinate, define

$$h(\boldsymbol{i}, \boldsymbol{j}) = \frac{\pi(\boldsymbol{j}) p(\boldsymbol{j}, \boldsymbol{i})}{\pi(\boldsymbol{i}) p(\boldsymbol{i}, \boldsymbol{j})} \wedge 1.$$

The mutation mechanism then makes a mutant of type j with probability $h(i, j)$, and otherwise makes a 'mutant' of type i, the original type.

When the mutation matrices M_l are identical at each site with $M_l \equiv M = (m(i, j))$, the stationary distribution of this model is π determined by (3.3). To check this, we need only observe that this construction is a variant of Hasting's algorithm [12], familiar to Markov chain Monte Carlo enthusiasts. The form of $h(i, j)$ simplifies considerably for the present example. Simple algebra shows that when i and j differ in just the lth coordinate

$$
h(i, j) = \begin{cases}
\dfrac{m(j_1, i_1)\mu(j_1)t(j_1, i_2)}{m(i_1, j_1)\mu(i_1)t(i_1, i_2)}, & l = 1 \\[2ex]
\dfrac{m(j_l, i_l)t(i_{l-1}, j_l)t(j_l, i_{l+1})}{m(i_l, j_l)t(i_{l-1}, i_l)t(i_l, i_{l+1})}, & 2 \le l \le s - 1 \\[2ex]
\dfrac{m(j_s, i_s)t(i_{s-1}, j_s)}{m(i_s, j_s)t(i_{s-1}, i_s)}, & l = s
\end{cases}
$$

Thus a mutation mechanism determined by the bases at sites adjacent to the target site can readily produce a Markov dependent stationary distribution. Clearly, this scheme can be generalized in many ways to produce stationary measures of great complexity, for example those with high order Markov dependence, and those with non-homogeneous Markov structure.

In the next section, we show how the distribution of the allelic configuration of a sample of genes undergoing such a mutation mechanism in a population of deterministically varying size can be calculated.

3.2. Sampling distributions. In this section, we return to our original labeling of types as $1, 2, \ldots, d$, with mutation matrix $P = (p_{ij})$. Let $q(t, n)$ be the probability that a sample of n genes taken at time t in the past has a type configuration of $n = (n_1, \ldots, n_d)$, where n_i is the number of copies of type i in the sample. The fundamental integro-recurrence relation for $q(t, n)$ is derived by considering the configuration of genes at the time of the first event (either a coalescence or a mutation) in the ancestry of the sample prior to time t. The time W_t of this first event has distribution determined by

$$
(3.4) \qquad \mathbb{P}(W_t > s) = \exp\left(-\int_t^s \gamma(u, n)\,du\right), \quad s \ge t,
$$

where

$$
\gamma(u, n) = \frac{n(n-1)}{2}\lambda(u) + \frac{n\theta}{2}.
$$

The integro-recurrence for the sampling formula takes the form

$$
q(t, n) = \int_t^\infty \left\{ \frac{n\theta}{2\gamma(s, n)} \sum_{\substack{i, j \in [d] \\ n_j > 0, i \ne j}} \frac{n_i + 1}{n} p_{ij} q(s, n + e_i - e_j) \right.
$$

$$+ \frac{n\theta}{2\gamma(s, \boldsymbol{n})} \sum_{i \in [d]} \frac{n_i}{n} p_{ii} q(s, \boldsymbol{n})$$

(3.5)

$$\left. + \frac{n(n-1)\lambda(s)}{2\gamma(s, \boldsymbol{n})} \sum_{j \in [d], n_j > 0} \frac{n_j - 1}{n - 1} q(s, \boldsymbol{n} - \boldsymbol{e}_j) \right\} g(t, \boldsymbol{n}; s) ds,$$

where $\{\boldsymbol{e}_i\}$ are the d unit vectors, and

$$g(t, \boldsymbol{n}; s) = \gamma(s, \boldsymbol{n}) \exp \left(- \int_t^s \gamma(u, \boldsymbol{n}) du \right)$$

is the density of W_t. Boundary conditions are required to determine the solution to (3.5). These have the form

(3.6) $$q(t, \boldsymbol{e}_i) = \pi_i^*, \ i = 1, \ldots, d,$$

where π_i^* is the probability that the most recent common ancestor is of type i. It is often assumed that

(3.7) $$\pi_i^* = \pi_i, \ i = 1, \ldots, d,$$

where $\boldsymbol{\pi} = (\pi_1, \ldots, \pi_d)$ is the stationary distribution of P. Of particular interest is the solution $q(\boldsymbol{n}) \equiv q(0, \boldsymbol{n})$.

To derive (3.4), we first calculate the probability of no coalescence events in a sample of size n in time (t, s). This event has probability

$$\exp \left(- \frac{n(n-1)}{2} (\Lambda(s) - \Lambda(t)) \right).$$

Given no coalescences in (t, s), the conditional probability of no mutations in (t, s) is just the chance that no mutations occur on the n branches of the ancestral tree in time (t, s). This is

$$\exp \left(- \frac{n\theta}{2} (s - t) \right).$$

To verify (3.5) suppose that the first event prior to time t occurred at time $W_t = s$. The relative rates of mutation and coalescence for the n genes are $n\theta/2 : n(n-1)\lambda(s)/2$, so the probability that the event at time s is a mutation is $n\theta/2\gamma(s, \boldsymbol{n})$. To obtain a configuration of $\boldsymbol{n}$ after a mutation the configuration at time $s+$ must be either $\boldsymbol{n}$, and a transition $i \to i$ takes place for some $i \in [d]$ (the mutation resulted in no observable change), or $\boldsymbol{n} + \boldsymbol{e}_i - \boldsymbol{e}_j, \ i, j \in [d], n_j > 0, i \neq j$ and a transition $i \to j$ takes place. On the other hand, the probability that the event at s is a coalescence is $n(n-1)/2\gamma(s, \boldsymbol{n})$. To obtain a configuration $\boldsymbol{n}$ the configuration must be $\boldsymbol{n} - \boldsymbol{e}_j$ for some $j \in [d]$ with $n_j > 0$ and the ancestral lines involved in the coalescence must be of type j. Averaging over the density of W_t produces (3.5).

An alternative (and equivalent) integro-recurrence relation for $q(t, \boldsymbol{n})$ can be derived by considering the configuration of genes at the time of the first event *that changes the configuration* in the ancestry of the sample prior to time t. The time W_t^* of this first event has distribution determined by

$$\mathbb{P}(W_t^* > s) = \exp\left(-\int_t^s \gamma^*(u, \boldsymbol{n})du\right), \quad s \geq t,$$

where

(3.8) $$\gamma^*(u, \boldsymbol{n}) = \frac{n(n-1)}{2}\lambda(u) + \frac{n\theta}{2}\left(1 - \sum_{i \in [d]} \frac{n_i}{n}p_{ii}\right).$$

The corresponding recurrence is

$$
\begin{aligned}
q(t, \boldsymbol{n}) \;=\; \int_t^\infty &\left\{ \frac{n\theta}{2\gamma^*(s, \boldsymbol{n})} \sum_{\substack{i, j \in [d] \\ n_j > 0, i \neq j}} \frac{n_i + 1}{n} p_{ij} q(s, \boldsymbol{n} + \boldsymbol{e}_i - \boldsymbol{e}_j) \right. \\
(3.9) \qquad &\left. + \frac{n(n-1)\lambda(s)}{2\gamma^*(s, \boldsymbol{n})} \sum_{j \in [d], n_j > 0} \frac{n_j - 1}{n - 1} q(s, \boldsymbol{n} - \boldsymbol{e}_j) \right\} g^*(t, \boldsymbol{n}; s)ds,
\end{aligned}
$$

where g^* is the density of W_t^*. This equation is somewhat simpler to solve than (3.5), and so may be preferred in practice.

To see that the solutions to (3.5) and (3.9) are the same, note that (3.5) is equivalent to the differential equation

$$
\begin{aligned}
\frac{dq(u, \boldsymbol{n})}{du} \;=\; & \gamma(u, \boldsymbol{n})q(u, \boldsymbol{n}) - \frac{n\theta}{2} \sum_{i \in [d]} \frac{n_i}{n} p_{ii} q(u, \boldsymbol{n}) \\
& - \frac{n\theta}{2} \sum_{\substack{i, j \in [d] \\ n_j > 0, i \neq j}} \frac{n_i + 1}{n} p_{ij} q(u, \boldsymbol{n} + \boldsymbol{e}_i - \boldsymbol{e}_j) \\
& - \frac{n(n-1)\lambda(u)}{2} \sum_{j \in [d], n_j > 0} \frac{n_j - 1}{n - 1} q(u, \boldsymbol{n} - \boldsymbol{e}_j).
\end{aligned}
$$

Both (3.5) and (3.9) can be found as the solution of this differential equation with integrating factors $\exp(-\int_0^s \gamma(u, \boldsymbol{n})du)$ and $\exp(-\int_0^s \gamma^*(u, \boldsymbol{n})du)$, respectively.

When the population size is constant through time, equations (3.5) and (3.9) reduce to a discrete recurrence for the configuration probability

$q(\boldsymbol{n}) \equiv q(t, \boldsymbol{n})$:

$$q(\boldsymbol{n}) = \frac{\theta}{n+\theta-1} \left\{ \sum_{\substack{i,j\in[d] \\ n_j>0, i\neq j}} \frac{n_i+1}{n} p_{ij} q(\boldsymbol{n}+\boldsymbol{e}_i-\boldsymbol{e}_j) + \sum_{i\in[d]} \frac{n_i}{n} p_{ii} q(\boldsymbol{n}) \right\}$$

$$(3.10) \qquad + \frac{(n-1)}{\theta+n-1} \sum_{j\in[d], n_j>0} \frac{n_j-1}{n-1} q(\boldsymbol{n}-\boldsymbol{e}_j).$$

This recursion has been studied in various forms by several authors, among them Sawyer, Dykhuizen, and Hartl (1987), and Lundstrom (1990). Given $\{q(\boldsymbol{m}); \boldsymbol{m} < \boldsymbol{n}\}$, simultaneous equations for the $\binom{n+d-1}{d-1}$ unknown probabilities $\{q(\boldsymbol{m}); \boldsymbol{m} = \boldsymbol{n}\}$ are non-singular, and in theory can be solved. In practice a numerical solution is difficult because of the large number of equations, and the situation is considerably more difficult for the recursion in (3.5). In the next section, we describe the Markov chain Monte Carlo approach that we have used to solve systems like (3.5), and describe some of the applications of the technique.

4. Markov chain Monte Carlo. The recursions for sampling probabilities, typified by (3.5) and (3.9), have a common structure. Let $\mathcal{X}$ denote the discrete set of states of the recursion. In (3.5) for instance, $\mathcal{X}$ is the set of d-dimensional vectors $\boldsymbol{n} = (n_1, \ldots, n_d)$ with nonnegative integer entries and sum $m = 1, 2, \ldots, n$, n being the size of the sample. The recursions may be written in the form

$$q(t, x) = \int_t^\infty \sum_{y\in\mathcal{A}} r(s; x, y) q(s, y) g(t, x; s) ds$$

$$(4.1) \qquad + \int_t^\infty \sum_{y\in\mathcal{B}} r(s; x, y) q(s, y) g(t, x; s) ds, \quad x \in \mathcal{B},$$

where $q(t, x)$ is known explicitly for $x \in \mathcal{A}$, $r(s; x, y) \geq 0$, $g(t, x; s)$ is a probability density satisfying $\int_0^\infty g(t, x; s) ds = 1$, and $\mathcal{X} = \mathcal{A} \cup \mathcal{B}$.

For the sampling probabilities determined by (3.9) we have $x = \boldsymbol{n}, n = \sum_{i=1}^d n_i, y = \boldsymbol{m}$, and the non-zero entries of the kernel r in (4.1) are

$$r(s; \boldsymbol{n}, \boldsymbol{m}) = \frac{\theta}{2\gamma^*(s, \boldsymbol{n})}(n_i+1)p_{ij}, \quad \boldsymbol{m} = \boldsymbol{n}+\boldsymbol{e}_i-\boldsymbol{e}_j, \; i,j \in [d],$$

$$n_j > 0, \; i \neq j$$

$$= \frac{n\lambda(s)}{2\gamma^*(s, \boldsymbol{n})}(n_j-1), \quad \boldsymbol{m} = \boldsymbol{n}-\boldsymbol{e}_j, \; j \in [d], \; n_j > 0,$$

where γ^* is given in (3.8).

These recursions are typically impossible to solve explicitly except perhaps for very small sample sizes. Standard numerical solutions are often

not practicable either because of the enormous dimension of $\mathcal{X}$ or the difficulty in evaluating multiple integrals. Instead, we have developed a Markov
chain Monte Carlo method that can be used to find approximants to $q(t, x)$.
The idea, as in all Monte Carlo methods, is to express $q(t, x)$ as the mean
of a functional of $X(\cdot)$, and then repeatedly simulate from $X(\cdot)$. This can
be achieved in the following way.

Let $P(s; x, y)$ be a transition probability kernel on the state space $\mathcal{X}$
satisfying $\sum_{y \in \mathcal{X}} P(s; x, y) = 1$ for all $s \geq 0, x \in \mathcal{X}$ and

$$P(s; x, y) > 0 \text{ if } r(s; x, y) > 0.$$

We use P and g to define a non-homogeneous Markov chain $X(\cdot)$ on $\mathcal{X}$ that
evolves as follows: Given that $X(t) = x \in \mathcal{B}$, the time of the next change
of state has density $g(t, x; s)$, and given that this change occurs at time s,
the probability that the next state is y is $P(s; x, y)$. We are interested in
the process up to the time τ that it reaches the set $\mathcal{A}$. We assume that P
has been chosen so that $\mathbb{P}_x(\tau < \infty) = 1$ for all $x \in \mathcal{B}$.

Now write (4.1) as

$$q(t, x) = \int_t^\infty \sum_{y \in \mathcal{A}} f(t, x; s, y) q(s, y) P(s; x, y) g(t, x; s) ds$$

$$(4.2) \qquad + \int_t^\infty \sum_{y \in \mathcal{B}} f(t, x; s, y) q(s, y) P(s; x, y) g(t, x; s) ds, \quad x \in \mathcal{B},$$

where

$$(4.3) \qquad f(t, x; s, y) = \frac{r(s; x, y)}{P(s; x, y)}.$$

Let $\tau_1 < \tau_2 \cdots < \tau_k = \tau$ be the jump times of $X(\cdot)$ and define $\tau_0 = t$.
It is shown in [10] that

$$(4.4) \qquad q(t, x) = \mathbb{E}_{(t,x)} q(\tau, X(\tau)) \prod_{j=1}^k f(\tau_{j-1}, X(\tau_{j-1}); \tau_j, X(\tau_j)),$$

where $\mathbb{E}_{(t,x)}$ denotes expectation with respect to $X(t) = x$. This representation provides a simple Markov chain Monte Carlo approximant to
$q(t, x)$: Simulate many independent copies of the process $X(\cdot)$ starting from
$X(t) = x$, and compute the observed value of the functional under the expectation sign in (4.4) for each of them. The average of these values is an
unbiased estimate of $q(t, x)$, and standard theory may be used to assess
how accurately $q(t, x)$ has been estimated.

There is a canonical candidate for P, obtained by setting

$$f(x; s) = \sum_y r(s; x, y),$$

and

$$(4.5) \qquad P(s; x, y) = \frac{r(s; x, y)}{f(x; s)}.$$

Equation (4.3) shows that $f(t, x; s, y)$ reduces to

$$f(t, x; s, y) = f(x; s).$$

For the special case in (3.9), if we define

$$w^*(s; \boldsymbol{n}) = \theta \sum_{\substack{i, j \in [d] \\ n_j > 0, i \neq j}} (n_i + 1)p_{ij} + n\lambda(s) \sum_{j \in [d], n_j > 0} (n_j - 1),$$

then

$$
\begin{aligned}
P(s; \boldsymbol{n}, \boldsymbol{m}) &= \frac{\theta}{w^*(s; \boldsymbol{n})}(n_i + 1)p_{ij}, \quad \boldsymbol{m} = \boldsymbol{n} + \boldsymbol{e}_i - \boldsymbol{e}_j, \ i, j \in [d], \\
&\qquad\qquad\qquad\qquad\qquad\qquad\qquad n_j > 0, \ i \neq j \\
&= \frac{n\lambda(s)}{w^*(s; \boldsymbol{n})}(n_j - 1), \quad \boldsymbol{m} = \boldsymbol{n} - \boldsymbol{e}_j, \ j \in [d], \ n_j > 0.
\end{aligned}
$$

We have found it important in practice, particularly in the context of variance reduction, to have some flexibility in choosing the stopping time τ, or, equivalently, the set $\mathcal{A}$. For instance, the natural choice for the case (3.9) has $\mathcal{A} = \{\boldsymbol{m} : \sum_{i=1}^d m_i = 1\}$. This corresponds to tracing the ancestry back to a single individual. However, it is sometimes possible to calculate sampling probabilities, either explicitly or perhaps numerically, when there are two or three distinct ancestors, rather than tracing the genealogy back to just a single individual. In this case we can take $\mathcal{A} = \{\boldsymbol{m} : \sum_{i=1}^d m_i = 2\}$ for example.

4.1. Surface simulation and Monte Carlo likelihoods. The sampling probability $q(t, x)$ is usually a function of some unknown parameters, denoted here by Γ; we write $q_\Gamma(t, x)$ to emphasize the dependence on Γ. Often we are interested in finding the solution q_Γ for a variety of values of Γ, for example when using q as a likelihood function. To compute q on a surface of Γ-values, we use the following approach based on importance sampling. We construct a single process $X(\cdot)$ with parameters Γ_0, from which estimates of $q_\Gamma(t, x)$ may be found for other values of Γ. Write (4.1) in the form

$$(4.6) \quad q_\Gamma(t, x) = \int_t^\infty \sum_{y \in \mathcal{X}} h_{\Gamma, \Gamma_0}(t, x; s, y) P_{\Gamma_0}(s; x, y) q_\Gamma(s, y) \ g_{\Gamma_0}(t, x; s) ds$$

where

$$h_{\Gamma, \Gamma_0}(t, x; s, y) = \frac{f_\Gamma(t, x; s, y) g_\Gamma(t, x; s) P_\Gamma(s; x, y)}{g_{\Gamma_0}(t, x; s) P_{\Gamma_0}(s; x, y)}.$$

The representation of $q_\Gamma(t, x)$ is, from [10],

$$(4.7) \quad q_\Gamma(t, x) = \mathbb{E}_{(t,x)} q_\Gamma(\tau, X(\tau)) \prod_{j=1}^{k} h_{\Gamma, \Gamma_0}(\tau_{j-1}, X(\tau_{j-1}); \tau_j, X(\tau_j)).$$

Estimates of $q_\Gamma(t, x)$ may be now obtained as described above. This method is faster than simulating independent runs at a variety of grid points when the cost of producing observations on the process $X(\cdot)$ outweighs the cost of calculating the functionals in (4.7). In exchange for this time saving, the estimates are no longer independent, but rather they are correlated because of the common generating process. This makes the analysis of the output somewhat more complicated than in the independent replicates case. In practice, several different values of the generating parameters Γ_0 are used, and the results combined to form a single estimate of $q_\Gamma(t, x)$ for several different values of Γ.

4.2. Non-homogeneous recursions. Another class of problems that arise in studying probabilistic aspects of the coalescent involves recursions that are non-homogeneous. These may be written in the form

$$(4.8) \qquad m(t, x) = w(t, x) + \int_t^\infty \sum_{y \in \mathcal{X}} r(s; x, y) m(s, y) g(t, x; s) ds,$$

where $w(t, x)$ is a known function, and where $m(t, x)$ is known on the set $\mathcal{A}$. Recursions of this form may be solved in a similar way to their homogeneous counterparts, as follows. For $x \in \mathcal{B}$, write (4.8) as

$$
\begin{aligned}
m(t, x) \;=\; & w(t, x) + \int_t^\infty \sum_{y \in \mathcal{A}} f(t, x; s, y) m(s, y) P(s; x, y) g(t, x; s) ds \\
(4.9) \qquad & + \int_t^\infty \sum_{y \in \mathcal{B}} f(t, x; s, y) m(s, y) P(s; x, y) g(t, x; s) ds,
\end{aligned}
$$

and iterate to obtain

$$
\begin{aligned}
m(t, x) = & \int_t^\infty \sum_{y_1 \in \mathcal{A}} f(t, x; s_1, y_1) m(s_1, y_1) P(s_1; x, y_1) g(t, x; s_1) ds_1 \\
& + \int_t^\infty \sum_{y_1 \in \mathcal{B}} \int_{s_1}^\infty \sum_{y_2 \in \mathcal{A}} f(t, x; s_1, y_1) f(s_1, y_1; s_2, y_2) m(s_2, y_2) \\
& \qquad\qquad P(s_1; x, y_1) P(s_2; y_1, y_2) g(t, x; s_1) g(s_1, y_1; s_2) ds_2 ds_1
\end{aligned}
$$

$$
\begin{aligned}
(4.10) \quad + \cdots \\
& + w(t, x) + \int_t^\infty \sum_{y_1 \in \mathcal{B}} f(t, x; s_1, y_1) w(s_1, y_1) P(s_1; x, y_1) g(t, x; s_1) ds_1 \\
& + \int_t^\infty \sum_{y_1 \in \mathcal{B}} \int_{s_1}^\infty \sum_{y_2 \in \mathcal{B}} f(t, x; s_1, y_1) f(s_1, y_1; s_2, y_2) w(s_2, y_2)
\end{aligned}
$$

$$P(s_1; x, y_1)P(s_2; y_1, y_2)g(t, x; s_1)g(s_1, y_1; s_2)ds_2 ds_1$$
$$+ \cdots$$

In terms of the Markov process $X(\cdot)$, $m(t, x)$ may be represented as follows. Let $\tau_1 < \tau_2 < \cdots < \tau_k = \tau$ be the times of the jumps $X(\cdot)$ makes until it reaches the set $\mathcal{A}$, and set $\tau_0 = t$. Then

$$m(t, x) = \mathbb{E}_{(t,x)} \sum_{l=0}^{k-1} w(\tau_l, X(\tau_l)) \prod_{j=1}^{l} f(\tau_{j-1}, X(\tau_{j-1}); \tau_j, X(\tau_j))$$

(4.11)
$$+ \mathbb{E}_{(t,x)} m(\tau, X(\tau)) \prod_{j=1}^{k} f(\tau_{j-1}, X(\tau_{j-1}); \tau_j, X(\tau_j)).$$

Once more, independent replicates of $X(\cdot)$ starting from $X(t) = x$ may be used to estimate the expectation in (4.11), values of the sum on the right being accumulated as each simulation progresses. It is straightforward to adapt this scheme to the surface simulation setting of the last section.

4.3. Other sampling properties: the distribution of the time to MRCA. Sampling distributions are not the only quantities that produce recursions to which the Markov chain Monte Carlo method can be applied. One example arises in studying the joint distribution of the sample configuration and the time to the most recent common ancestor. Let $q(t, x, w)$ be the probability that a sample taken at time t has configuration x, and the (further) time to the MRCA is at most w. Of particular interest is the distribution function

(4.12)
$$\mathbb{P}(T_{MRCA} \le w | x) = \frac{q(t, x, w)}{q(t, x)}.$$

It is shown in [9] and [10] that $q(t, x, w)$ satisfies a recursion of the form

(4.13)
$$q(t, x, w) = \int_t^\infty \sum_{y \in \mathcal{X}} r(s; x, y) q(s, y, t + w - s) g(t, x; s) ds,$$

and that for $w > 0$

(4.14)
$$q(t, x, w) = \mathbb{E}_{(t,x)} q(\tau, X(\tau), t + w - \tau) \prod_{j=1}^{k} f(\tau_{j-1}, X(\tau_{j-1}); \tau_j, X(\tau_j)).$$

Under the initial condition

$$q(t, x, w) = I\{w \ge 0\}, \quad x \in \mathcal{A}$$

the term $q(\tau, X(\tau), t + w - \tau)$ in (4.14) reduces to $I\{\tau \le t + w\}$. If we simulate the process $X(\cdot)$ R times, and define

$$F_l = \prod_{j=1}^{k_l} f(\tau_{j-1}, X(\tau_{j-1}); \tau_j, X(\tau_j)),$$

the value of the functional under the expectation sign in (4.4) for the lth simulation, then the distribution in (4.12) can be approximated by the ratio

$$\frac{\sum_{l=1}^{R} F_l I\{\tau^{(l)} \le t + w\}}{\sum_{l=1}^{R} F_l},$$

where $\tau^{(l)}$ is the time the lth simulation hits $\mathcal{A}$. Conditional moments can be computed in a similar way.

4.4. Applications. In this section, we review briefly some of the applications of this computational approach. Further details may be found in the original papers. Computer code is available from the authors on request.

4.4.1. The infinitely-many-sites model. The simplest mutation structure is the infinitely-many-sites model of sequence evolution, in which every mutation in the ancestral tree of the sample produces a new segregating site in the sample. Hudson [13] [14] gives a variety of applications. The sample may be described by a collection of sequences of zeros and ones. If the labeling of the ancestral base at each site is known, we can suppose that the ones denote mutant bases at a site, while the zeros denote sites at which the ancestral type is still present. Typically, this labeling is unknown. The distribution of the sample is determined by certain rooted and unrooted genealogical trees that are embedded in the process; rooted trees correspond to known labeling of sites, unrooted trees to unknown labeling of sites. The theory of these trees is developed in [7]. Markov chain Monte Carlo methods are used to estimate parameters in the varying population size model in [8]. population Inference about the distribution of the time to the most recent common ancestor, conditional on the structure of a sample, is discussed in [9], where applications to mitochondrial sequence data are given. These computer-intensive methods are sometimes time-consuming, and it is therefore of some interest to know how inferences based on simpler summary statistics of the data (for example, the number of segregating sites and alleles) compare to inferences based on the full data. Inference about θ and the time to the MRCA are addressed in [10].

If distinct sequences in the sample are identified as alleles, the sampling theory of the allele frequencies in the constant population size case is given by the Ewens sampling formula [4]. The analogous theory for the variable population size case appears in [8], where the Markov chain Monte Carlo method is also explored. See also [10].

4.4.2. The finitely-many-sites model. Of central interest in the analysis of DNA sequence data is the development of methods for estimating parameters of the substitution process. In the population genetics setting, this can be thought of as the problem of estimating the parameters of the mutation rate matrix R in (3.1). One method, developed by Lundstrom [19] and extended in [20], uses a method of moments approach. In the simplest mutation model determined by (3.2) with identical substitution matrices $M_l \equiv M$ and equal rates $h_l = 1/s$ at each site, the vectors the count the number of each type of base observed at each site are exchangeable. In particular, they have the same distribution (but of course they are not independent). This observation provides a simple moment method for estimating the entries of the rate matrix θM: equate observed and expected counts, and minimize the sum of squares of the differences. A detailed study of the behavior of this method appears in [19] and [20]; the extension to hypervariable sites is described in [21].

Our development of the Markov chain Monte Carlo method for coalescents was motivated in part by trying to assess whether the estimation methods described above had good statistical properties. The simulation method, together with the surface simulation for likelihoods, is developed for the sampling distribution (3.10) in [6]. Among the issues addressed is the effect on variance reduction of choosing the stopping time τ, and a variety of suggestions for speeding up the method. Note that it is simple to use the same Monte Carlo approach to estimate parameters for the more complicated sequence models described in Section 3.1, and the effects of variable size can be accommodated simply as well [8].

Notice that in the model of sequence evolution determined by (3.2), the mutation processes at different sites are conditionally independent given the genealogy. This means that if the genealogical tree is known, the probability of a set of sequences may be computed by, in effect, reducing the problem to the computation of sampling probabilities at a single site. For simple models for M_l, the mutation matrix at the lth site, it is possible to compute the probability that a base that is of type i at time 0 is type j at time t, and so compute the probability that a site has a particular set of types at the tips of the ancestral tree.

Kuhner, Yamato and Felsenstein [17] [18] have developed an alternative approach to maximum likelihood estimation of θ in this constant population size model. They use a Metropolis-Hastings sampler to sample genealogies, and compute the probability of the set of sequences by using the conditional independence property.

4.4.3. The effects of recombination. The previous examples have been concerned with samples in which the effects of recombination can be ignored. However, the same principles can be applied to study recombination as well. The simplest case is the one with completely unlinked loci, for which computational aspects of the sampling theory can be found in [24], [20], and [21]. For the linked case, think of two finitely-many-alleles loci, A and B, with K alleles at the first locus, L at the second, and mutation rate matrices

$$R_A = \frac{\theta_A}{2}(P^A - I), \quad R_B = \frac{\theta_B}{2}(P^B - I).$$

The analog of the sampling equation (3.10) is a linear system satisfied by the probability $q(\boldsymbol{a}, \boldsymbol{b}, \boldsymbol{c})$ of ordered configurations of the form $(\boldsymbol{a}, \boldsymbol{b}, \boldsymbol{c})$, where $\boldsymbol{a} = (a_1, \ldots, a_K), \boldsymbol{b} = (b_1, \ldots, b_L)$, and $\boldsymbol{c} = (c_{ij}, i \in [K], j \in [L])$. Here, a_i gametes have type i at the A locus and unspecified alleles at the B locus, b_j gametes have type j at the B locus and unspecified alleles at the A locus, and c_{ij} gametes have allele i at the A locus and allele j at the B locus, for $i \in [K], j \in [L]$. The linear system can be derived from a simple coalescent argument, and the sampling formula $q(\boldsymbol{0}, \boldsymbol{0}, \boldsymbol{c})$ of the gamete configuration $\boldsymbol{c}$ found by the Markov chain Monte Carlo approach. The same method works to solve the analogous linear system for two infinitely-many-alleles loci that is discussed by Ethier and Griffiths [2], [3]. The methods can also be extended to allow for variable population size, more loci and more complex mutation schemes.

4.4.4. The effects of migration. Nath and Griffiths (1993) derive a recursion analogous to (3.10) in an island model with migration among L islands. $q(\boldsymbol{n})$ is then replaced by $q(\boldsymbol{n}_1, \ldots, \boldsymbol{n}_L)$ the configuration probability of samples of sizes $n_1, \ldots, n_L$ taken from the L islands.

The Markov chain Monte Carlo technique in Section 4 is developed, and the estimated surface of probabilities with the migration rate varying is used to study likelihood estimation of the migration rate in the case of $L = 2$ islands with $d = 2$ possible alleles.

5. Discussion. In this paper, we have reviewed one computational approach for calculating sampling probabilities and related quantities for models arising from versions of the coalescent. The progenitor of this approach dates back at least to the late 1940s, where it was used to solve matrix equations of the form $Ax = b$; see Forsythe and Leibler [5] and Halton [11] for example. The techniques advocated here are similar in spirit to the Hastings- Metropolis method [22], [12], where the quantity of interest is represented as the mean (under the stationary distribution) of a function of an ergodic Markov chain, and this mean is estimated by computing an ergodic average. This uses a single run of the chain to produce estimates, the observations within the run being correlated. In the present approach we use independent runs of random lengths, which in principle makes the subsequent analysis of the output somewhat simpler.

These techniques may also be applied to other variants on the population genetics theme. The models are described here in terms of 'alleles' and 'mutations', but these may be interpreted in other ways as well. For example, imagine a population of individuals reproducing according to the coalescent, but now think of the 'alleles' as describing the structure of a population of mitochondria within each individual. For example, the parameter θ may be interpreted as the birth-and-death rate of the individual mitochondrial populations, and the transition matrix P describes how a given population reproduces at the birth-and-death times. If each of the mitochondria is labeled as type A or type B, then a plausible model for the evolution of the individual populations is the two-type Moran model with mutation. More complicated within-individual reproduction mechanisms could of course be used. This provides a simple model for the evolution of a mitochondrial lineage within a reproducing human population. These methods also work for other models in which the branching structure of the coalescent is replaced by other branching processes, such as the binary splitting, or Yule, process. In this case, all that changes in equation (3.10) is the relative rate of 'splits' and 'mutations'.

REFERENCES

[1] BORODOVSKY, M.Y., SPRIZHITSKY, Y., GOLOVANOV, E. and ALEXANDROV, A. *Statistical patterns in the primary structures of functional regions in the genome of E. coli: II Nonuniform Markov models*, Mol. Biol., **20**, 1024-1033, 1986.

[2] ETHIER, S.N. and GRIFFITHS, R.C., *The neutral two-locus model as a measure-valued diffusion*, Adv. Appl. Prob., **22**, 773-786, 1990.

[3] ETHIER, S.N. and GRIFFITHS, R.C., *On the two-locus sampling distribution*, J. Math. Biol., **29**, 131-159, 1990.

[4] EWENS, W.J., *The sampling theory of selectively neutral alleles*, Theoret. Popul. Biol., **3**, 87-112, 1972.

[5] FORSYTHE, G.E. and LEIBLER, R.A, *Matrix inversion by the Monte Carlo method*, Math. Comp., **26**, 127-129, 1950.

[6] GRIFFITHS, R.C. and TAVARÉ, S., *Simulating probability distributions in the coalescent*, Theoret. Popul. Biol., **46**, 131-159, 1994.

[7] GRIFFITHS, R.C. and TAVARÉ, S., *Unrooted genealogical tree probabilities in the infinitely-many-sites model*, Math. Biosci., **127**, 77-98, 1995.

[8] GRIFFITHS, R.C. and TAVARÉ, S., *Sampling theory for neutral alleles in a varying environment*, Phil Trans. R. Soc. Lond. B, **344**, 403-410, 1994.

[9] GRIFFITHS, R.C. and TAVARÉ, S., *Ancestral inference in population genetics*, Statistical Science, **9**, 307-319, 1994.

[10] GRIFFITHS, R.C. and TAVARÉ, S., *Monte Carlo inference methods in population genetics*, Mathl. and Comput. Modelling, in press, 1996.

[11] HALTON, J.H., *A retrospective and prospective study of the Monte Carlo method*, SIAM Review, **12**, 1-63, 1970.

[12] HASTINGS, W.K., *Monte Carlo sampling methods using Markov chains and their applications*, Biometrika, **57**, 97-109, 1970.

[13] HUDSON, R.R., *Gene genealogies and the coalescent process*, In: Oxford Surveys in Evolutionary Biology, Volume 7. Edited by D. Futuyma and J. Antonovics, 1-44, 1991.

[14] HUDSON, R.R., *The how and why of generating gene genealogies*, In: Mechanisms of molecular evolution, N. Takahata and A.G. Clark (editors), 23-36, 1992. Sinauer.

[15] KINGMAN, J.F.C., *On the genealogy of large populations*, J. Appl. Prob., **19A**, 27-43, 1982.

[16] KINGMAN, J.F.C., *Exchangeability and the evolution of large populations*, In: Exchangeability in probability and statistics, G. Koch and F. Spizzichino (editors), 97-112. North-Holland Publishing Company, 1982.

[17] KUHNER, M.K., YAMATO, J. and FELSENSTEIN, J., *Estimating effective population size from sequence data using Metropolis-Hastings sampling*, Genetics, submitted, 1994.

[18] KUHNER, M.K., YAMATO, J. and FELSENSTEIN, J., *Applications of Metropolis-Hastings genealogy sampling*, IMA volume, in press, 1994.

[19] LUNDSTROM, R. *Stochastic models and statistical methods for DNA sequence data.*

Ph.D. thesis, Mathematics Department, University of Utah, 1990.

[20] LUNDSTROM, R., TAVARÉ, S. and WARD, R.H., *Estimating mutation rates from molecular data using the coalescent*, Proc. Natl. Acad. Sci. USA, **89**, 5961-5965, 1992.

[21] LUNDSTROM, R., TAVARÉ, S. and WARD, R.H., *Modelling the evolution of the human mitochondrial genome*, Math. Biosci, **112**, 319-335, 1992.

[22] METROPOLIS, N., ROSENBLUTH, A.W., ROSENBLUTH, M.N., TELLER, A.H., and TELLER, E., *Equations of state calculations by fast computing machines*, J. Chem. Phys., **21**, 1087-1092, 1953.

[23] NATH, H.B. and GRIFFITHS, R.C., *Estimation in an island model undergoing a multidimensional coalescent process*, Statistics Research Report 226, Monash University, 1993.

[24] SAWYER, S., DYKHUIZEN, D. and HARTL, D., *Confidence interval for the number of selectively neutral amino acid polymorphisms*, Proc. Natl. Acad. Sci. USA, **84**, 6225-6228, 1987.

[25] SLATKIN, M. and HUDSON, R.R., *Pairwise comparisons of mitochondrial DNA sequences in stable and exponentially growing populations*, Genetics, **129**, 555-562, 1991.

[26] TAVARÉ, S., *Calibrating the clock: using stochastic processes to measure the rate of evolution*, Chapter 5 in "Calculating the secrets of life", E.S. Lander and M.S. Waterman (editors). National Academy Press, Washington DC, pp. 114-152, 1995.

[27] TAVARÉ, S., *The effects of site dependence on estimating the topology of a tree from DNA sequence data*, in preparation, 1995.

[28] WATTERSON, G.A., *A stochastic analysis of three viral sequences*, Mol. Biol. Evol., **9**, 666-677, 1992.

APPLICATIONS OF METROPOLIS-HASTINGS GENEALOGY SAMPLING

MARY K. KUHNER*, JON YAMATO[†] , AND JOSEPH FELSENSTEIN[†]

Abstract. The genealogy (scaled in units of mutations per unit time) of a random sample of sequences from a population provides information on the parameter Θ (neutral mutation rate times effective population size). When the genealogy is unknown, Θ could in principle be estimated by summing over possible genealogies, but there are too many of them. We use a Metropolis-Hastings sampler to concentrate attention on the genealogies of high posterior probability for a given value of Θ; a likelihood curve for nearby values of Θ can be constructed from the sampled genealogies, and its maximum is a maximum likelihood estimate of Θ. This method can in principle be extended to populations with migration and population growth, and to sequences with recombination and selection. It can also be applied to types of data other than nucleotide sequences in cases where an appropriate likelihood model is available.

Introduction. The relationships among individuals from a population can be summarized by a genealogy showing the common ancestors of the individuals and the times at which those ancestors lived. If the individuals were sampled at random from a single population, the times back to the common ancestors have an approximate expected distribution called the *coalescent* which was derived by Kingman (1982a, b). The coalescent distribution depends only on the number of individuals sampled and the parameter $4N_e\mu$, also called Θ, which combines the effects of effective population size N_e and the neutral mutation rate μ. (We use capital Θ rather than the conventional lowercase θ in this paper because we are considering mutation rate per site, not per locus as in previous studies.) Using the coalescent approximation, Θ can be estimated from the genealogy. Such an estimate is potentially more efficient than one made from the sequence data without reference to the genealogy (Felsenstein 1992a). Unfortunately the genealogy is generally unknown.

At least three approaches have been previously proposed for estimates of Θ incorporating genealogical information. A method proposed by Fu (1994) makes an estimate based on a single reconstructed genealogy, combined with a correction factor. The method of Griffiths and Tavaré (1994a, b) uses recursion over the sampling probabilities to construct a Markov chain from which likelihood estimation of parameter values is possible, without explicitly attempting to reconstruct genealogies. The bootstrap Monte Carlo method of Felsenstein (1992b) makes an estimate based on genealogies produced by bootstrapping the original data set. (We have recently found the bootstrap Monte Carlo method to be biased [Kuhner,

---

* Corresponding Author: University of Washington GENETICS Box 357360 Seattle, WA 98195-7360 Phone (206) 543-8751, FAX (206) 543-0754, Internet: *mkkuhner@genetics.washington.edu*

[†] Department of Genetics, University of Washington.

Yamato and Felsenstein 1995] and do not recommend its use.)

The approach proposed in the current paper is importance sampling: we sample among all genealogies, but concentrate on those which are expected to make substantial contributions to the estimate. This can be done by means of a Metropolis-Hastings sampler. New genealogies are created by small modifications of an initial one, and accepted or rejected based on their fit to the data. This creates a sequence of genealogies which is a Markov chain. Samples taken from this chain of genealogies can be used to construct a likelihood curve for Θ and to find its maximum. This approach avoids the bias which may be caused by using a single reconstruction of the genealogy, since such reconstructions unavoidably involve some degree of error.

We have designed such a sampler, which can be used for non-recombining nucleotide sequences (such as mitochondrial DNA) from a single population of constant size. The details of the sampling procedure are presented in another paper (Kuhner, Yamato, and Felsenstein 1995): this paper will briefly review the sampler, and discuss ways in which it can be extended to other types of data.

Methods. The Metropolis-Hastings sampler on genealogies is a form of importance sampling: we sample genealogies from a known distribution which we hope will be similar to the unknown true distribution, in order to gain as much information about Θ per sampled genealogy as possible. In evaluating the genealogies, we must then correct for the known distribution from which we sampled.

The known distribution we will use incorporates the probability $P(D|G)$ of the sequence data given the genealogy, and the probability $P(G|\Theta_0)$ of the genealogy given a chosen value of Θ which we will call Θ_0. The true distribution which we are trying to estimate is identical to this except that the unknown true Θ replaces Θ_0. The importance sampling function we use is:

$$\frac{P(D|G)P(G|\Theta_0)}{P(D|\Theta_0)}$$

which is the posterior probability of the genealogy. Since only ratios of posterior probabilities are needed, the unknown constant $P(D|\Theta_0)$ need not be considered.

The prior probability, $P(G|\Theta_0))$, can be readily computed using the coalescent approximation of Kingman (1983a, b). The probability of the data, $P(D|G)$, is the quantity maximized by ML methods of phylogeny estimation, and can be computed for a variety of models: we chose the nucleotide sequence model of Felsenstein (described by Kishino and Hasegawa 1989) which allows the specification of the ratio of transitions to transversions and of the frequencies of the four bases. Under Extensions we describe some alternative models which could also be used.

To carry out this sampling, we construct a new genealogy by modifying the current genealogy, drawing the new coalescence times from a distribution proportional to $P(G|\Theta_0)$. (This procedure by itself, if repeated many times, would eventually sample all genealogies in proportion to coalescent expectations for the given Θ_0.) Having constructed a new genealogy, we decide whether to accept or reject it by comparing the relative probability of the sequence data, $P(D|G)$, on the old and new genealogies. The outcome of this process, repeated many times, is a Markov chain of genealogies where each possible genealogy is represented in proportion to the product $P(G|\Theta_0)P(D|G)$, its posterior probability. Some details of the rearrangement procedure follow.

To make a local rearrangement, we choose a random node in the tree and consider a neighborhood around it: its two descendent nodes, its parent node, and its parent's other descendent. ("Parent" and "child" here are relationships among nodes in the genealogy, not literal parentage.) A series of neighborhood rearrangements of this type will eventually transform any topology into any other.

The interior of the neighborhood is erased and replaced with a new set of coalescences chosen based on a conditional coalescent distribution. Two factors must be considered: (1) the probability of coalescence at a given moment depends on the number of lineages present in the entire genealogy at that moment; (2) we require the three lineages within the neighborhood to coalesce with one another, and not with any other lineage, by the bottom of the neighborhood (this condition limits the process to local rearrangements). We use a modification of the state-array approach of Viterbi (1967). Briefly, we construct a lattice containing the conditional probabilities that three, two or one lineages are present in the neighborhood at each moment, then trace a weighted random path through this lattice beginning with one lineage at the bottom of the neighborhood and ending with three lineages at the top. This path defines the intervals in which the coalescences are to occur: the lineages coalescing are then chosen at random from the neighborhood lineages available at the given times, and the exact coalescence times within each interval are also chosen. (Complete details of this algorithm are given in Kuhner, Yamato, and Felsenstein 1995.)

Now that the modified genealogy has been constructed, the probability of the data on the new and old genealogies is evaluated under the chosen model. The decision to accept or reject the new genealogy depends on the ratio r of these probabilities; if $r \geq 1$ the new genealogy is accepted with probability 1, whereas if $r < 1$ it is accepted with probability r (Metropolis et al. 1953). The method of lineage erasure and redrawing does not introduce any bias towards particular topologies, so the corrective terms described by Hastings (1970) need not be computed in this form of the sampler. However, more complex future extensions of the algorithm may require these corrective terms to compensate for any tendency to over-propose certain genealogies.

At intervals, genealogies can be sampled from this Markov chain, and the set of sampled genealogies can be used to construct a likelihood curve for Θ via the following relationship, which correctly compensates for the use of the importance sampling function:

$$(1) \qquad L(\Theta) = \sum_G \frac{P(G|\Theta)}{P(G|\Theta_0)}$$

The Metropolis-Hastings sampler should produce the correct value of $L(\Theta)$ for any value of Θ_0 asymptotically as the number of steps along the Markov chain approaches infinity. In practice, for finite numbers of steps there is a bias towards Θ_0 if it is distant from Θ. A useful approach is to run several chains, each using as its Θ_0 the maximum of the curve for the previous chain. This brings Θ_0 close to the final estimate of Θ, so that the likelihood curve is well-estimated in that region. The genealogies from all chains can be used to make a single estimate by treating them as if they were sampled from a mixture distribution (Geyer 1991).

Applications. The Metropolis-Hastings genealogy sampler was originally designed in order to estimate Θ, but the ability to sample genealogies in proportion to their posterior probability has other potential uses:

Phylogeny estimation. The maximum likelihood method of phylogeny estimation searches for the phylogeny G maximizing $P(D|G)$. Maximum likelihood is one of the most effective phylogeny methods (Kuhner and Felsenstein 1994), but it is computationally intensive. Current heuristics such as the one implemented in Felsenstein's DNAML program from PHYLIP are not guaranteed to find the maximum likelihood phylogeny, and they are very slow on large data sets.

The Metropolis-Hastings sampler is not attempting to maximize $P(D|G)$. However, in the course of its search the sampler can easily make note of the highest value encountered, and the genealogy producing it. In practice, this genealogy often has the same topology as the maximum likelihood genealogy. Its branch lengths can then be optimized by standard methods. Thus, the Metropolis-Hastings sampler can be used as a heuristic method for finding the genealogy of maximum likelihod with respect to the data. It is much faster on large data sets (50 sequences and up) than DNAML. We plan to compare their performance on simulated data. Gary Olsen (pers. comm.) reports that DNAML's heuristic frequently becomes trapped in local maxima when analyzing large data sets. Preliminary results suggest that the sampler is less prone to become trapped.

For data drawn from a single population (as opposed to data representing multiple populations or species) the genealogy of maximum posterior probability at the best value of Θ may in fact be a superior reconstruction of the true genealogy, since the coalescent distribution contributes some information about expected branch lengths. However, this approach is probably not appropriate for interspecific data, since speciation may not follow coalescent dynamics.

Bootstrap-like resampling. The genealogies sampled during a Metropolis-Hastings run with the best value of Θ represent a weighted random sample from the space of all genealogies, and can be used to test hypotheses about that space. For example, they can be used in the same way as a set of bootstrap reconstructions to test the strength of support for a particular branch or rooting. However, since successive steps of the Markov chain are not independent, care must be taken to sample at sufficiently long intervals that successive sampled genealogies are close to independence. Even with a very long sampling interval, the Metropolis-Hastings sampler is much faster than extant likelihood bootstrapping approaches for medium to large numbers of sequences. It should be feasible to apply it to questions such as the degree of support for an African root in the human mitochondrial DNA data of Vigilant et al. (1991).

Extensions.

Other likelihood models. Any likelihood model can be substituted into this algorithm, leaving the genealogy-rearrangement machinery intact. For example, the sequence likelihood model of Yang (1993), which assumes that mutation rates follow a gamma distribution, could be used. Types of data other than nucleotide sequences can be accomodated: for example, the protein likelihood model of Kishino et al. (1990), or the restriction site likelihood model of Felsenstein (1992c). These models will, however, be more computationally intensive than the current one.

Variable population size. Change in population size changes the expected shape of the coalescent, and thus reconstructions of the genealogy can be used to infer past population size changes. We are currently exploring estimation of an exponential growth rate g. The approach is to assume an initial estimate of g (called g_0) and an initial estimate of the present-day Θ (called Θ_0). Modification of genealogies is then done in much the same way as the original algorithm, substituting terms involving g_0 and Θ_0 for the original terms involving Θ_0. From the collection of sampled genealogies it is possible to make a joint estimate of g and Θ, though there will be bias unless g_0 and Θ_0 are fairly close to g and Θ.

Preliminary results indicate that this method produces an upward bias in estimation of both Θ and g (i.e. the population is estimated to be bigger than it really is, and to have grown more rapidly), but that use of multiple unlinked loci can decrease this bias.

A further possibility in this direction is to allow several different episodes of exponential growth or decline, and attempt to estimate the entire collection of growth rates and time periods. This is complicated by the fact that the model with the largest number of episodes will be able to fit the data best unless the model explicitly penalizes rate changes. It is difficult to suggest a likelihood model incorporating changes in growth rate, since little is known about the statistical distribution of such changes. Another approach would be to run independent Markov chains with different numbers of episodes, and then accept the results of the first chain for which adding a given number of additional episodes does not significantly improve the results. Alternatively, model-selection criteria such as Akaike's Information Criterion (AIC) could be used (Akaike 1973).

Recombination. The coalescent distribution is very stochastic, and a single realization of it (such as is provided by mitochrondrial DNA) cannot provide much certainty about the history of the population. Multiple independent samples are desirable, but in order to analyze nuclear DNA we will need to deal with recombination.

Recombinant genealogies can be thought of as collections of ordinary genealogies, each applying to a part of the sequence (Figure 1). There is no difficulty in evaluating the likelihood of sequence data on such a genealogy, since each individual site "sees" a normal genealogy. The major issue that must be resolved is how to make rearrangements that sample among recombinant as well as non-recombinant genealogies in proportion to $P(G|\Theta_0, c_0)$ (where c is the recombination rate per site). The concept of the neighborhood of rearrangement, used in the non-recombinant algorithm, no longer applies, since any lineage can potentially recombine with any other, including lineages not currently represented in the genealogy.

One approach, which we are currently investigating, is to allow two different types of modifications: one type which modifies branching order and length (analogous to the modifications in the non-recombinant version), and one which adds and removes recombinations. Some simplification of the full recombinant coalescence process will probably be required. The non-recombinant coalescent need only consider the probability that three lineages will coalesce into two or one during an interval of time. In a recombinant genealogy, three lineages could split into an unlimited number of lineages via recombination. An explicit solution for this infinite family of probabilities (three lineages become four, five, ...) is probably too difficult. A truncated version that allows only a limited number of new recombinations to occur during a time interval may be an adequate substitute, especially as c is likely to be small.

Migration. Takahata (1985) explored the coalescent theory of populations that exchange migrants, and found it intractable except for very small cases. This difficulty arises from allowing for all possible combinations of migration events. As an alternative approach, a Metropolis-Hastings sampler which explicitly included migration events as well as coalescences in its genealogies could be used. Such a sampler would allow two types of modification to a genealogy: a change of branching order or length (as in the

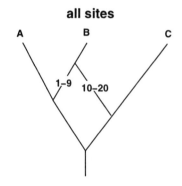

all sites

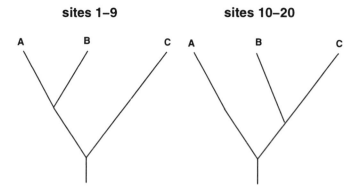

sites 1–9   **sites 10–20**

FIG. 1. *A recombinant genealogy. Recombination has occured between sites 9 and 10. The upper genealogy represents the entire sequence; the two lower genealogies show its decomposition into multiple non-recombinant genealogies, each applying to a subset of the sites.*

original algorithm), and a modification of the putative migration events. Such a search could be made fairly efficient by a modification scheme which did not propose migration events inconsistent with the present-day subpopulation structure, although it would still need to search through a very large state space of genealogies with migration.

Selection. General models of selection are probably too complex for the Metropolis-Hastings approach, but some special cases can be analyzed in this fashion. Consider a mutation at a single site which is strongly favored and thus rapidly goes to fixation in the population. Sites which are closely linked to the favored site will also tend to be dragged to fixation with it ("hitchhiking"), while more distant sites will retain polymorphism due to recombination. (In non-recombining DNA a single selected site will draw the entire molecule to fixation, leaving no information about which site was selected.) In essence, the selected site in a recombining sequence sees an extremely small Θ (because the only individuals who contributed to the current population were those carrying the site destined to be fixed) while adjacent sites see a larger and larger Θ approaching the neutral Θ of the genome as a whole. This type of pattern should be detectable with Metropolis-Hastings sampling of recombinant genealogies that explicitly include instances of hitchhiking.

Another potentially tractable type of selection is balancing selection at a specific site. The population could be considered to consist of two subgroups, each with a specific base at the selected site: the two subgroups would be maintained at near-constant size by natural selection. Mutation at the key site could move an entire sequence from one subgroup to another; recombination could move the part of the sequence not containing the key site. It should be possible to detect which site is defining the subgroups, although this will require using the machinery of both recombination and migration.

Disequilibrium mapping. The Metropolis-Hastings approach could be used for a type of disequilibrium linkage mapping. Conventional linkage mapping uses recombinations observed in family data to locate a gene of interest with respect to marker loci. The resolution of conventional mapping is limited: the closer one is to the gene of interest, the rarer are the informative recombinations. Disequilibrium mapping attempts to use information from recombinations that have occured throughout the evolutionary history of the region, which should potentially make many more recombinations available for study.

Current methods of disequilibrium mapping are based on pairwise combinations of loci, losing some of the available information. A Metropolis-Hastings method with recombination could be used for mapping by computing a likelihood function across the sampled genealogies for each possible location of the gene to be mapped. This will presumably require a likelihood model for non-sequence data such as mapped restriction sites (Felsenstein 1992c), since sequence data across the entire area to be mapped are unlikely to be available.

One problem to be overcome is that a sample with many copies of a rare trait (such as a genetic disease) is likely to be quite different from a random population sample, and this ascertainment bias must be allowed for in constructing the Metropolis-Hastings sampler. Furthermore, if the trait in question is not recessive allowance must be made for uncertainty about which of the affected individual's two chromosomes carries it. However, even if the estimates of recombination rate and population size are biased by such ascertainment issues, the assignment of the trait to a particular chromosome region may still escape bias and be made correctly.

Discussion. The Metropolis-Hastings sampler is a relatively efficient way to explore the space of possible genealogies for a given data set. In preliminary simulations (Kuhner, Yamato, and Felsenstein 1995) we used samples of 100 sequences of length 1000, performing a total of 105,500 steps divided among 7 Markov chains: the entire estimation took approximately 181 minutes on a DECstation 5000/125, a mid-speed workstation. The run time does not increase quickly with increasing numbers of sequences or sites. Samples of over 800 sequences are feasible on a workstation, although in such cases many steps along the chain will be necessary in order to adequately survey the space of plausible genealogies.

In principle the Metropolis-Hastings genealogy sampling approach can be extended to a number of key problems, including variable population size, recombination, migration and selection. The more complex models will probably demand large numbers of relatively long sequences, and possibly multiple loci, in order to simultaneously estimate several parameters, but the recent explosion in sequencing technology should soon make such data sets available. It is interesting to contemplate a "grand unified model" of sequence evolution incorporating all the listed processes, and perhaps insertion, deletion, rearrangment, and interactions among sites as well. Only time will tell whether such a unified model is feasible.

Software availability. We are making available a new package, LAMARC, containing our Metropolis-Hastings programs. The first program in LAMARC, COALESCE, implementing the Metropolis-Hastings genealogy sampler for the simplest case (neutral non-recombining sequences in a constant-size, randomly mating population) is available by anonymous ftp from *evolution.genetics.washington.edu* in directory pub/lamarc. It is written in C. Future programs will also be available in the same directory.

Acknowledgements. We thank the Institute for Mathematics and its Applications for inviting us to the Workshop on Mathematical Population Genetics, which two of us were able to attend supported by funds provided by the National Science Foundation. The research presented here was supported by National Science Foundation grants BSR-8918333 and DEB-9207558 and National Institute of Health grant 2-R55GM41716-04 (all to J. F.). We thank Charles Geyer, Elizabeth Thompson, Richard Hudson and Ellen Wijsman for helpful suggestions that have contributed to the progress of this research.

REFERENCES

AKAIKE, H, 1973 *Information theory and an extension of the maximum likelihood principle.* In: Petrov, B. N., and F. Csaki (eds) Second international symposium on information theory. Akademiai Kiado, Budapest, pp. 267–281.

FELSENSTEIN, J., 1992a *Estimating effective population size from samples of sequences: inefficiency of pairwise and segregating sites as compared to phylogenetic estimates.* Genet. Res. **59:** 139–147.

FELSENSTEIN, J., 1992b *Estimating effective population size from samples of sequences: a bootstrap Monte Carlo integration method.* Genet. Res. **60:** 209–220.

FELSENSTEIN, J., 1992c *Phylogenies from restriction sites, a maximum likelihood approach.* Evolution **46:** 159–173.

FU, Y.-X., 1994 *A phylogenetic estimator of effective population size or mutation rate.* Genetics **136**: 685–692.

GEYER, C.J., 1991 *Estimating normalizing constants and reweighting mixtures in Markov chain Monte Carlo.* Technical Report No. 568, School of Statistics, University of Minnesota.

GRIFFITHS, R.C., and S. TAVARÉ, 1994a *Sampling theory for neutral alleles in a varying environment.* Proc. R. Soc. Lond. B **344**: 403–410.

GRIFFITHS, R.C., and S. TAVARÉ, 1994b *Simulating probability distributions in the coalescent.* Theoret. Popul. Biol. **46**: 131–159.

KISHINO, H., T. MIYATA and M. HASEGAWA, 1990 *Maximum likelihood inference of protein phylogeny and the origin of chloroplasts.* J. Mol. Evol. **31**: 151–160.

HASTINGS, W.K., 1970 *Monte Carlo sampling methods using Markov chains and their applications.* Biometrika **57**: 97–109.

KINGMAN, J.F.C., 1982a *The coalescent. Stochastic Processes and Their Applications* **13**: 235–248.

KINGMAN, J.F.C., 1982b *On the genealogy of large populations.* J. Applied Prob. **19A**: 27–43.

KISHINO, H., and M. HASEGAWA, 1989 *Evaluation of the maximum likelihood estimate of the evolutionary tree topologies from DNA sequence data, and the branching order in Hominoidea.* J. Mol. Evol. **29**: 170–1790.

KUHNER, M.K., and J. FELSENSTEIN, 1994 *A simulation comparison of phylogeny algorithms under equal and unequal evolutionary rates.* Mol. Biol. Evol. **11**: 459–468.

KUHNER, M.K., J. YAMATO, and J. FELSENSTEIN, 1995 *Estimating effective population size and mutation rate from sequence data using Metropolis-Hastings sampling.* Submitted for publication.

METROPOLIS, N., A.W. ROSENBLUTH, M.N. ROSENBLUTH, A.H. TELLER, and E. TELLER, 1953 *Equations of state calculations by fast computing machines.* J. Chem. Phys. **21**: 1087–1092.

TAKAHATA, N., 1985 *The coalescent in two partially isolated diffusion populations.* Genet. Res. **52**: 213-222.

VIGILANT, L., M. STONEKING, H. HARPENDING, K. HAWKES, and A.C. WILSON, 1991 *African populations and the evolution of human mitochondrial DNA.* Science **253**: 1503–1507.

VITERBI, A.J., 1967 *Error bounds for convolutional codes and an asymptotically optimum decoding algorithm.* IEEE Trans. Inform. Theory **IT-13**: 260–269.

YANG, Z., 1993 *Maximum-likelihood estimation of phylogeny from DNA sequences when substitution rates differ over sites.* Mol. Biol. Evol. **10**: 1396–1401.

ESTIMATING SELECTION AND MUTATION RATES FROM A RANDOM FIELD MODEL FOR POLYMORPHIC SITES*

STANLEY A. SAWYER†

Abstract. Selection and mutation rates are estimated from aligned DNA sequences by using a Poisson random field model for base frequencies at polymorphic sites. This approach is applied at amino acid polymorphic codon positions to estimate selection rates against observed amino acid polymorphisms and also the proportion of codon positions that admit a weakly selected replacement. The technique can also be used at silent sites to obtain a numerical estimate of codon bias, which is shown in two examples to be of comparable strength to the selection against observed variant amino acids. A nonparametric test that can detect selection against amino acids using silent sites as a control is also discussed. Finally, a technique that estimates silent mutation rates assuming selective neutrality of silent polymorphisms, but which is not sensitive to saturation, is used as a consistency check.

1. Introduction. The basic genetic material or DNA of plants and animals is composed of one or more chromosomes, where the actual number of chromosomes depends on the plant or animal. Each chromosome can be thought of as a long string of letters from the alphabet T, C, A, G, where each letter corresponds to a specific nucleotide. In this sense, a mouse is the same as a tomato to a geneticist, since both have about the same amount of DNA.[1] A gene or genetic locus is a segment of a chromosome that affects some trait. Typically, a gene contains one or more coding regions for a protein or RNA type along with some recognition sites for regulatory molecules. Some of the most important proteins are enzymes, which are proteins that catalyze biochemical reactions.

Beginning in the 1960's, many experiments found that enzymes in human and natural populations were often polymorphic; i.e., had multiple forms in the population (HARRIS, 1966; LEWONTIN and HUBBY, 1966). Some biologists felt that most of this variation is selectively neutral, with at most negligible effects on fitness (KIMURA, 1983). That is, whichever form of the enzyme that a creature possessed made no difference to its survival or lifestyle. Other biologists felt that different enzymes were unlikely to be selectively equivalent, and so various forms of selection must be involved (WILLS, 1973; LEWONTIN, 1974; *see also* HARTL, 1989; HARTL and SAWYER, 1991).

The coding regions of genes in DNA are directly translated into strings of amino acids by various RNA enzymes. Proteins and protein enzymes are built up from one or more strings of amino acids. Triples of nucleotides from a coding region of a gene are called *codons*. There are $4^3 = 64$

---

* This work was partially supported by National Science Foundation Grant DMS-9108262 and National Institutes of Health research grant GM-44889.

† Washington University, St. Louis, MO 63130, USA.

[1] Attributed to Eric Lander.

possible codons, of which 61 code for the 20 different amino acids and the remaining 3 indicate the end of a string of amino acids. Most amino acids are coded for by either two or four distinct codons, with the variability in the third nucleotide position. Thus the two codons CAT and CAC (and no other codons) code for the amino acid histine, and the four codons GTT, GTC, GTA, and GTG code for valine. A change in the DNA in a coding region that does not change the corresponding amino acid (for example, a change from CAT to CAC or from GTT to GTA) is called a *silent* or *synonymous* change. If a change in the DNA causes a change in the amino acid, it is called a *replacement* change. (The term "replacement" in this context refers only to amino acids.) As one might expect, DNA from natural populations shows a great deal of silent as well as replacement variation.

Curiously, the different codons for the same amino acid tend to occur in different frequencies (HARTL and CLARK, 1989; LI and GRAUR, 1991). This tendency is called *codon bias*. One possible reason for codon bias could be biased mutation rates at the DNA level. That is, some of the nucleotides T, C, A, G that make up DNA may be more mutable than others, or else mutation may preferentially result in some nucleotides as opposed to others. On the other hand, selection may also be involved. Some codons may be necessary for the proper configuration of DNA. The transfer RNA's that are necessary for each codon to translate DNA occur in different frequencies for different codons for the same amino acid, which may slow the translation of some synonymous codons. However, rates of nucleotide substitution at silent codon sites are similar to those in pseudogenes (LI *et al.* 1985; WOLFE *et al.* 1989). (A pseudogene is a DNA segment that resembles a gene but is not expressed.) This suggests that most silent substitutions are selectively neutral or nearly selectively neutral. While in general silent DNA changes appear to be under weaker selective constraints than changes that cause amino acid replacement, some silent changes are known that have a selective effect.

The purpose here is to discuss methods for detecting and estimating selection based on an aligned set of DNA sequences. The main technique that we discuss is a random field model for the frequencies of variant nucleotides at the various polymorphic sites that allows quantitative estimates of both selection and mutation rates (Section 4). This approach can be applied independently to silent polymorphisms and to amino acid variation, and provides independent estimates for both the selection against replacement substitutions and the selection involved in codon bias.

We begin with a nonparametric test for selection against replacement substitutions that uses silent substitutions as a selectively neutral "control" (Section 2). While this is the simplest of the three methods that we discuss, it does not give quantitative estimates of selection or mutation, and cannot detect selection against replacement substitutions if selection against silent substitutions is equally strong, which appears to be the case in one of our two examples. Section 3 is devoted to an independent method for estimating mutation rates at silent sites under different assumptions that provides a consistency check for the random field model.

A fringe benefit of the random field analysis is that you can estimate separate gene locus-wide mutation rates for silent changes and for amino-acid replacements. Many changes to a functioning protein or enzyme are presumably lethal or nearly lethal. Amino-acid variation that is common enough to be detected in a sample of DNA sequences must be subject to relatively weak selection. An estimate of the ratio of the amino-acid replacement mutation rate to the silent mutation rate should give an estimate of the proportion of amino acids in a protein that can be replaced without lethal effects on the host.

The models discussed below all assume that changes to an ancestral base or amino acid are either strongly deleterious (and so will never be seen in a natural population) or else change the fitness of the host by an equal amount, equal both for changes to different bases at the same site and for changes at different sites. Mutations at different sites have multiplicative effects on fitness. In particular, these models are not designed to detect "balancing" selection, in which rare enzymes have a selective advantage, or selection which is nonmultiplicative across sites. Of course, no statistical test can detect *arbitrary* forms of selection, since *anything* that you observe could be the result of selection for exactly that configuration.

2. A contingency-table test for unidirectional selection. Suppose that you have n aligned DNA sequences from a coding region. Most aligned sites will have a single most common base with at most a few sequences with different bases at that site. If most changes from this consensus base are selectively deleterious, then you would expect relatively few sequences at any particular site to be different from the consensus base. Similarly, if changes from the consensus base were advantageous, then you would also expect relatively few differences from the consensus, since otherwise the consensus would be quickly driven from the population.

If polymorphic replacement sites are observed to be more bunched (i.e., have fewer deviations from the consensus) than polymorphic silent sites, then one possible explanation is unidirectional selection against amino acid replacements.

Specifically, given n aligned DNA sequences, we say that a site is *simply polymorphic* if $n - 1$ of the sequences have one base at this site and one sequence has a second base. All other polymorphic sites are called *multiply polymorphic*. A three-site codon is called *regular* if the first two nucleotide positions are nondegenerate (i.e., any replacement changes the amino acid). This is equivalent to saying that the codon corresponds to an amino acid other than leucine or arginine, which are the only two amino acids for which a single change in the first or second nucleotide position of a codon will not change the amino acid.

About half of regular amino acids are fourfold degenerate at the third position, which means that any base can be substituted at the third codon site without changing the amino acid. Most of the other amino acids are twofold degenerate at the third position. Two-fold degenerate amino acids are of two types. The first type can have either T or C at the third position, but any other change at the third site changes the amino acid. For the second type, the third base is either A or G. There is one threefold degenerate amino acid (isoleucine), which corresponds to the three codons ATT, ATC, and ATA. Since the codon ATA is extremely rare in most natural populations, we treat isoleucine as twofold degenerate and the rare codon positions with an ATA as irregular.

Table 2.1: 2 × 2 tables for selection

14 strains of *Escherichia coli* at the *gnd* locus (1407bp): [a,b]

	All silent[c]		Two-fold silent[c]	
	simple poly	multiple poly	simple poly	multiple poly
Silent (regular)[c]	60	83	27	31
Replacements (1,2 pos'n regular)	20	7	20	7
	$P = 0.003^f$		$P = 0.021^f$	

8 strains of *Salmonella typhimurium* at the *PutP* locus (1467bp): [d,e]

Silent (regular)	93	59	43	26
Replacements (1,2 pos'n regular)	12	4	12	4
	$P = 0.416^f$		$P = 0.398^f$	

a – The *gnd* locus transcribes the enzyme 6-phosphogluconate dehydrogenase.

b – DYKHUIZEN and GREEN (1991); BISERCIC, FEUTRIER, and REEVES (1991).

c – See text for definitions.

d – The *PutP* locus transcribes the enzyme proline permease.

e – NELSON and SELANDER (1992).

f – Two-sided Fisher exact test.

Now, consider a 2×2 contingency table with the numbers of silent simply and multiply polymorphic sites at amino-acid monomorphic regular codon positions in the first row, and the numbers of simply and multiply polymorphic sites at the first and second positions of regular codons in the second row (SAWYER, DYKHUIZEN, and HARTL, 1987; see Table 2.1). Regular codon positions have the potential of supplying two replacement polymorphisms, but this is rare, and historically may have been the result of two distinct amino-acid replacements.

The contingency table in Table 2.1 is highly significant for 14 strains of *Escherichia coli* at the *gnd* locus when all silent sites are used, but is not significant for 8 strains of *Salmonella typhimurium* at the *PutP* locus. Thus the selective forces affecting variant bases appear to be different for silent and replacement sites in *gnd*, but not in *PutP*.

Most replacement variation may be lethal or at least subject to highly deleterious selection, so that there may be at most two weakly-selected bases at any amino-acid varying site. Thus it may be fairer to compare replacement polymorphisms with twofold degenerate silent polymorphisms, which are more likely to be simply polymorphic than fourfold degenerate sites. When twofold degenerate sites are used, the contingency table in Table 2.1 is significant for *gnd* but is not highly significant.

Note this test cannot detect selection against amino acid replacements if there is the same amount of selection against silent differences due to codon bias. Similarly, a positive result for this test might even be due to positive selection for silent nucleotide variants in combination with selectively neutral amino acid variation.

3. Estimating mutation rates assuming silent sites are neutral. There are many different ways to estimate the amount of mutation at silent sites in an aligned set of DNA sequences (see e.g. FU and LI, 1993). The following approach has the advantage that it automatically allows for parallel or repeated mutations at the same site, and can also be adapted to estimate the divergence time between two species (SAWYER and HARTL, 1992; *see also* SAWYER, DYKHUIZEN, and HARTL, 1987).

Assume that a fourfold degenerate site (for example) has mutation rates μ_T, μ_C, μ_A, and μ_G *to that base* per chromosome per generation. Thus the mutation rate depends on the base, but depends only on the end-product base (TAJIMA and NEI, 1982). Note that "mutations" of e.g. T to T that do not change the base are permitted here, but will not be counted below when estimating the locus-wide silent mutation rate.

Under these conditions, the *population* frequencies of the four bases at that site will have the joint probability density

$$(3.1) \qquad C_A \, p_T^{\alpha_T - 1} \, p_C^{\alpha_C - 1} \, p_A^{\alpha_A - 1} \, p_G^{\alpha_G - 1} \, dp_T \, dp_C \, dp_A \, dp_G$$

where $\alpha_T = 2N_e\mu_T$, $\alpha_C = 2N_e\mu_C$, ..., where N_e is the haploid effective population size and $C_A = C(\alpha_T, \alpha_C, \ldots)$, under the usual conditions for diffusion approximations (WRIGHT, 1949; KINGMAN, 1980). The density in equation 3.1 is called a Dirichlet density. The corresponding density for TC-twofold degenerate sites is the beta density $C'_A \, p_T^{\alpha_T - 1} \, p_C^{\alpha_C - 1} \, dp_T dp_C$ for $p_T + p_C = 1$, with a similar expression for AG-twofold degenerate sites.

Now assume that the site is part of an aligned sample of n DNA sequences, and consider the probability that the sample has n_T sequences with the base T at that site, n_C sequences with C, n_A with A, ..., where $n = n_T + n_C + n_A + n_G$. This probability can be obtained by integrating the density in equation 3.1, and is

$$(3.2) \qquad C_N \, \frac{\alpha_T^{(n_T)} \alpha_C^{(n_C)} \alpha_A^{(n_A)} \alpha_G^{(n_G)}}{\alpha^{(n)}}, \qquad \alpha = \alpha_T + \alpha_C + \alpha_A + \alpha_G$$

where $x^{(k)} = x(x+1)\ldots(x+k-1)$ and $C_N = n!/(n_T! \, n_C! \, n_A! \, n_G!)$ (WATTERSON, 1977). The corresponding probability for TC-twofold degenerate sites is $C'_N \alpha_T^{(n_T)} \alpha_C^{(n_C)}/(\alpha_T + \alpha_C)^{(n)}$. The probabilities in equation 3.2 for fourfold degenerate sites, and the corresponding probabilities at the two different types of twofold degenerate sites, can be combined to obtain maximum likelihood estimators for $\alpha_T, \alpha_C, \alpha_A,$ and α_G (Table 3.1).

Table 3.1: Maximum likelihood estimates of $\alpha_T, \alpha_C, \alpha_A, \alpha_G$ from the probabilities 3.2 at silent sites

gnd[a]		ADH[b]	
alpha's	4-fold[c]	alpha's	4-fold[c]
$\alpha_T = 0.128$	0.407	$\alpha_T = 0.0080$	0.155
$\alpha_C = 0.109$	0.288	$\alpha_C = 0.0300$	0.610
$\alpha_A = 0.063$	0.106	$\alpha_A = 0.0023$	0.066
$\alpha_G = 0.057$	0.199	$\alpha_G = 0.0097$	0.169
$\mu_{\text{sil}} = 30.82$[d]		$\mu_{\text{sil}} = 2.05$[d]	

a – Likelihoods for 14 strains of *E. coli* (1407bp; see Table 2.1).

b – Pooled likelihoods for 6 *Drosophila simulans* and 12 *D. yakuba* strains (771bp; McDONALD and KREITMAN, 1991; pooling means that within-species log likelihoods are summed).

c – Base frequencies at 4-fold degenerate regular silent sites.

d – Locus-wide silent mutation rate scaled by N_e (see text).

Given equation 3.1, the mean frequency of the base T at fourfold degenerate sites is $E(p_T) = \alpha_T/\alpha$ for α in equation 3.2. Thus the expected rate of transitions $T \to C$ at fourfold degenerate sites in a genetic locus is $N_4 \alpha_T \alpha_C/(2\alpha)$, where N_4 is the number of fourfold degenerate regular codon positions in the locus. (The factor of two is because μ_C is the mutation rate to the base C per N_e generations, while $\alpha_C = 2N_e\mu_C$ in equation 3.1.) Similarly, the mutation rate at pyrimidine twofold degenerate sites in the locus is $N_{2,TC}\,\alpha_T \alpha_C/(\alpha_T + \alpha_C)$, where $N_{2,TC}$ is the number of pyrimidine twofold degenerate regular codon positions. These considerations lead to a formula for the locus-wide silent mutation rate μ_{sil} (Table 3.1).

Remarks. The maximum likelihood method assumes that the distributions at different silent sites can be treated as independent. Recombination and gene conversion both help to insure the independence of site distributions. Independence can be tested by computing the significance of autocorrelations of the events monomorphic/polymorphic for adjacent silent sites. The first three autocorrelations are not significant for either the *E. coli* data in Table 2.1 nor the two ADH data sets in Table 3.1. Maximum likelihood theory uses independence only for the central limit theorem for the log likelihoods for the various terms, and so can tolerate some deviation from joint statistical independence.

The methods that are used in this paper assume that each sample is a random sample from a panmictic population. If a sample contains some strains that are significantly different from the others, then model parameters will be estimated incorrectly. For example, the references quoted in Table 2.1 have 16 strains of *E. coli* at the *gnd* locus, but two strains (labeled r4 and r16) are as distant from the other *E. coli* strains as they are from *Salmonella*. The remaining *E. coli* strains have an estimated phylogeny with a more coalescent-like appearance. These two aberrant *E. coli* strains were excluded from the analysis. Similarly, the 13 *Salmonella* strains at the *PutP* locus in Table 2.1 were reduced to 8 strains.

4. Estimating mutation *and* selection rates. We now discuss a model that will allow us to estimate the mutation rate μ and the relative selection rate γ for mutants, both scaled by the haploid effective population size N_e (SAWYER, 1994; HARTL, MORIYAMA, and SAWYER, 1994; *see also* SAWYER and HARTL, 1992).

This model is sensitive to saturation (repeated mutations at the same site), but should give reliable results if the estimate for μ_{sil} (the parameter μ for regular silent sites) is comparable to or greater than the more accurate estimate of μ_{sil} based on the Dirichlet density 3.1 of the previous section (which, however, assumes selective neutrality at silent sites). This model will be applied both for bases at regular silent sites and for amino acids at codon positions.

Consider a flux of mutations at the rate of μ per generation into the population. Each mutation changes one base at one site in one individual, and each new mutant base confers a relative selective advantage of $s = \gamma/N_e$ with respect to the current base. Most of the resulting new mutant alleles quickly go extinct by chance, but some survive to have appreciable base frequencies in the population:

Subsequent mutations are ignored at that site. Since we are assuming that all new mutant bases (or amino acids) are selectively equivalent, we can ignore mutation between mutant bases at the same site.

We now view the *population frequencies at polymorphic sites* for the surviving mutant bases as a *point process* of frequencies on $[0, 1]$. Under the usual diffusion approximation conditions (EWENS, 1979; ETHIER and KURTZ, 1986), this will be a Poisson point process with the expected density

(4.1)
$$2\mu \frac{1-e^{-2\gamma(1-p)}}{1-e^{-2\gamma}} \frac{dp}{p(1-p)} \qquad \text{for} \quad 0 < p < 1$$
$$= 2\mu \frac{dp}{p} \qquad \text{if } \gamma = 0 \text{ (i.e., no selection)}$$

(See SAWYER and HARTL, 1992, for a sketch of the proof.) Note that the densities in 4.1 are *not* integrable at $p = 0$. This corresponds to the fact that the population contains a large number of rare mutants at any one time. The formula 4.1 was first derived by WRIGHT (1938) as the transient distribution of the frequency of a single allele under selection and irreversible mutation.

Now suppose that we have a sample of n aligned DNA sequences from this population. Let N_k be the number of polymorphic sites that have k bases different from the ancestral base at that site. Then the N_k $(1 \leq k \leq n-1)$ are independent Poisson random variables with means

(4.2)
$$N(k, \mu, \gamma) = 2\mu \int_0^1 \frac{1-e^{-2\gamma(1-p)}}{1-e^{-2\gamma}} \binom{n}{k} p^k (1-p)^{n-k} \frac{dp}{p(1-p)}$$
$$= 2\mu \frac{1}{k} \qquad \text{if } \gamma = 0 \text{ (i.e., no selection)}$$

(SAWYER and HARTL, 1992).

In practice, we will not be able to tell which base is the ancestral base by looking at the various bases at a polymorphic site. However, any polymorphic site has at most two bases in our model, of which *one* must be the ancestral base. Thus the number X_k of polymorphic sites that have *either* k or $n - k$ bases different from the ancestral base—or, equivalently, have k of one type of base and $n - k$ bases of a second type—is observable. Here

(4.3)
$$X_k = N_k + N_{n-k}, \qquad 1 \leq k < n/2$$
$$= N_{n/2}, \qquad k = n/2$$

The variables X_k are independent Poisson random variables with means

(4.4)
$$G(k,\mu,\gamma) = N(k,\mu,\gamma) + N(n-k,\mu,\gamma), \qquad 1 \le k < n/2$$
$$= N(n/2,\mu,\gamma), \qquad\qquad\qquad k = n/2,$$

for $N(k,\mu,\gamma)$ in (4.2). Since the X_k are independent Poisson, the joint likelihood is

(4.5)
$$L(\mu,\gamma) = \prod_{k=1}^{(n+1)/2} e^{-G(k,\mu,\gamma)} \frac{G(k,\mu,\gamma)^{X_k}}{X_k!}$$

If a site actually has three or more different bases in a sample, choose k so that $n - k$ is the number of representatives of one of the bases with the largest number of representatives in the sample. This is equivalent to assuming that either that base or the common ancestor of all the other bases is the ancestral base. Since there are likely to be relatively few sites with three or more different bases, this treatment is not likely to bias the analysis.

Since the expression $G(k,\mu,\gamma)$ in (4.4) can be written in the form $G(k,\mu,\gamma) = 2\mu J(k,\gamma)$ where $J(k,\gamma)$ does not depend on μ, maximizing (4.5) is equivalent to the following procedure. First, set

$$\widehat{\mu}_{MLE} = \widehat{\mu}_{MLE}(\gamma) = \frac{X_{TOT}}{2J(\gamma)} \qquad \text{where} \qquad X_{TOT} = \sum_{k=1}^{(n+1)/2} X_k$$

is the total number of polymorphic sites and $J(\gamma) = \sum_{k=1}^{(n+1)/2} J(k,\gamma)$. Then maximize

(4.6)
$$L(\gamma) = L(\widehat{\mu}_{MLE}(\gamma),\gamma) = C(X) \prod_{k=1}^{(n+1)/2} \left(\frac{J(k,\gamma)}{J(\gamma)} \right)^{X_k}$$

as a function of γ. In particular, finding $\widehat{\mu}_{MLE}$ and $\widehat{\gamma}_{MLE}$ can be reduced to a one-dimensional maximization.

Given a data set composed of n aligned DNA sequences, we use the counts for polymorphic silent sites to estimate parameters μ_{sil} and γ_{sil} for silent sites, and the counts for amino-acid polymorphic codon positions to estimate parameters μ_{rep} and γ_{rep} for replacement amino acids. The conclusions for the two data sets of Table 2.1 are given in Table 4.1.

The scaled selection rate $\gamma_{rep} = -3.66$ for *E. coli* in Table 4.1 corresponds to a selection rate of $s = -\gamma_{rep}/N_e$ per generation against replacements, where N_e is the effective population size of *E. coli*. We can estimate N_e as follows. The value $\mu_{sil} = 30.82$ in Table 3.1 corresponds to $N_{sil} \times \mu N_e$, where μ is the mutation rate per site per generation and N_{sil} is the number of amino-acid monomorphic codon positions with twofold or

Table 4.1: Joint estimates of the locus-wide mutation rate μ and the selection rate γ

14 *E. coli* strains, *gnd* locus (1407p):

$\mu_{\text{sil}} = 30.82$ (neutral Wright model)

$\mu_{\text{sil}} = 33.57 \pm 5.50^{a}$

$\gamma_{\text{sil}} = -1.34 \pm 0.83^{**\ ac}$

$\mu_{\text{rep}} = 12.51 \pm 4.47^{b}$ ($\gamma_{\text{sil}} \neq \gamma_{\text{rep}} : P = 0.029^{*}$)

$\gamma_{\text{rep}} = -3.66 \pm 2.24^{***\ bc}$

8 *S. typhimurium* strains, *PutP* locus (1467bp):

$\mu_{\text{sil}} = 41.05$ (neutral Wright model)

$\mu_{\text{sil}} = 65.41 \pm 10.40$

$\gamma_{\text{sil}} = -2.43 \pm 0.96^{***}$

$\mu_{\text{rep}} = 7.63 \pm 3.34$ ($\gamma_{\text{sil}} \neq \gamma_{\text{rep}} : P = 0.77$)

$\gamma_{\text{rep}} = -2.04 \pm 2.44$

$*\ P < 0.05$ $**\ P < 0.01$ $***\ P < 0.001$

a – Estimated from base distributions at polymorphic regular silent sites.

b – Estimated from amino acid distributions at amino-acid polymorphic codon positions.

c – The ranges $\pm$ are 95% normal-theory confidence intervals, while P-values are for likelihood ratio tests against $\gamma = 0$.

---

fourfold degenerate regular silent sites. The 14 strains of *E. coli* in Table 3.1 have $N_{\text{sil}} = 367$, and the estimate $\mu = 5 \times 10^{-10}$ per generation (OCHMAN and WILSON, 1987) implies $N_e = 1.7 \times 10^{8}$.

This estimate of N_e leads to $s = -\gamma_{\text{rep}}/N_e = 2.2 \times 10^{-8}$ per generation against replacements. Thus the average magnitude of selection per generation that acts against observed amino acid substitutions is quite small. One way in which such a small selection coefficient could be realized is if a substitution is selectively neutral in most environments, but disadvantageous in some rarely-encountered environments (HARTL, 1989).

The estimates $\mu_{\text{rep}} = 12.51$ (for 469 codons, or $2 * 469 = 938$ first and second codon position sites) but $\mu_{\text{sil}} = 30.82$ (for 367 codons) in Table 4.1 suggests that about one sixth of amino acid positions in *E. coli* are susceptible to a weakly-selected replacement. SAWYER, DYKHUIZEN, and HARTL (1987) estimated $s = 1.6 \times 10^{-7}$ against replacement amino acids in a similar model assuming that *all* codon positions in *E. coli* in *gnd* were vulnerable to a weakly-selected amino-acid replacement. The earlier estimate is about seven times as large as the value $s = 2.2 \times 10^{-8}$ obtained above (with selection acting on about six times more amino acids), and the two estimates for s are remarkably consistent.

Table 4.2: Observed versus fitted values for the counts (4.3) for silent polymorphic sites

14 *E. coli* strains, *gnd* locus (143 silent polymorphic sites): *

k	1	2	3	4	5	6	7
Obs. X_k	60	31	16	10	14	7	5
Est. X_k	60.97	28.08	17.59	12.76	10.26	9.02	4.32

8 *S. typhimurium* strains, *PutP* locus (152 silent poly. sites): **

k	1	2	3	4
Obs. X_k	93	33	17	9
Est. X_k	92.23	33.77	18.54	7.45

---

* $P = 0.712$ (5 d.f.; $\mu_{\text{sil}} = 33.57$, $\gamma_{\text{sil}} = -1.34$)

** $P = 0.796$ (2 d.f.; $\mu_{\text{sil}} = 65.41$, $\gamma_{\text{sil}} = -2.43$)

The closeness of the estimates of μ_{sil} from the Poisson random field model to the estimates from Wright's Formula 3.2 suggests that saturation or repeated mutations at the same site do not have a significant effect on the estimates in Table 4.1. The fitted values for the numbers of polymorphic silent sites are quite close in both cases (Table 4.2). The fitted values for the counts for replacement amino acids resembled the observed counts in both cases, but had too many empty or near-empty cells to carry out a chi-square goodness-of-fit test.

The goodness-of-fit test in Table 4.2 is a nested hypothesis test for r Poisson variables $X_1, X_2, \ldots, X_r$. The means $\mu_k = E(X_k)$ are arbitrary in the larger model but satisfy $\mu_k = G(k, \mu_{\text{sil}}, \gamma_{\text{sil}})$ in the restricted model.

If the restricted model is true, then twice the logarithm of the ratio of the maximum likelihoods of the data under the two models has a χ^2-distribution with $r - 2$ degrees of freedom (RAO, 1973, p418). Note that we cannot use a standard χ^2 cell test here since the sum of the counts is not constrained to have a preassigned value.

REFERENCES

[1] BISERCIC, M., J. Y. FEUTRIER, and P. R. REEVES (1991) Nucleotide sequences of the *gnd* genes from nine natural isolates of *Escherichia coli*: evidence of intragenic recombination as a contributing factor in the evolution of the polymorphic *gnd* locus. *J. Bacteriol.* 173, 3894–3900.

[2] DYKHUIZEN, D. E., and L. GREEN (1991) Recombination in *Escherichia coli* and the definition of biological species. *J. Bacteriol.* 173, 7257–7268.

[3] ETHIER, S. N., and T. G. KURTZ (1986) *Markov Processes*. Wiley and Sons, New York.

[4] EWENS, W. J. (1979) *Mathematical Population Genetics*. Springer-Verlag, New York.

[5] FU, Y.-X., and W.-HS. LI (1993) Maximum likelihood estimation of population parameters. *Genetics* 134, 1261–1270.

[6] HARRIS, H. (1966) Enzyme polymorphisms in man. *Proc. Royal Soc. London Ser. B* 164, 298–310.

[7] HARTL, D. L. (1989) Evolving theories of enzyme evolution. *Genetics* 122, 1–6.

[8] HARTL, D. L., and A. CLARK (1989) *Principles of population genetics*, 2nd Ed. Sinauer Associates, Sunderland, MA.

[9] HARTL, D. L., E. MORIYAMA, and S. A. SAWYER (1994) Selection intensity for codon bias. *Genetics* 138, 227–234.

[10] HARTL, D. L., and S. A. SAWYER (1991) Inference of selection and recombination from nucleotide sequence data. *J. Evol. Biol.* 4, 519–532.

[11] KIMURA, M. (1983) *The Neutral Theory of Molecular Evolution*. Cambridge University Press.

[12] KINGMAN, J. (1980) *Mathematics of Genetic Diversity*. CBMS-NSF Regional Conf. Ser. Appl. Math 34, Soc. Ind. Appl. Math., Philadelphia.

[13] LEWONTIN, R. C. (1974) *The Genetic Basis of Evolutionary Change*. Columbia University Press.

[14] LEWONTIN, R. C., and J. L. HUBBY (1966) A molecular approach to the study of genic heterozygosity in natural populations. II. Amount of variation and degree of heterozygosity in natural populations of *Drosophila pseudoobscura*. *Genetics* 54, 595–609.

[15] LI, W.-H., and D. GRAUR (1991) *Fundamentals of molecular evolution*. Sinauer Associates, Sunderland, MA.

[16] LI, W.-H., C.-I. WU, and C.-C. LUO (1985) A new method of estimating synonymous and nonsynonymous rates of nucleotide substitution considering the relative likelihood of nucleotide and codon changes. *Mol. Biol. Evol.* 2, 150–174.

[17] MCDONALD, J. H., and M. KREITMAN (1991) Adaptive protein evolution at the *Adh* locus in *Drosophila*. *Nature* 351, 652–654.

[18] NELSON, K., and R. K. SELANDER (1992) Evolutionary genetics of the proline permease gene (*PutP*) and the control region of the proline utilization operon in populations of *Salmonella* and *Escherichia coli*. *J. Bacteriol.* 174, 6886–6895.

[19] OCHMAN, H., and A. C. WILSON (1987) in *Escherichia coli and Salmonella typhimurium: Cellular and Molecular Biology* eds Ingraham, J. L., Low, K. B., Magasanik, B., Neidhardt, F. C., Schaechter, M. and Umbarger, H. E. Amer-

ican Society of Microbiology Pubs.

[20] RAO, C. R. (1973) *Linear statistical inference and its applications*, 2nd ed. John Wiley & Sons, New York.

[21] SAWYER, S. A. (1994) Inferring selection and mutation from DNA sequences: The McDonald-Kreitman test revisited. *In* G. B. Golding (Ed.) *Non-Neutral Evolution: Theories and Data*. Chapman & Hall, New York, 77–87.

[22] SAWYER, S. A., D. E. DYKHUIZEN, and D. L. HARTL (1987) Confidence interval for the number of selectively neutral amino acid polymorphisms. *Proc. Nat. Acad. Sci. USA* 84, 6225–6228.

[23] SAWYER, S. A. and D. L. HARTL (1992) Population genetics of polymorphism and divergence. *Genetics* 132, 1161–1176.

[24] TAJIMA, F. and M. NEI (1982) Biases of the estimates of DNA divergence obtain by the restriction enzyme technique. *J. Mol. Evol.* 18, 115–120.

[25] WATTERSON, G. (1977) Heterosis or neutrality? *Genetics* 85, 789–814.

[26] WILLS, C. (1973) In defense of naïve pan-selectionism. *Amer. Naturalist* 107, 23–34.

[27] WOLFE, K., P. SHARP, and W.-H. LI (1989) Mutation rates differ among regions of the mammalian genome. *Nature* 337, 283–285.

[28] WRIGHT, S. (1938) The distribution of gene frequencies under irreversible mutation. *Proc. Nat. Acad. Sci. USA* 24, 253–259.

[29] WRIGHT, S. (1949) Adaption and selection, pp365–389 in *Genetics, Paleontology, and Evolution*, edited by G. JEPSON, G. SIMPSON, and E. MAYR. Princeton Univ. Press, Princeton, N.J.

THE USE OF LINKAGE DISEQUILIBRIUM FOR ESTIMATING THE RECOMBINATION FRACTION BETWEEN A MARKER AND A DISEASE GENE

N.L. KAPLAN[*] AND B.S. WEIR[†]

Abstract. Linkage analysis, which requires informative pedigrees, can rarely detect markers that are within a centimorgan of the target gene. Linkage disequilibrium analysis, which is based on population data, has been used to further localize the gene of interest. A drawback of this approach is that with these type of data it is difficult to estimate the recombination fraction between the marker and the disease gene. The estimation techniques that have been used are based on theoretical population genetic models. Among other things, these models assume that the disease is sufficiently ancient that equilibrium has been achieved. If, however, the disease is not very old, then equilibrium models are not appropriate. For a Wright-Fisher model the initial growth of a mutant gene can be described by a Poisson branching process. Using this result, a simulation based procedure is developed for obtaining confidence bounds on the recombination fraction between a marker and the disease gene. Properties of the procedure are described and recent marker data for cystic fibrosis are discussed.

1. Introduction. Linkage analysis has been very successful in localizing a disease gene to a chromosomal region. This method requires recombinationally informative pedigrees, and so linkage between the disease gene and markers that are very close, e.g., within a centiMorgan, is rarely detected. Allelic association (also referred to as linkage disequilibrium), which is based on population data rather than family data, has been used to further localize the gene of interest. With this approach one compares the marker allele distributions in samples of disease and normal chromosomes, and concludes that the two loci are probably close if the distributions are judged different. This conclusion is based on the observations that the rate of recombination tends to correlate with physical distance, and if the rate of recombination is low, then it is likely that the allele distributions in the normal and disease samples will be different for markers in a neighborhood of the disease locus.

Although there are many evolutionary forces that can cause an association between alleles at different loci, e. g., lack of recombination, selection, migration, founder effect or drift, only recombination is correlated with physical distance between the loci. It is reasonable to assume that for diseases that arose within the past few hundred generations, allelic association results from a lack of recombination in a neighborhood of the disease locus. The size of the neighborhood varies inversely with the age of a disease mutation, and the rate of recombination between the marker and disease locus. If there is no unusual reduction in the rate of recombination, then

---

[*] Statistics and Biomathematics Branch, National Institute of Environmental Health Sciences, Research Triangle Park, NC 27709.

[†] Program in Statistical Genetics, Department of Statistics, North Carolina State University, Raleigh, NC 27695–8203.

for older diseases, e.g. those that have been in the population thousands
of generations, it is likely that disease chromosomes will carry many less
ancestral marker alleles because of the many more intervening recombination events. Hence, assuming normal recombination rates, diseases for
which linkage disequilibrium to several markers in a chromosomal region is
observed are most likely of recent origin.

If the disease is not very old and an association is detected between
the marker and disease, then one would expect that at least one mutation
causing the disease has a high frequency in the disease population. If all
disease mutations were in low frequency, then, assuming that mutations
occur at random, the frequencies of ancestral disease haplotypes would reflect those in the normal population, and no association should be detected.
There are at least two examples that support the prediction. Cystic fibrosis, which is an autosomal recessive disorder, has a high frequency mutation
that is responsible for about 70% of the disease population (Kerem et al.
1989). Also there is evidence suggesting that Huntington disease, a late
onset dominant disorder, may have a single mutation responsible for as
much as one third of the disease population (MacDonald et al. 1992).

A drawback with population association studies is that it is difficult to
quantify the relation between recombination and measures of association.
Since recombination events are not observed, the recombination fraction,
c, between the marker and disease loci must be estimated on the basis of
a population genetic model. The models traditionally invoked assume the
population is at equilibrium and so the parameter estimated is not c but
$N_e c$, where N_e is the effective population size. In studies such as those
of Chakravarti et al. (1984) or Estivill et al. (1987), the expected value
of a statistic based on squared linkage disequilibrium was approximated
by $1/(1 + 4N_e c)$, and $N_e c$ estimated by equating this expression with the
observed value of the statistic. In addition to the complication that c is
confounded with N_e, there is the further problem that the estimates have
large variances because of the stochastic nature of the evolutionary forces
that shaped the population (Weir and Hill, 1986). Empirical information on
this genetic sampling would require information from replicate populations,
which is unlikely to be possible. We have also previously noted (Hill and
Weir, 1980) that there is no simple algebraic expression for the equilibrium
value of the expectation of squared linkage disequilibrium, and that account
must be taken of the sampling framework of normal and disease genotypes
(Kaplan and Weir, 1992).

The probability that a disease chromosome is identical by descent at
the marker locus is $(1 - c)^G \approx e^{-cG}$, where G is the number of generations
since the disease mutation occurred. A disease chromosome is identical
by descent at the marker if the chromosomal segment between disease and
marker has remained intact in all of its ancestors since the time of the
mutation. This equation confounds c with G instead of with N_e as in
the equilibrium models. If the association is due to an excess of disease

chromosomes that have an ancestral allele at the marker locus, then we would expect that e^{-cG} is not small, suggesting that cG is not large. At a minimum, we might expect that $cG < 1$ since $e^{-1} = .37$ and so less than 40% of the disease chromosomes are expected to have the ancestral marker allele. The conclusion that $cG < 1$ has important implications. If, for example, G was as large as 5000 generations, then the condition would require that $c < .0002$ meaning that every marker was fortuitously within 20 kb of the disease gene. (This assumes that $1cM = 1,000$ kb). It follows that there is no basis for inferring ancestral haplotypes (Kerem et al. 1989, MacDonald et al. 1992) suggesting that a larger c and smaller G is probably more appropriate. Hence we need to focus on modeling the initial growth phase of the disease rather than its behavior under an equilibrium assumption.

The major difficulty in studying the statistical properties of any estimate of c is taking into account the variability associated with the evolutionary history of the disease population. We addressed these concerns when we considered equilibrium population models (e.g. Hill and Weir 1988, Kaplan and Weir 1992). One way to deal with this problem is to assume an evolutionary model and simulate its dynamics. Hill and Weir (1994) recently used this approach in the equilibrium case to estimate the likelihood of $N_e c$. In this paper we will use the same approach except that we assume that the disease population is young rather than old. It is well known that the stochastic behavior of the initial growth phase of a mutation can be modeled with a Poisson branching process (Ewens, pg. 24 1979). This model is easy to simulate and so for any choice of parameters we can determine the sampling properties of any estimate of c.

In this study we propose a method for constructing confidence bounds on the recombination fraction between a disease and a marker locus that relies on simulating the evolution of the disease population to estimate the likelihood of the recombination fraction. We illustrate the method with cystic fibrosis data.

2. Methods. Suppose a marker M with A alleles $(A \geq 2)$ is determined to be in linkage disequilibrium with the disease locus. For a restriction length polymorphism $A = 2$, but for a microsatellites marker A can be greater than 2. To simplify the discussion we assume that A is an RFLP with marker alleles M_1 and M_2. The most frequent allele in the disease sample is denoted by M_1 and is assumed to be the marker allele on the ancestral chromosome on which the disease mutation occurred. For a multiallele marker the choice of ancestral allele may not be so clear if there are several alleles of comparable frequency in the disease sample. If the disease is young, then it is reasonable to assume that the marker polymorphism existed in the population when the disease mutations occurred. Even if there are different mutations resulting in the same disease phenotype, detecting linkage disequilibrium suggests that at least one disease mutation is in high frequency in the disease population, (e.g. the ΔF_{508} mutation causing cystic fibrosis, Kerem et al. 1989). Therefore, it is assumed that there is a single disease mutation that occurred at some time in the past on a chromosome carrying an M_1 allele which is identified as the high frequency allele in the disease sample.

In view of the low frequency of the disease, it is assumed that all individuals are either heterozygous for the disease (carrier) or are homozygous normal (non-carrier). Since the age of the disease is small compared to the population size, p_{n1} and p_{n2}, the proportions of the normal population carrying marker alleles M_1 and M_2 respectively, change so slowly that they are assumed constant in time. The proportions in the disease population, p_{d1} and p_{d2}, are changing, and it is the stochastic process dictating this change that we need to model. Sample sizes are denoted by k_n and k_d for the normal and disease chromosomes.

The stochastic rules for going from one generation to the next are based on a Wright Fisher sampling scheme. All carriers are assumed to be selectively equivalent. If the disease is recessive, then this assumption is reasonable. However, if the disease is dominant, then the disease is assumed to occur after the reproductive years, such as is generally the case for Huntington disease. Time will be measured in generations with $t = 0$ the generation in which the disease mutation occurred and $t = G$, the current generation from which the samples are taken. Suppose in generation t there are $X_T(t) = X_1(t) + X_2(t)$ carriers, where $X_i(t)$ is the number of disease chromosomes that carry marker allele M_i, $i = 1, 2$. The proportion of nonrecombinant disease gametes in generation t that carry marker allele M_i, $i = 1, 2$, is $(1-c)X_i(t)/2N$, where N is the total number of individuals in the population in generation t. Since the human population is growing, N is not assumed constant. A recombinant disease gamete from generation t carries marker allele M_1 or M_2, depending on whether the normal chromosome involved in the recombination carried marker allele M_1 or M_2. Hence, the proportion of recombinant disease gametes with marker allele M_i is $cX_T(t)p_{ni}$, $i = 1, 2$. The proportion of the gamete pool formed by generation t that have marker allele $M_i, i = 1, 2$ is therefore

$$g_i = \frac{(1 - c)X_i(t) + cX_T(t)p_{ni}}{2N}.$$

So long as g_1 and g_2 are small, $X_1(t+1)$ and $X_2(t+1)$ can be modeled as independent Poisson random variables with means $2N(1 + \lambda)g_1$ and $2N(1 + \lambda)g_2$, respectively (Ewens, pg. 24 1979). The parameter λ is small compared to one and can be interpreted as the sum of two quantities; ρ, the rate of growth of the overall population and s, the possible selective advantage of the carrier over the non-carrier. Since the population is large, we assume $\lambda > 0$. The population size N cancels in the expressions for the mean and we end up with the stochastic recursive relationship:

$$(2.1) \quad X_i(t+1) = \text{Poisson} \left[(1+\lambda)\{(1-c)X_i(t) + cX_T(t)p_{ni}\} \right]$$
$$i = 1, 2,$$

where *Poisson*$[\Gamma]$ denotes a Poisson variable with mean Γ. The initial values are $X_T(0) = X_1(0) = 1$ and $X_2(0) = 0$. If there are more than two marker alleles, then (1) holds for each allele.

The stochastic recursion (1) can be used to simulate the evolution of the disease population for any set of values of λ, c, and p_{n1}. The parameter of interest is c and so its value will vary, but will typically be less than a few centiMorgans. The obvious estimate of p_{ni} is f_{ni}, the frequency of marker allele i in the normal sample. A more conservative estimate, taking into account binomial sampling, is

$$f_{ni} + 2\sqrt{\frac{f_{ni}(1 - f_{ni})}{k_n}},$$

where k_n is the normal sample size. Specifying the values of the remaining parameters, λ and G, is more of a problem. The following is one possible approach for doing this which exploits available information about the disease population.

The current size of the disease population, $X_T(G)$, contains information about λ and G, and so the first step is to estimate $X_T(G)$. Very often an estimate is available of the incidence of the disease in the population under study, e.g., the incidence of cystic fibrosis is about $1/2000=0.0005$ (Kerem et al. 1989). From the incidence data and the genetic characteristics of the disease, (dominant or recessive), the frequency of disease chromosomes in the population can be estimated. For example, since cystic fibrosis is recessive, an estimate of its frequency is $\sqrt{0.0005} \approx 0.02$. To estimate $X_T(G)$ requires an estimate of the current population size which we take to be the total Caucasian population of about 500 million people or 10^9 chromosomes. Hence $X_T(G)$ is of the order of about 20 million for cystic fibrosis. An alternative approach for estimating $X_T(G)$ which might be useful for isolated populations is to use historical records. For example, Hastbacka et al. (1992) estimated the number of chromosomes carrying the mutation for diastrophic dysplasia to be about 85,000 for the Finnish population.

It seems reasonable in any application to consider only evolutionary histories that lead to values of $X_T(G)$ that are close to the estimated value. In the Finnish example, one might condition on histories that have values of $X_T(G)$ that fall in the interval (50,000-125,000). A strength of simulation is that this very natural conditioning is easy to implement.

Demanding that $X_T(G)$ be in a specified interval puts constraints on λ and G since the process describing the growth of the disease process, X_T, is a Poisson branching process whose offspring distribution has mean

$1 + \lambda$. For a given value of G, we need to choose λ so that $X_T(G)$ will fall in the specified interval a reasonable amount of the time. A bad choice of λ will result in very inefficient simulations. The mean of $X_T(G)$ equals $(1 + \lambda)^G$. Also the probability that the X_T process does not go extinct is approximately 2λ (Ewens 1979). Hence if we imagine that all the mass of $X_T(G)$ is at the estimated value, then we can estimate λ using the equation

$$\text{estimated value of } X_T(G) = \frac{(1 + \lambda)^G}{2\lambda}$$

The determination of $X_T(G)$ is not very precise and so it is important that the results be reasonably robust with respect to this quantity. For a fixed value of G, large changes in $X_T(G)$ cause small changes in the estimate of λ. For example, if $G = 200$, and $X_T(G) = 10^5$, then the corresponding value of λ is .047, while $X_T(G) = 10^7$ results in $\lambda = .074$. Since the behavior of a branching process is governed by the offspring distribution, and since the Poisson distribution is completely determined by its mean, small changes in $1 + \lambda$ will not materially change the behavior of the X_T process. In the example a hundred fold change in $X_T(G)$ results in less than a 3% change in the mean.

If $X_T(G)$ is specified, then the sole remaining parameter that needs to be specified is G. In the absence of historical records suggesting a value of G, other methods are needed. For example, suppose we have two markers in the region of interest and an estimate of the recombination fraction between them. If the assumed value of G is too large, then the model-based estimates of c will be too small, and so an upper bound on G can be determined.

Fortunately, the simulations show that the age of the population acts essentially as a scaling factor, and so confidence bounds need to be computed only for a single reference value of G. The reason for this is that the critical parameter of the model is the product cG. Although we cannot prove this property about G analytically, evidence of the importance of the product cG comes from the formula for the expectation of p_{d1},

$$\begin{aligned} E(p_{d1}) &= (1 - c)^G + (1 - (1 - c)^G)p_{n1} \\ &\approx e^{-cG} + (1 - e^{-cG})p_{n1}. \end{aligned}$$

Once values for p_{ni}, λ and G have been decided, the simulations can be used to make inferences about c from the data. To do this we estimate the likelihood of c for specified data. This is easy to do since conditional on p_{d1}, the number of chromosomes in the disease sample carrying marker allele M_1 has a binomial distribution with parameters k_d and p_{d1}. We denote the binomial probability by $B(\tilde{k}, k_d, p_{d1})$, where $\tilde{k}$ is the vector of allele counts in the disease sample (data will be denoted with a tilde). To calculate the likelihood we repeatedly simulate the population frequency p_{d1}, always requiring that $X_T(G)$ be near the estimated value, and then average the associated binomial probabilities. The only numerical difficulty is that the likelihood must be scaled in order to obtain nonnegligible values. In all cases we scale by $B(\tilde{k}, k_d, f_{d1})$, where f_{d1} the frequency of M_1 in the disease sample. Unless stated otherwise, the likelihood will always be scaled by this term. By repeating this process for a number of values of c, we estimate the likelihood as a function of c. Following standard procedure, we drop down 2 units from the maximum to obtain the support interval for c which is approximately a 95% confidence interval. There is no problem considering multiple allele markers. The only difference is that the binomial distribution is replaced by the multinomial distribution.

3. Results. In this section we apply the proposed method to marker data for cystic fibrosis. The markers discussed are in linkage disequilibrium with the disease locus, and M_1 is the high frequency allele in the disease sample which is assumed to be the allele on the ancestral chromosome.

Cystic fibrosis (CF) is an autosomal recessive disease that to date is the best example of the usefulness of linkage disequilibrium in mapping a disease gene. Extensive linkage analysis localized the CF gene to chromosome 7 (region q31), (Eiberg et al. 1985) and linkage disequilibrium data proved useful in further localizing the gene (Kerem et al. 1989). Approximately two percent of the Caucasian population are carriers of the disease gene, implying that the carrier population is in the millions (Estivill et al. 1987). As already noted, it is not critical what value is assigned to $X_T(G)$ so long as it is large. For the calculations we will be somewhat conservative and assume that $X_T(G) = 10^7$. To guarantee that the size of the simulated disease populations is in a neighborhood of 10^7, we consider only realizations that are in the interval $(.9x10^7, 1.1x10^7)$. The size of the interval is not that critical, but it can affect the length of the simulation if the interval is very small. If G, the age of the disease, is specified, then the growth rate, λ, is the solution of the equation

$$\frac{(1+\lambda)^G}{2\lambda} = 10^7$$

For G of 200, which is close to the estimates of the age of the major CF mutation, ΔF_{508}, given by Serre et al. (1990), the corresponding value of λ is 0.074. To simplify the analysis, the estimate of p_{n1} is the frequency of M_1 in the normal sample. For each likelihood calculation 1000 repetitions were carried out.

The two loci considered are the flanking markers MET and D7S8. In Table 3.1 are the data for three two allele polymorphisms: $pmetD/TaqI$, $pmetH/TaqI$ and $pJ311/MspI$. Since $pmetD/TaqI$ and $pmetH/TaqI$ are so close together, we also considered the two polymorphisms as a single one with three alleles. The double recombinant was rare and so it was ignored. All of the data except that for $pmetH/TaqI$ came from the paper of Beaudet et al. (1986) which was the first extensive set of haplotype data

TABLE 3.1

Data for the cystic fibrosis flanking markers MET and D7S8

Polymorphism	Sample size		Allele frequencies			Chi square
				CF	N	
$pmetD/TaqI^a$	CF	402	M_1	.88	.77	17.4
	N	384	M_2	.12	.23	
$pmetH/TaqI^b$	CF	115	M_1	.73	.47	15.0
	N	105	M_2	.27	.53	
$pJ3:11/MspI^a$	CF	448	M_1	.56	.41	18.5
	N	429	M_2	.44	.59	
$pmetH/pmetD^a$	CF	348	M_1	.62	.42	29.0
	N	330	M_2	.12	.21	
			M_3	.26	.37	

[a]Beaudet et al 1986
[b]Cutting et al. 1989

published for MET and *D7S8*. The remaining data were published by Cutting et al. (1989). In Figure 3.1 we plot the ln(likelihood) curves for four different polymorphisms for five different values of G : $100, 150, 200, 300$ and 400. To make the four curves with different values of G comparable, we scaled c in each case by the factor $G/200$. For each of the four polymorphisms, the curves are virtually identical, showing that G does indeed act as a scaling factor. In Figure 3.2, we plotted the ln(likelihood) curves for different values of $X_T(G)$ for the polymorphism $pmetD/TaqI$ ($G = 200$). In the three cases, $X_T(G) = 10^6$, 10^7 and $2x10^7$, there is virtually no difference in the curves showing that the value of $X_T(G)$ is not that critical so long as it is large.

In Table 3.2 we give confidence bounds for the four polymorphisms of Table 3.1 for $G = 200$. To calculate the bounds we evaluated the likelihood at a series of c values at intervals of .001. The results suggest an upper bound for MET of .010 and .012 for D7S8. The lower bounds are .002 and .003 respectively. Also included in Table 3.2 are the maximum likelihood estimates of c.

Beaudet et al.(1986) used linkage analysis to obtain upper confidence bounds on MET and D7S8 of .012 and .011 respectively. These bounds are comparable to those in Table 3.2 and so for these markers little additional information is gained from the linkage disequilibrium analysis. Beaudet et al. reported lower bounds to only one decimal place and so their bounds are 0.0.

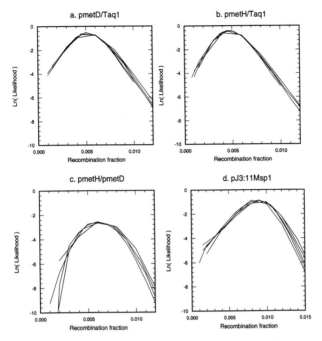

FIG. 3.1. *For each of the four polymorphisms in Table 3.1 the ln(likelihood) is plotted as a function of c for five different values of G : 100, 150, 200, 300 and 400 generations. The reference value of G is 200, and so for each curve c is scaled by G/200. The curves are expressed relative to ln(B($\tilde{k}, k_d, f_{d1}$)).*

TABLE 3.2

Upper and lower endpoints of the support interval for the recombination fraction ($G = 200$)

Polymorphism	Upper bound	Lower bound	Maximum likelihood estimate
$pmetD/TaqI$	.009	.002	.005
$pmetH/TaqI$	.008	.002	.005
$pJ3 : 11/MspI$	.012	.005	.008
$pmetH/pmetD$	.010	.003	.005

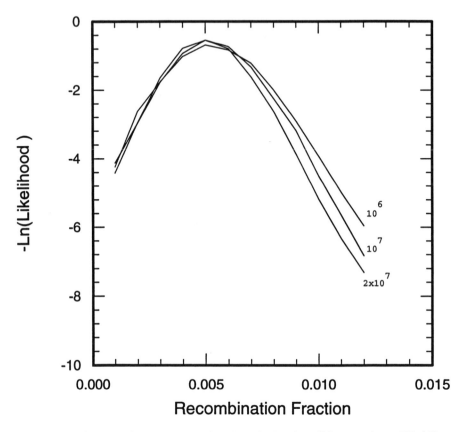

FIG. 3.2. *ln(likelihood) is plotted as a function of c for three different values of $X_T(G)$:* 10^6, 10^7 *and* 2×10^7. *The polymorphism is pmetD/TaqI and G = 200.*

4. Discussion. It is not easy to estimate the recombination fraction between a marker and a disease gene from linkage disequilibrium data. Previous estimates have been criticized for their large variance (Hill and Weir 1988), and for failing to take into account the sampling scheme (Kaplan and Weir 1992). Another problem which has not received much attention is that the underlying equilibrium model is probably not appropriate for describing the evolution of rare human diseases. The use of linkage disequilibrium analysis for localizing a disease gene is based on the simple observation that descendants of a disease mutation tend to share the haplotype of the ancestral chromosome in a neighborhood of the disease gene. The size of the neighborhood depends on the recombination rate (c) and the age of the mutation (G). The product, cG, is the critical parameter, and a simple argument shows that this parameter cannot be very large if one hopes to reconstruct ancestral haplotypes from samples of disease chromosomes in the current population. If G is large, as would be necessary in the equilibrium case, then c must be small, and so any marker that shows a significant association with the disease because of an excess of ancestral haplotypes in the disease population, is necessarily very close to the disease gene. This seems unlikely, and therefore we are forced to consider the nonequilibrium case. Hence, if one believes that ancestral haplotypes are present in the current disease population, then the disease cannot be too old.

Under reasonable assumptions a Poisson branching process gives an adequate description of the growth of the disease population. There is no simple procedure for obtaining bounds on the recombination fraction and so a simulation approach is used to estimate the likelihood of c. There are several parameters that must be specified in order to carry out the simulation. The results are fairly robust to all the parameters except the age of the disease. This is not surprising since the product cG is the critical parameter. As Figure 3.1 clearly shows G acts a scaling factor, and so results for a given value of G can be scaled to obtain results for any other value of G. One advantage of using simulations is that it is easy to incorporate additional information into the analysis. For example, if we have some idea as to how large the disease population is, then we can easily require that the simulated populations have about the same size.

The method in this paper was applied to marker data for cystic fibrosis. The bounds obtained for markers near the CF gene in Table 3.2 are reasonably consistent with the physical map given by Kerem et al. (1989), assuming that $1\ cM = 1000kb$. Their estimates for MET and D7S8 are approximately 900 kb and 800 kb, respectively, which convert to recombination estimates of .009 and .008. Our results are for the reference value of $G = 200$, and so if a different value of G was used, the bounds would change accordingly. This dependence on the value of G is a nuisance, but it appears to be the price that one pays for using population data rather than family data.

It is instructive to point out why the method works so well for CF. First, approximately 70% of the CF chromosomes descend from a single mutation, ΔF_{508}. Secondly, on the ancestral chromosome most of the markers in the vicinity of the disease gene had the less frequent allele. Finally the mutation was recent enough so that recombination did not totally degrade the ancestral haplotype in the neighborhood of the CF gene. The success of CF is due to fortuitous evolutionary events and so there is no reason to expect that all diseases will be so lucky, evolutionarily speaking.

There are several areas of future research. First, the scaling property of G needs to be proven analytically. For two allele markers the likelihood of c can be expressed as moments of p_{d1}, and so it would suffice to show that the moments of p_{d1} only depend on cG. Second, the method needs to be applied to data sets for different diseases. Some examples of diseases where there is linkage disequilibrium data are Huntington disease, Friedreich ataxia and diastrophic dysplasia. Third, the method should be compared with other proposed methods for estimating confidence bounds on c such as the one suggested by Hastbacka et al. (1992). Finally, the method can easily be extended to study the simultaneous behavior of two marker loci. In this case there is an added complication since the position of the disease locus relative to the marker loci also has to be specified. One would expect that the likelihood approach could be used to help decide on the appropriate topology. Work on these problems is currently underway.

REFERENCES

[1] Chakravarti A., Buetow K.H., Antonarakis S.E., Waber P.G., Boehm C.D., Kazazian H.H., *Nonuniform recombination within the human b-globin gene cluster*, Am. J. Hum. Genet. 36 (1984), pp. 1239-1258.

[2] Cutting G.R., Antonarakis S.E., Buetow K.H., Kadch L.M., Rosenstein B.J., Kazazian H.H., *Analysis of DNA polymorphism haplotypes linked to the cystic fibrosis locus in North American black and Caucasian families supports the existence of multiple mutations of the cystic fibrosis gene*, Am. J. Hum. Genet. 14 (1989), pp. 307-318.

[3] Eiberg H., Mohr J., Schmiegelow K., Nielson L.S., Williamson R., *Linkage relationships of paraoxonase (PON) with other markers: indication of PON-cystic fibrosis synteny*, Clin. Genet. 28 (1985), pp. 265-271.

[4] Estivill X., Farrall M., Scambler P.J., Bell G.M., Hawley K. M.F., Lench N.J., Bates G.P., Kruyer H.C., Frederick P.A., Stanier P., Watson E.K., Williamson R., Wainwright B.J., *A candidate for the cystic fibrosis locus isolated by selection for methylation-free islands*, Nature 326 (1987), pp. 840-845.

[5] Ewens W.J., *Mathematical Population Genetics Springer-Verlag*, New York (1979).

[6] Hastbacka J., de la Chappella A., Kaitila I., Sistonen P., Weaver A., Lander E., *Linkage disequilibrium mapping in isolated founder populations: diastrophic dysplasia in Finland*, Nature Genet. 2 (1992), pp. 204-211.

[7] Hill W.G., Weir B.S., *Variances and covariances of squared linkage disequilibria in finite populations*, Theor. Pop. Biol. 33 (1988), pp. 54-78.

[8] Hill W.G., Weir B.S., *Maximum likelihood estimation of gene location with linkage disequilibrium*, Am. J. Hum. Gen. 54 (1994), pp. 705-714.

[9] Kaplan N., Weir B.S., *Expected behavior of conditional linkage disequilibrium*, Am. J. Hum. Genet. 51 (1992), pp. 333-343.

[10] Kerem B., Rommens J.M., Buchanana J.A., Markiewicz D., Cox T.K., Chakravarti A., Buchwald M., Tsui L.C., *Identification of the cystic fibrosis gene:genetic analysis*, Science 245 (1989), pp. 1073-1080.

[11] MacDonald M.E., Novelletto A., Lin C., Tagle D., Barnes G., Bates G., Taylor S., Allitto B., Altherr M., Myers R., Lehrach H., Collins F.S., Wasmuth J.J., Frontali M., Gusella J.F., *The Huntington's disease candidate region habits many different haplotypes*, Nature Genet. 1 (1992), pp. 99-103.

[12] Serre J.L., Simon-Bouy B., Mornet E., Jaume-Roig B., Balassopoulou A., Schwartz M., Taillander A., *Studies of RFLP closely linked to the cystic fibrosis locus throughout Europe lead to new considerations of population genetics*, Hum. Genet. 84 (1990), pp. 449-454.

[13] Weir B.S., Hill W.G., *Nonuniform recombination within the human b-globin gene cluster*, Am. J. Hum. Genet. 38 (1986), pp. 776-778.

THEORY AND APPLICATIONS OF RAPD-PCR:
MISPRIMING

ANDREW G. CLARK*

Abstract. By applying the polymerase chain reaction (PCR) to random 10-nucleotide DNA primers, one can obtain Random Amplification of Polymorphic DNA (RAPDs). The information provided by these banding patterns has proven useful for mapping and for verification of identity of bacterial strains. A simple renewal process provides expected distributions for the number and length of fragments obtained when genomic DNA is amplified by RAPD-PCR. The observation of more than the expected number of bands from RAPD analysis of bacterial DNA suggests that a frequent problem is mispriming (amplification at sites where the match to the primer is not perfect). Expected numbers of fragments and fragment length distributions with mispriming are examined.

Key words. DNA, RAPD, PCR, renewal process.

1. Introduction. Random amplification of polymorphic DNA by the polymerase chain reaction (RAPD-PCR) has proven to be a useful technique for detecting polymorphisms in genetic mapping and strain identification (Williams *et al.* [13]; Welsh and McClelland [12]; Wang *et al.* [9]). The method is generally faster and less expensive than any previous method for detecting DNA sequence variation. In principle it should be possible to use RAPD banding patterns to estimate statistics that quantify the level of DNA polymorphism. Any such procedure would require a model to establish a quantitative relationship between DNA sequence polymorphism and expected levels of RAPD band polymorphism. Clark and Lanigan [2] made some progress in this area, but they assumed that RAPD amplification occurred only if there was an exact match (exact complementarity) between at all ten nucleotides of the primer and the genomic DNA being tested.

RAPD-PCR fragment patterns can be highly repeatable, and by recombination mapping, the patterns can be interpreted as being caused by clear genetic segregation of DNA sequence differences. But the observation that priming can occur at sites that do not exactly match the primer has sullied the reputation of the RAPD technique because different laboratories can obtain different banding patterns (Smith *et al.* [8]). Here we look at some of the consequences of mispriming on expected products of RAPD-PCR in terms of the numbers of bands and their size distribution. This analysis follows that of Clark and Lanigan [2] and Weissing and Velterop [11].

2. How RAPD patterns are scored. RAPD-PCR applies the polymerase chain reaction with a single short oligonucleotide primer, randomly amplifying fragments of genomic DNA which are size-fractionated by agarose

* Department of Biology, Institute of Molecular Evolutionary Genetics, Pennsylvania State University, University Park, PA 16802

gel electrophoresis. For each reaction, a single 10-nucleotide random primer is added to a rather standard PCR reaction containing the template DNA, a buffer, and the DNA polymerase of *Thermus aquaticus* (also known as *Taq* polymerase), or the DNA polymerase of some other thermophilic bacterium (*AmpliTaq* is the trade name that Perkin Elmer uses). PCR works by a series of cycles of denaturing the DNA (making it single-stranded), annealing (allowing the primer molecules to find complementary sequences on the target DNA and to form a base-paired double helix), followed by elongation (when the *Taq* polymerase actually synthesizes the new DNA strand).

In the first cycle, DNA is synthesized starting at the position on the target DNA where the primer hybridizes and going in the 5′ to 3′ direction. After elongation, the cycle repeats, except that in the second and subsequent rounds, primers hybridize not only to the original DNA, but also to the newly synthesized fragments. In the case of a target DNA sequence with two primer sites in opposite orientation (and "pointing" toward each other), the synthesis results in nearly a doubling in the amount of DNA for the fragment located between the two primers, provided they are between about 100 and 5000 nucleotides apart. PCR can also be performed with a polymerase that has much better error correction than does *Taq* polymerase, and DNA molecules up to 35,000 nt have been amplified ([1]). In principle such error-correcting polymerases could be used in RAPD-PCR, but the very long fragments will still probably not be seen because they will be out-competed by the shorter fragments in the PCR reaction. Typically 30–40 cycles of denaturation-annealing-elongation are performed, resulting in vast amplification of a particular fragment or set of fragments. Standard PCR uses a programmable heat block to carry out the whole operation, and temperatures for the three phases are typically 94° for denaturation, 56° for annealing and 72° for synthesis. RAPD-PCR uses an unusually low temperature of $35 - -42°$ for annealing, and it is this low stringency that makes mispriming a frequent occurrence. After the PCR cycles are complete, a portion of the solution is loaded on an agarose gel and subjected to electrophoresis. Typically, fragments in the range from about 200–10,000 nucleotides in length can be separated and scored.

The RAPD method is useful in genetic analysis only if the variation in banding patterns accurately represent allelic segregation at independent loci. Polymorphism is detected as band presence vs. absence, and may be caused either by failure of the primer to hybridize to sites on the target DNA in some individuals due to nucleotide sequence differences or by insertions or deletions in the fragment between two conserved primer sites. True allelic segregation may be confused with intermittent PCR artifacts (Riedy *et al.* [7]), unless additional genetic analysis is performed. Other technical problems and a number of experimental solutions are presented by Hadrys *et al.* [4] and Lanigan [5]. Pedigree-structured data makes it possible to identify bands that exhibit Mendelian segregation.

3. Expected distribution of RAPD fragment lengths. Let us begin with the assumption that the primers hybridize to all sites in the genome that precisely match in sequence, and one or more base mismatches at the primer site precludes hybridization. Hypothetically, assume that a fragment is amplified in all cases in which two successive primer sites are located on complementary strands in opposite orientation. We can label the orientations as $-$ and $+$, and suppose that only the $-+$ order produces a PCR product. This is equivalent to assuming that when primer sites are nested (as in $---+++$) only the smallest of the possible fragments is amplified from the innermost pair of primers, and in practice this appears to be true (Williams *et al.* [14]). Under the assumption that the nucleotide sequence is random and unstructured, the probability of matching a primer at a site is independent of the proximity of other matches, provided the primer sequence is non-self-overlapping. This means that the occurrence of matches can be approximately modeled as a discrete renewal process, with an geometric distribution of distances between matches and a bionomial distribution of the number of matches (Feller [3]).

Following the notation and derivation of Weissing and Velterop [11], let a be the probability that any given n-nucleotide run in the genomic DNA matches the n-nucleotide primer. If successive bases are statistically independent, then a is the product of the genomic frequencies of the bases in the primer. If the bases were equally frequent, $a = (1/4)^n$. If we assume that a is small and the target DNA length is large, then the expected distribution of the interval between such primer sites on one strand of the DNA is approximately exponential with parameter a. The same distribution applies to the opposite strand. Amplification occurs only when successive primer sites are of opposite orientation. To get a PCR product of length L nucleotides between the primers, we need L consecutive sites that do not correspond to a match with the primer in either orientation (which has probability $(1 - 2a)^L$), and two independent sites which match the primer (which has probability a^2). Thus we get

Prob(fragment of length L with fixed starting site) $= a^2(1 - 2a)^L$

The probability of a fragment of any length is the sum of this quantity over all possible lengths (including zero), or

$$\text{Prob(fragment)} = \sum_L a^2(1 - 2a)^L = \frac{a}{2}$$

The probability density function of RAPD fragment lengths (p_L) is therefore the probability that a fragment of length L is obtained, normalized by the probability of obtaining any fragment:

$$p_L = \frac{a^2(1 - 2a)^L}{a/2} = 2a(1 - 2a)^L \approx 2ae^{-2aL}$$

So the distribution of fragment lengths expected from RAPD PCR is exponential with a mean $1/(2a)$, and a variance of $1/(2a)^2$. In the case where all four nucleotides are equally frequent, a 10 nt primer has $a = (1/4)^{10}$, and the mean fragment length expected after PCR is 524,288 nt. Because typical agarose gel electrophoresis only identifies fragments less than 10,000 nt, it is clear that the useful bands are in the tail of this distribution.

As described by Clark and Lanigan [2], RAPD-PCR generally produces more fragments than is expected based on the theory presented in the next section. Some bands may also be generated when mispriming results in amplification at sites that do not perfectly match the primer sequence. The observation that many 10 nt primers produce five or more RAPD bands from bacterial genomic DNA suggests that either mispriming occurs frequently or that bacterial genomes are highly structured (Wang *et al.* [9]). Moya *et al.* [6] did RAPD-PCR with 20 different random primers of length 10 nt on phage lambda, whose genome is 48,502 nt in length. The observed average number of RAPD fragments was 0.7 per primer, a number nearly 100-fold greater than expected. They conclude that only 7 of the 10 nucleotides need to match to effectively prime the PCR reaction. Additional experiments showed that the 3 nucleotides at the 3′ end of the primer must always match exactly, and that approximately 4 of the remaining 7 must also match. If all four nucleotides are equally frequent, the chance of such a match is $\frac{1}{4}^7 \binom{7}{4} = 0.002136$. This is 2240 times as frequent as matches requiring a perfect 10 nt match. The discrepancy between the observed 100-fold excess and the expected 2240-fold excess is probably due to competitive interactions in the PCR reaction, so that only a fraction of possible misprimed PCR products are seen. With this level of mispriming, the mean fragment length is expected to be $1/2240$ that expected in the case of a perfect 10 nt match, or 234 nt.

4. Expected number of RAPD bands per primer. The expected number of locations on the template DNA where the primer will hybridize in one orientation is Ca, where C is the number of nucleotides in the template DNA. We need to consider primers hybridizing in both orientations, where the expected number of matches in either orientation is $2Ca$, and the distribution of the number of matches is Poisson. Because adjacent primers must be of opposite orientation, and are therefore not independent, the number of RAPD fragments that are amplified does not have a Poisson distribution.

Amplification occurs only if two adjacent primer matching sites are in opposite orientation and in the right order. Suppose that only a $-+$ transition results in amplification. In order to derive the probability that m fragments are amplified given that there are M primer sites, we need to calculate the number of ways m $-+$ transitions can occur. For any given sign sequence of length M, it turns out there are $\binom{M+1}{2m+1}$ ways to get m $-+$

transitions. Because there are 2^M sequences of $+$ and $-$ of length M, we get the conditional probability:

$$\text{Prob}(m \text{ fragments} \mid M \text{ matches}) = \frac{1}{2^M}\binom{M+1}{2m+1}$$

For example, there are 16 possible sign sequences of length $M = 4$ matches, and of these 5 have 0 $-+$ transitions, 10 have 1 $-+$ transition and 1 sequence has 2 $-+$ transitions. To obtain the density function for the probability of obtaining m fragments, we need to take the sum over M of the probability of obtaining M primer matches times the conditional probability of obtaining m fragments given M matches. As Weissing and Velterop [11] show:

$$
\begin{aligned}
q_m &= \sum_M e^{-2aC}\frac{(2aC)^M}{M!}\frac{1}{2^M}\binom{M+1}{2m+1} \\
&= e^{-aC}\left(\frac{(aC)^{2m}}{(2m)!} + \frac{(aC)^{2m+1}}{(2m+1)!}\right)
\end{aligned}
$$

This distribution has mean $\mu = \frac{aC}{2} - \frac{1}{4}(1 - e^{-2aC})$ and variance $\sigma^2 = \frac{aC}{4}(1 - 2e^{-2aC}) + \frac{1}{16}(1 - e^{-4aC})$. Note that the effect of mispriming can be entirely accounted for by its effect on a, the probability of the primer hybridizing at a site, assuming the primer is nonrepetitive and non-selfcomplementary (including partial self-complementarity). This means that a doubling in the probability of a match results in slightly less than a doubling in the expected number of fragments and slightly less than a doubling in the variance. The effect of mispriming on fragment number is expected to be enormous in the case of multiple mismatches being allowed. We noted above that allowing 3 mismatches in the first 7 of 10 nucleotides results in 2240 times as many fragments as the case of exact matching. The fact that mammalian genomes rarely produce more than 20 bands suggests that the degeneracy is not as great as Moya *et al.* [6] found it to be for phage lambda. If the primer sequence can overlap with itself, a needs to be corrected in a manner analogous to that described by Waterman [10].

5. Computer simulations. Twenty random sequences of length $C = 25,000$ nt were generated having all four nucleotides equally frequent. For each template sequence, 16 nonrepetitive nonselfcomplementary primer sequences of length 5 were generated.

In each case, the computer scanned the template sequence for $-+$ transitions and recorded the number of fragments generated and their sizes. This process was done in two ways, first requiring perfect 5 nt matches and second allowing a mismatch at the $5'$ most nucleotide of the primer.

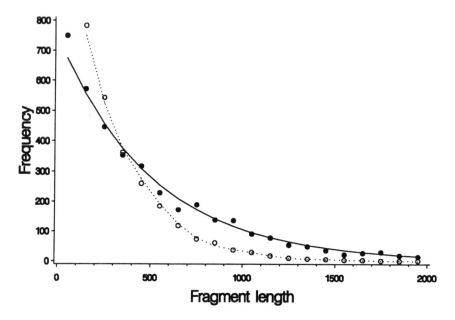

FIG. 1. *Distribution of fragment lengths. 20 random sequences of length* $C = 25,000$
nt were generated, and each template sequence, 16 nonrepetitive nonselfcomplementary
primer sequences of length 5 were generated. The computer scanned each template
sequence and recorded the number of fragments generated and their sizes. The solid
curve represents the expected distribution of fragment lengths when a perfect match of
all five nucleotides of the primer is required. The dots represent observed frequencies
from the simulation. The dotted line represents the theoretical expectation in the case
where one of the 5 sites is degenerate, so that only a perfect match at four sites is
necessary. Open circles represent simulations of this situation.

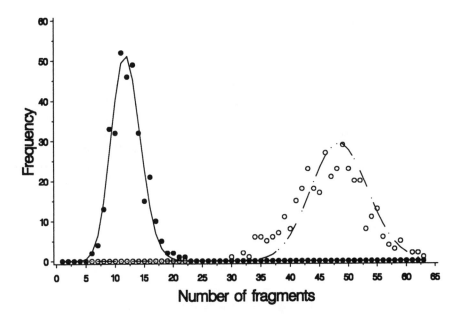

FIG. 2. *Distribution of fragment number. The same simulations as described in Figure 1 also had fragment numbers tallied. As in Figure 1, the solid line is the theoretical expectation, and the dots are the observed simulation when a perfect match of all 5 nt is required. Similarly, the dashed line is the theoretical prediction and the open circles are the observed numbers from the simulations.*

Figure 1 shows the observed and expected distributions of fragment lengths for both cases of perfect matching and mispriming. Mispriming produces an exponential distribution of fragment lengths whose exponential parameter is 1/4 that of the case of perfect match, and the mean fragment length is 1/4 as long. The observed and expected distributions of fragment numbers are plotted in Figure 2 for both perfect matching and mispriming. The case with mispriming is expected to have four times as many primer sites, and this produces four times as many fragments in a distribution having four times the variance. The agreement with the theory looks acceptable.

6. Conclusions. Discrete renewal theory provides the tools needed to infer the expected distributions of fragment lengths and fragment numbers generated by RAPD-PCR. The distribution of fragment lengths is exponential, and the effect of mispriming (allowing amplification at primer sites whose sequence does not exactly match the primer) is to change the parameter of this distribution (giving it a shorter mean fragment) but mispriming does not change the form of the distribution. Similarly, the expected number of fragments generated is close to a Poisson distribution and the mean is increased by the same proportion as the change in the exponential parameter. In going from a perfect matching 10-mer to having 7 matches constrained to have three be at the 3' end, the latter case leads one to expect 2240 times as many fragments. The ability of mismatching sites to prime RAPD-PCR synthesis depends on many factors including the annealing temperature.

REFERENCES

[1] Barnes, W.M. 1994. PCR amplification of up to 35-kb DNA with high fidelity and high yield form lambda bacteriophage templates. Proc. Natl. Acad. Sci. USA 91:2216-2220.

[2] Clark, A.G. and C.M.S. Lanigan. 1993. Prospects for estimating nucleotide divergence with RAPDs. Mol. Biol. Evol. 10:1096-1111.

[3] Feller, W. 1968. *An Introduction to Probability Theory and its Applications.* Third edition. Wiley, New York.

[4] Hadrys, H., M. Balick, and B. Schierwater. 1992. Applications of random amplified polymorphic DNA (RAPD) in molecular ecology. Mol. Ecol. 1:55-63.

[5] Lanigan, C.M.S. 1992. RAPD analysis of primates: phylogenetic and genealogical considerations. Ph.D. dissertation, Genetics Graduate Group, University of California, Davis.

[6] Moya, A., R. Carrio, D. Martinez-Torres, and A. Latorre. 1994. Mathematical model for studying DNA polymorphisms amplified by arbitrary primers. Molec. Ecol., *in press.*

[7] Riedy, M.F., W.J. Hamilton, and C.F. Aquadro. 1992. Excess of nonparental bands in offspring from known primate pedigrees assayed using RAPD PCR. Nuc. Acids Res. 20:918.

[8] Smith, J.J., J.S. Scott-Craig, J.R. Leadbetter, G.L. Bush, D.L. Roberts, and D.W. Fulbright. 1994. Characterization of random amplified polymorphic DNA (RAPD) products from Xanthomonas campestris and

some comments on the use of RAPD products in phylogenetic analysis. Mol. Phyl. Evol. 3:135-142.

[9] Wang, G., T.S. Whittam, C.M. Berg, and D.E. Berg. 1993. RAPD (arbitrary primer) PCR is more sensitive than multilocus enzyme electrophoresis for distinguishing related bacterial strains. Nuc. Acids Res. 21:5930-5933.

[10] Waterman, M.S. 1983. Frequencies of restriction sites. Nuc. Acids Res. 11:8951-8956.

[11] Weissing, F.J. and O. Velterop. 1993. The expected distribution of RAPD bands. Mol. Biol. Evol. 10:1107-1110.

[12] Welsh, J. and M. McClelland. 1990. Fingerprinting genomes using PCR with arbitrary primers. Nuc. Acids Res. 18:7213-7218.

[13] Williams, J.G.K., A.R. Kubelik, K.J. Livak, J.A. Rafalski, and S.V. Tingey. 1990. DNA polymorphisms amplified by arbitrary primers are useful as genetic markers. Nuc. Acids Res. 18:6531-6535.

[14] Williams, J.G.K., J.A. Rafalski, and S.V. Tingey. 1992. Genetic analysis using RAPD markers. Methods in Enzymol. 218:704-740.

THE STRUCTURED COALESCENT

HILDE M. HERBOTS*

Abstract. The genealogy of a sample from a subdivided population is, under reasonable assumptions about reproduction and migration, well described by the structured coalescent, which is the extension to structured populations of Kingman's coalescent. As an example of an application of the structured coalescent, the distribution of the time since the most recent common ancestor of a pair of individuals from the same subpopulation in a symmetric island model of population structure is compared to that of two individuals from a panmictic population of the same total size. For a simple model for reproduction and migration, a rigorous proof is given that, measuring time in diffusion time-scale, the ancestral process converges (weakly) to the structured coalescent, as the subpopulation sizes tend to infinity.

1. Introduction. The genealogical history of a population is often of interest in its own right. More importantly, it is central to an understanding of the population's evolution. The intimate relationship between genealogy and genetic composition is, under a variety of idealized models, well understood (see for example Tavaré [20], Hudson [6] and references therein). A major development in the study of genealogy has been the so-called "coalescent approach", the basic idea of which is to draw a sample from the population and to trace back the ancestry of the individuals in the sample, focusing on the times in the past when two or more individuals in the sample have a common ancestor. For a wide range of reproductive models, the family relationships in such a sample from the population are well described by a Markov chain called the "coalescent" (Kingman [7,8,9]), which for $t > 0$, partitions the sample into equivalence classes, individuals belonging to the same class if they had the same ancestor time t ago. As t increases (i.e. as we look further back into the past), the number of distinct ancestors of the sample decreases and the equivalence classes coalesce, until ultimately the whole sample forms one equivalence class sharing a single common ancestor. A simpler version of this process is the "ancestral process" which only keeps account of the number of distinct ancestors of the sample at each time in the past.

One advantage of this approach backward in time is that only those individuals in the past who have descendants in the current sample need to be considered, which considerably reduces the complexity of simulations. In addition, the coalescent is a simple and elegant time-homogeneous Markov chain which lends itself very well to analytical study.

Whereas the coalescent, in its original formulation, assumes a panmictic population, most real populations have geographic (or other) structure.

---

* (Present address: Department of Statistical Science, University College London, Gower Street, London WC1E 6BT, U.K.) School of Mathematical Sciences, Queen Mary and Westfield College, London, U.K. and Department of Mathematics, Katholieke Universiteit te Leuven, Belgium.

231

When modelling genealogy in a structured population, it is not sufficient
to trace the ancestry of the individuals in the sample under study, but one
also needs to keep track of the *locations* of the ancestors of the sample at
each time in the past, if the Markov character of the coalescent is to be
maintained. The resulting coalescent-like process for structured populations is called the "structured coalescent", introduced by Takahata [18] for
a population consisting of only two subpopulations, and formulated in its
general form by Notohara [13].

This paper is concerned with the structured coalescent. In the next
section, the structured coalescent is described in detail and references are
given to applications. As an example of an application, we examine the
distribution of the time since the most recent common ancestor of two individuals from a single subpopulation under the symmetric island model of
population structure (Wright [21]). This distribution is a mixture of two
exponential distributions (Griffiths [3]). The third section provides a rigorous proof that, under a reasonable (discrete-time) model for reproduction
and migration, the genealogy of a sample from a subdivided population is
indeed, in the diffusion time-scale approximation, described by the structured coalescent. This result was proved by Notohara [13] for a different
model for migration and reproduction. However, certain steps in Notohara's proof are not entirely clear to us. More details will be given in
Section 3.

2. The structured coalescent. We consider a haploid population
divided into a finite or infinite number of subpopulations which are all
large and panmictic and which are partially isolated from each other. We
denote by $\mathcal{S}$ the set of the subpopulation labels ($\mathcal{S}$ is countable). The
number of haploid individuals in subpopulation i is $N_i = 2c_i N$, where
c_i is a positive integer constant and N is large. (In diploid applications,
$2c_i N$ is the number of homologous genes in subpopulation i rather than
the number of individuals.) At a particular generation which we call time
zero, we draw a sample of n individuals from the total population (where
n is finite), and we trace the ancestry of the individuals in the sample. At
each time in the past, we count how many distinct ancestors the n sampled
individuals have in each subpopulation. We denote by $\alpha_N^i(\tau)$ the number
of distinct ancestors the sample has in subpopulation i, τ generations ago
($i \in \mathcal{S}$), and by $\boldsymbol{\alpha}_N(\tau)$ the ordered set $\left(\alpha_N^i(\tau)\right)_{i \in \mathcal{S}}$, with a component for
each subpopulation. If there are K subpopulations and K is finite, then
$\boldsymbol{\alpha}_N(\tau)$ is a K-tuple. If the number of subpopulations is infinite, $\boldsymbol{\alpha}_N(\tau)$ is
a sequence with index set $\mathcal{S}$. In standard mathematical notation, we write
that $\boldsymbol{\alpha}_N(\tau) \in \mathbb{N}^{\mathcal{S}}$, which is the set of all functions from $\mathcal{S}$ to $\mathbb{N}$, where $\mathbb{N}$
is the set of the natural numbers, including zero: $\mathbb{N} = \{0, 1, 2, \ldots\}$. The
process $\boldsymbol{\alpha}_N = \{\boldsymbol{\alpha}_N(\tau); \tau \in \mathbb{N}\}$ will be called the ancestral process.

Tracing the ancestral lineages of the individuals in the sample, two
types of events can occur. Two particular lineages can coalesce at the most

recent common ancestor of the corresponding individuals in the sample (this can only occur when these lineages reside in the same subpopulation), in which case the number of distinct ancestors in that subpopulation (i.e. the value of $\alpha_N^i(\tau)$) decreases by one. The rate at which such a coalescence event occurs is, for the exchangeable models of reproduction (Cannings [1]), inversely proportional to the size of the subpopulation. If an ancestor in subpopulation i is an immigrant from subpopulation j (which we also describe as a "migration" of the ancestor from subpopulation i to subpopulation j backward in time), the number of distinct ancestors in subpopulation i decreases by one, while that in subpopulation j increases by one. We denote by ϵ^i the element of $I\!\!N^S$ with components

$$\left(\epsilon^i\right)_j = \delta_{ij} = \left\{ \begin{array}{ll} 1 & \text{if } j = i \\ 0 & \text{otherwise} \end{array} \right.$$

and we define addition and subtraction of elements of $I\!\!N^S$ to be component-wise, i.e. the sum or difference of two sequences (or K-tuples) in $I\!\!N^S$ is obtained by adding or subtracting their corresponding components. If $\alpha_N(\tau) = \alpha$ and two lineages in subpopulation i coalesce, the value of $\alpha_N(\tau)$ is changed to $\alpha - \epsilon^i$; the migration of an ancestral lineage from subpopulation i to subpopulation j (backward in time) changes the value of $\alpha_N(\tau)$ from α to $\alpha - \epsilon^i + \epsilon^j$.

Under reasonable assumptions about reproduction and migration, the ancestral process α_N is, with an appropriate re-scaling of time, well approximated by the "*ancestral process associated with the structured coalescent*", which is the continuous-time Markov chain $\{\alpha(t); t \geq 0\}$ with **Q**-matrix **Q** whose entries are

$$(1) \quad Q_{\alpha,\beta} = \left\{ \begin{array}{ll} -\sum\limits_{i \in S}\left\{\alpha_i \frac{M_i}{2} + \frac{1}{c_i}\left(\begin{array}{c}\alpha_i \\ 2\end{array}\right)\right\} & \text{if } \beta = \alpha \\[3mm] \alpha_i \frac{M_{ij}}{2} & \text{if } \beta = \alpha - \epsilon^i + \epsilon^j \ (j \neq i) \\[3mm] \frac{1}{c_i}\left(\begin{array}{c}\alpha_i \\ 2\end{array}\right) & \text{if } \beta = \alpha - \epsilon^i \\[3mm] 0 & \text{otherwise} \end{array} \right.$$

(Notohara [13]), where M_{ij} is the scaled migration rate (forward in time) from subpopulation j to subpopulation i and $M_i = \sum_{j \neq i} M_{ij}$. For example, if reproduction in each subpopulation follows the neutral Wright-Fisher model and a constant proportion q_{ji} of the individuals born in subpopulation j migrate to subpopulation i every generation ($i, j \in S$), time-scaling is in units of $2N$ generations and $M_{ij} = \lim_{N \to \infty}\left(4N\frac{c_j}{c_i}q_{ji}\right)$. Intuitively the matrix **Q** is understood as follows: tracing the ancestral lineages of a sample from the population, any two lineages in subpopulation i have coalescence rate $1/c_i$, while the rate at which a lineage moves from subpopulation i to subpopulation j is $M_{ij}/2$ (the factor of $1/2$ is standard and is convenient in applications involving pairs of genes). For a range of models for reproduction and migration and using diffusion time-scale, the ancestral process α_N converges weakly, as N tends to infinity (i.e. as the subpopulations become infinitely large), to the ancestral process associated with the structured coalescent, which we will call the "structured coalescent" for simplicity. We will prove this in section 3 for a particular model of reproduction and migration.

The structured coalescent has proved valuable in understanding and modelling the genetic and demographic processes of interest in a variety of applications concerning structured populations. Applications of the coalescent approach to subdivided populations include those by Griffiths [3], Takahata [18,19], Tajima [17], Notohara [13], Hudson [6] and references in the latter. Marjoram and Donnelly [10,11] used the coalescent approach to simulate structured populations of variable size, in a study of pairwise comparisons of mitochondrial DNA sequences and the time since Eve, the most recent common ancestor of human mitochondrial DNA. Studying Wright's coefficient F_{ST} (Wright [22]), which serves as a measure of the genetic differentiation among subpopulations and which is used to estimate the effective level of gene flow between subpopulations, Slatkin [14] related F_{ST} in an approximate way to the mean coalescence times of pairs of genes drawn within and among subpopulations. Slatkin [15] showed these results provide a way of testing for isolation by distance in a natural population. In fact, F_{ST} can be expressed in an exact way in terms of the Laplace transforms of the distributions of the coalescence times of pairs of genes. This was done in Chapter 4 of Herbots [4], where the dependence of F_{ST} on the parameters of population structure and on the mutation rate was examined.

To illustrate the use of the structured coalescent, we examine the time since the most recent common ancestor of a pair of individuals under the island model of population structure (Wright [21]). Under this model, the population is divided into K subpopulations (K finite) of equal size $2N$ with the same migration rate between any two subpopulations. With the appropriate assumptions (see, for example, the next section) and time-scale, the genealogy of a sample of individuals from this population is well described by the structured coalescent $\{\alpha(t); t \geq 0\}$ with infinitesimal generator $\mathbf{Q}$, given by equation (1), where $\mathcal{S} = \{1, \ldots, K\}$, $c_i = 1$ for $i = 1, \ldots, K$, where all M_i are equal, say $M_i = M$ ($i = 1, \ldots, K$), and where $M_{ij} = M/(K-1)$ for $j \neq i$. For example, if reproduction within each subpopulation follows the neutral Wright-Fisher model, time-scaling is in units of $2N$ generations and $M = 4Nm$, where m is the proportion of each subpopulation that is replaced by immigrants every generation. The time since the most recent common ancestor of a sample of individuals from this population is (in the coalescent approximation and corresponding time-scale) the time until the structured coalescent, starting from that sample, enters into the absorbing set of states $\{\alpha \in \mathbb{N}^{\mathcal{S}}; \sum_{i \in \mathcal{S}} \alpha_i = 1\}$.

The distribution of this time can be calculated from the structured coalescent by a first step argument. We restrict attention here to samples of size $n = 2$, and we denote by T_0 the coalescence time (i.e. the time since the most recent common ancestor) of two individuals from the same subpopulation, and by T_1 that of two individuals from different subpopulations. Starting from two individuals in a single subpopulation (i.e. $\alpha(0) = 2\epsilon^i$ for some $i \in \{1, \ldots, K\}$), the structured coalescent remains in state $\alpha(0)$ during a time X_0 which is exponentially distributed with mean $|Q_{\alpha(0),\alpha(0)}|^{-1} = (1 + M)^{-1}$ (see equation (1)). The first move of the structured coalescent corresponds with probability $1/(1 + M)$ to the coalescence of the two individuals' lineages, in which case the coalescence time of these individuals is precisely the waiting time X_0; with probability $M/(1 + M)$, the first step corresponds to one of the individuals migrating to another subpopulation in which case the remaining coalescence time of the two individuals has the same distribution as T_1, the coalescence time of two individuals from different subpopulations. This yields the following equation for the Laplace transform of the distribution of T_0:

$$(2) \qquad E[e^{-sT_0}] = E[e^{-sX_0}]\left(\frac{1}{1 + M} + \frac{M}{1 + M}E[e^{-sT_1}]\right).$$

Two individuals in different subpopulations have, when migrating, probability $1/(K - 1)$ of choosing the subpopulation where the other one is resident. Hence we find similarly that

$$(3) \qquad E[e^{-sT_1}] = E[e^{-sX_1}]\left\{\frac{1}{K - 1}E[e^{-sT_0}] + \left(1 - \frac{1}{K - 1}\right)E[e^{-sT_1}]\right\},$$

where the waiting time X_1 is exponentially distributed with mean M^{-1}. Re-writing equations (2) and (3), the Laplace transforms of the distributions of T_0 and T_1 satisfy the following system of linear equations:

$$\begin{cases} (1 + M + s)E[e^{-sT_0}] - ME[e^{-sT_1}] = 1 \\ \{M + (K - 1)s\}E[e^{-sT_1}] - ME[e^{-sT_0}] = 0. \end{cases}$$

The solutions of these equations are given by

$$(4) \qquad E[e^{-sT_0}] = \frac{M + (K - 1)s}{M + (KM + K - 1)s + (K - 1)s^2}$$

$$E[e^{-sT_1}] = \frac{M}{M + (KM + K - 1)s + (K - 1)s^2}$$

(Griffiths [3], Hudson [6]). A partial fraction expansion of (4) shows that the distribution of T_0 is a mixture of two exponential distributions:

$$(5) \qquad E[e^{-sT_0}] = A_1 E[e^{-sZ_1}] + A_2 E[e^{-sZ_2}],$$

where Z_i is exponentially distributed with mean $1/\lambda_i$, with

$$\lambda_1 = \frac{KM + K - 1 - \sqrt{D}}{2(K-1)}$$

$$\lambda_2 = \frac{KM + K - 1 + \sqrt{D}}{2(K-1)}$$

and where

$$D = (KM + K - 1)^2 - 4(K-1)M$$

$$A_1 = \frac{\lambda_2 - 1}{\lambda_2 - \lambda_1}$$

$$A_2 = \frac{\lambda_1 - 1}{\lambda_1 - \lambda_2}$$

(Griffiths [3]). From (5), the density of T_0 is easily found:

$$(6) \quad f_{T_0}(t) = e^{-\left(1+\frac{KM}{K-1}\right)\frac{t}{2}} \left\{ \cosh\left(\frac{t\sqrt{D}}{2(K-1)}\right) - \frac{(K-2)M+K-1}{\sqrt{D}} \sinh\left(\frac{t\sqrt{D}}{2(K-1)}\right) \right\}.$$

To calculate the density of T_1, we note that the coalescence time of two individuals in different subpopulations is the time T_r until these two individuals are present in a single subpopulation for the first time, plus the coalescence time of two individuals in the same subpopulation:

$$T_1 \overset{\text{d}}{=} T_r + T_0.$$

Because of the Markov character of the structured coalescent, both times are independent, so that their probability density functions satisfy

$$f_{T_1} = f_{T_r} * f_{T_0},$$

where $*$ denotes the convolution product. As T_r is exponentially distributed with mean $\left(\frac{M}{K-1}\right)^{-1}$, a straightforward calculation yields

$$(7) \qquad f_{T_1}(t) = \frac{2M}{\sqrt{D}} e^{-\left(1+\frac{KM}{K-1}\right)\frac{t}{2}} \sinh\left(\frac{t\sqrt{D}}{2(K-1)}\right).$$

Results (6) and (7) were obtained earlier by Takahata [18] and Nath and Griffiths [12] in the case of $K = 2$ (two subpopulations), but appear to be new in their general form (K colonies).

In figure 1, we compare the density of the coalescence time T_0 of two individuals from a single subpopulation in a symmetric island model with $K = 4$ subpopulations with the density of the coalescence time T_{pan} of

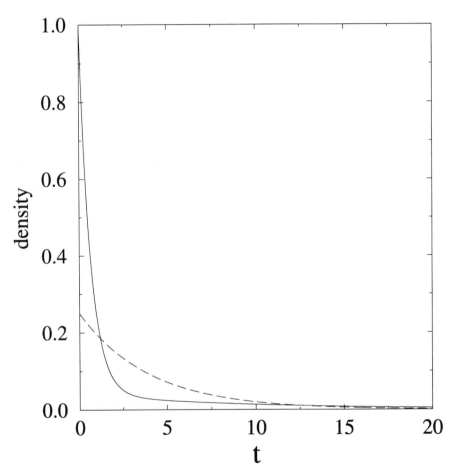

FIG. 1. *The coalescence time of two individuals in a single subpopulation of a subdivided population, compared to that of two individuals in a panmictic population of the same total size. The solid line is the density function $f_{T_0}(t)$, given by equation (6), of the coalescence time T_0 of two individuals in a single subpopulation under a symmetric island model with $K = 4$ subpopulations and a scaled migration rate of $M = 0.5$. The dashed line is the density function $f_{T_{pan}}(t)$ of the coalescence time T_{pan} of two individuals in a panmictic population of the same total size (and with the same time-scaling). The distribution of T_{pan} is exponential with mean K. The distribution of T_0 is a mixture of two exponential distributions.*

two individuals in a panmictic population of the same total size $(2KN)$. From (1) or from the coalescent for a panmictic population (Kingman [7]), T_{pan} is, in the same time-scaling as used above (in units of $2N$ generations if reproduction follows the neutral Wright-Fisher model), exponentially distributed with mean K. Figure 1 shows that for a small migration rate, two individuals from the same subpopulation in the subdivided population (solid line) are more likely to have a very short coalescence time than two individuals in the panmictic population (dashed line). This is because the subpopulation is much smaller than the total population, so that the probability that two individuals share their parent in the previous generation is larger in a single subpopulation of a structured population than it is in a large panmictic population. Once one of the individuals' lineages moves to a different subpopulation, however, coalescence cannot occur until the two lineages are present in a single subpopulation again, which is likely to take a long time if the migration rate is small. This explains the longer and thicker tail of the density of T_0, compared to that of T_{pan}. In fact, $f_{T_0}(t)/f_{T_{\mathrm{pan}}}(t) \to \infty$ as $t \to \infty$. These observations also explain that the variance of T_0 (easily calculated by differentiation of equation (4)) is much larger than that of T_{pan} and increases as the migration rate decreases:

$$\mathrm{Var}(T_0) = K^2 + 2\frac{(K-1)^2}{M} > K^2 = \mathrm{Var}(T_{\mathrm{pan}})$$

(Hey [5]). However, both distributions have the same mean:

$$ET_0 = ET_{\mathrm{pan}} = K$$

(Notohara [13], Hudson [6] and references therein). Strobeck [16] proved that in *isotropic*[1] models of population structure, the mean coalescence time of two individuals from any single subpopulation is independent of the migration rate and equal to that of two individuals in a panmictic population of the same total size. This is not true if the population structure is not isotropic. In Chapter 3 of Herbots [4] some examples are given of (non-isotropic) population structures in which the mean coalescence time of two individuals from a particular single subpopulation depends on the migration rate.

3. Convergence to the structured coalescent. Considering a population divided into a finite or infinite number of subpopulations and evolving in discrete generations, Notohara [13] showed that the genealogy of a sample from this population is approximately described by the structured coalescent. In Notohara's model, each generation is made up of two discrete steps, the first one due to migration and the second one due to reproduction.

---

[1] An isotropic model of population structure is one in which all subpopulations are identical with respect to size, migration pattern and migration rates. For a formal definition of isotropy, see Strobeck [16].

In the migration step the individuals migrate independently between the subpopulations. After the migration step, Wright-Fisher type reproduction brings the size of each subpopulation back to its size before migration. Although under this model the individuals migrate independently forward in time, their migration is not independent backward in time, as migration backward in time necessarily brings the size of each subpopulation to a fixed number, constant in time. Therefore, it is not entirely clear to us that Notohara's argument, which inappropriately relies on individuals migrating independently backward in time, is valid. Some minor additional complications may arise from the fact that under Notohara's model, the mechanisms of migration and reproduction are not independent, since at reproduction, the offspring distribution of the individuals in a particular subpopulation depends on the number of individuals present in that subpopulation after the migration step, from the fact that the offspring distributions (and hence coalescence events when we look backward in time) in different subpopulations are not independent, as the subpopulation sizes after migration are not independent, and from the possibility that migration empties a subpopulation. It is plausible (but not obvious) that these problems disappear in the limit as the subpopulations become infinitely large. In this section it is proved in detail that, for a slightly simpler model than Notohara's, the ancestral process converges to the structured coalescent.

As in the last section, the population considered is haploid and divided into a finite or infinite number of subpopulations. The set of the subpopulation labels is countable and is denoted by $\mathcal{S}$. The subpopulation sizes, migration pattern and migration rates are assumed to be constant in time. The size of subpopulation i is $N_i = 2c_i N$, where c_i is a positive integer constant, while we will let N become large. Denoting $c = \sum_{i=1}^n c_i$, the total population size is $2cN$ (where $c = \infty$ if the number of subpopulations is infinite). The population evolves in discrete non-overlapping generations. Every generation is made up of two discrete steps, the first one due to reproduction, the second one due to migration.

Reproduction is haploid (each member of a certain generation is the child of exactly one member of the previous generation) and takes place within each subpopulation according to the neutral Wright-Fisher model. This means that in each subpopulation, the joint distribution of the offspring numbers of the different individuals at any particular generation is symmetric multinomial, maintaining the size of that subpopulation: denoting by Y_{ij} the number of offspring of the jth individual in subpopulation i ($i \in \mathcal{S}, j = 1, \ldots, N_i$),

$$P\{Y_{i1} = y_1, \ldots, Y_{i,N_i} = y_{N_i}\} = \begin{cases} \frac{N_i!}{y_1! \cdots y_{N_i}!} N_i^{-N_i} & \text{if } \sum_{j=1}^{N_i} y_j = N_i \\ 0 & \text{otherwise.} \end{cases}$$

Offspring numbers in different subpopulations or at different generations are independent. This description forward in time is equivalent to the following

backward description: each individual in subpopulation i immediately after the reproduction step chooses its parent at random, independently and
uniformly from among the N_i individuals in subpopulation i just before
reproduction. This simple structure backward in time is the major advantage of the Wright-Fisher model for reproduction. While the convergence
result proved in this section explicitly assumes the neutral Wright-Fisher
model, it is likely to be valid (under mild moment conditions) for any exchangeable model of reproduction (Cannings [1]), provided an appropriate
time-scale is used; if in the limit as $N \to \infty$, the variance of the number of
offspring of an individual in a subpopulation varies among subpopulations,
the coalescence rates $1/c_i$ in (1) also need to be adapted.

At each generation, the reproduction step is followed by a migration
step. We assume that in every generation, a fixed proportion q_{ij} ($q_{ij} \geq 0$,
$\sum_{j \neq i} q_{ij} \leq 1$) of the individuals born in subpopulation i migrate to subpopulation j (these migrants are chosen at random, independently and
uniformly, without replacement, from subpopulation i), where q_{ij} is constant in time ($i, j \in \mathcal{S}$). It is assumed that the size of each subpopulation
is maintained under migration, which requires that

$$\forall i \in \mathcal{S} : c_i \sum_{j \neq i} q_{ij} = \sum_{j \neq i} c_j q_{ji}.$$

While each individual in subpopulation i has probability q_{ij} of migrating to subpopulation j, the requirement of constant subpopulation sizes
implies that the individuals do not migrate independently of each other.
Among the N_i individuals making up subpopulation i just after the migration step at any particular generation, there are $N_j q_{ji}$ who are immigrants
from subpopulation j. The "backward migration rate" m_{ij} from subpopulation i to subpopulation j, defined as the proportion of the individuals
in subpopulation i immediately after the migration step who were born in
subpopulation j, is therefore given by

$$m_{ij} = \frac{N_j q_{ji}}{N_i} = \frac{c_j}{c_i} q_{ji}$$

($i, j \in \mathcal{S}$). We also denote $m_i = \sum_{j \neq i} m_{ij}$, which is the proportion of the
individuals in subpopulation i after migration who were born in another
subpopulation.

Assume that the population has been evolving in this way indefinitely.
At a particular generation which we call time zero, n individuals are sampled from the total population. We trace back their ancestry, generation by
generation. As in the previous section, the ancestral process is the process
$\alpha_N = \{\alpha_N(\tau); \tau = 0, 1, 2, \ldots\}$, where $\forall \tau \in \mathbb{N} : \alpha_N(\tau) \in \mathbb{N}^{\mathcal{S}}$ with ith
component $\alpha_N^i(\tau)$ denoting the number of distinct ancestors the sample
has in subpopulation i, τ generations ago. As an initial sample of fixed size
n can have at most n distinct ancestors at any time, the state space of this

ancestral process is the set

$$E := \{\boldsymbol{\alpha} \in \mathbb{N}^{\mathcal{S}} : \sum_{i \in \mathcal{S}} \alpha_i \leq n\}.$$

This set is infinite when the number of subpopulations is infinite. Each element $\boldsymbol{\alpha}$ of E can be obtained by assigning $a = \sum_{i \in \mathcal{S}} \alpha_i$ individuals to the subpopulations in $\mathcal{S}$ and counting the number of individuals thus assigned to each subpopulation. As different elements of E require different assignments of individuals, E is in one-to-one correspondence to a subset of $\bigcup_{a=0}^{n} \mathcal{S}^a$. As $\mathcal{S}$ is countable, $\bigcup_{a=0}^{n} \mathcal{S}^a$ is countable and hence E is countable. We denote by $D_E[0, \infty)$ the space of right-continuous functions from $[0, \infty)$ into E having limits from the left. In this section it is proved that, measuring time in units of $2N$ generations (and making the appropriate assumptions), the ancestral process converges weakly to the structured coalescent:

THEOREM . *Assume that*
(i) $\forall i, j \in \mathcal{S}$ *with* $j \neq i$, $4N m_{ij}$ *increases monotonically with increasing*
   N, *with* $\lim_{N \to \infty} (4N m_{ij}) =: M_{ij}$,
and, denoting $M_i := \sum_{j \neq i} M_{ij}$, *assume that*
(ii) $\sup_{i \in \mathcal{S}} M_i < \infty$.
Then as $N \to \infty$, *the process* $\{\boldsymbol{\alpha}_N([2Nt]) : t \geq 0\}$ *converges weakly in*
$D_E[0, \infty)$ *to the structured coalescent* $\{\boldsymbol{\alpha}(t) : t \geq 0\}$ *defined by equation (1).*

Note that the definition of the scaled migration rate M_{ij} in (i) is a consequence of the time-scaling (in units of $2N$ generations) rather than an assumption. The quantity $4N m_{ij}$ is, up to the constant c_i, twice the number of immigrants into subpopulation i from subpopulation j per generation. The assumption that $4N m_{ij}$ $(j \neq i)$ monotonically increases as a function of N is equivalent to assuming that the number of migrants between any two subpopulations per generation increases as the subpopulation sizes increase. This assumption ensures, by the monotone convergence theorem, that $\forall i \in \mathcal{S}$: $M_i = \lim_{N \to \infty} (4N m_i)$, so that $c_i M_i$ is twice the limiting number of migrants from (or into) subpopulation i per generation. Assumption (ii) requires that the number of migrants from each subpopulation per generation remains bounded as the subpopulations become large.

   Proof. In order to prove the theorem, we show that the finite-dimensional distributions of the process $\boldsymbol{\alpha}_N([2N \cdot])$ converge to those of the structured coalescent as $N \to \infty$ and that the family of processes $\{\boldsymbol{\alpha}_N([2N \cdot])\}_{N \in \mathbb{N}_0}$ is relatively compact. As E is countable, and hence separable, Theorem 7.8 in Ethier and Kurtz [2] then yields the weak convergence of $\boldsymbol{\alpha}_N([2N \cdot])$ to the structured coalescent, as $N \to \infty$.

   Convergence of the finite-dimensional distributions. From the description of the model it is clear that the ancestral process $\boldsymbol{\alpha}_N$ is a (multi-dimensional) discrete-time Markov chain. We will calculate its transition matrix $\mathbf{P}_N$ and we will show that, for fixed $t \geq 0$, the transition matrix of $\boldsymbol{\alpha}_N$ over $[2Nt]$ generations, $\mathbf{P}_N^{[2Nt]}$, converges entry-wise to the transition matrix of the structured coalescent over time t, $e^{t\mathbf{Q}}$ (where $\mathbf{Q}$ is the infinitesimal generator of the structured coalescent, given by equation (1)), as $N \to \infty$. This means that the one-dimensional distributions of $\boldsymbol{\alpha}_N([2N\cdot])$ converge to those of the structured coalescent, as $N \to \infty$. Because of the Markov character of both the ancestral process and the structured coalescent and as E is countable, this is easily seen to be equivalent to the convergence of the finite-dimensional distributions.

Taking a step of one generation backward in time, we first have to take a migration step and subsequently a reproduction step. In order to find the transition probability of the ancestral process from state $\boldsymbol{\alpha} \in E$ to state $\boldsymbol{\beta} \in E$ in one generation,

$$(8) \qquad P_N(\boldsymbol{\beta}|\boldsymbol{\alpha}) := P\{\boldsymbol{\alpha}_N(\tau+1) = \boldsymbol{\beta}|\boldsymbol{\alpha}_N(\tau) = \boldsymbol{\alpha}\},$$

we calculate separately the probabilities $P_N^{(m)}(\boldsymbol{\beta}|\boldsymbol{\alpha})$ and $P_N^{(r)}(\boldsymbol{\beta}|\boldsymbol{\alpha})$ of a transition of the ancestral process from $\boldsymbol{\alpha}$ to $\boldsymbol{\beta}$ in, respectively, one backward migration step or one backward reproduction step.

In the migration step, $m_k N_k$ individuals from subpopulation k ($k \in \mathcal{S}$) move, backward in time, to another subpopulation. These migrants are a random sample without replacement from subpopulation k. The number of possible ways to draw $m_k N_k$ migrants without replacement from among the N_k individuals in subpopulation k is scriptsize $\begin{pmatrix} N_k \\ m_k N_k \end{pmatrix}$. The individuals counted in $\boldsymbol{\alpha}$ (the present value of the ancestral process) are the ancestors of our initial sample of n individuals; we call them individuals "belonging to" $\boldsymbol{\alpha}$. If an individual belonging to $\boldsymbol{\alpha}$ is a migrant, the ancestral process may change value. Migrants drawn from outside $\boldsymbol{\alpha}$ do not affect the ancestral process. The probability $R_N^{(m)}(\boldsymbol{\alpha})$ that more than one ancestor in $\boldsymbol{\alpha}$ is a migrant is the probability that at least two of the migrants are drawn from $\boldsymbol{\alpha}$ and, counting the number of possible ways to do this, satisfies

$$
\begin{aligned}
R_N^{(m)}(\boldsymbol{\alpha}) \;\leq\;& \sum_{k \in \mathcal{S}} \frac{\begin{pmatrix} \alpha_k \\ 2 \end{pmatrix} \begin{pmatrix} N_k - 2 \\ m_k N_k - 2 \end{pmatrix}}{\begin{pmatrix} N_k \\ m_k N_k \end{pmatrix}} \\
&+ \sum_{k \in \mathcal{S}} \frac{\begin{pmatrix} \alpha_k \\ 1 \end{pmatrix} \begin{pmatrix} N_k - 1 \\ m_k N_k - 1 \end{pmatrix}}{\begin{pmatrix} N_k \\ m_k N_k \end{pmatrix}} \sum_{l \neq k} \frac{\begin{pmatrix} \alpha_l \\ 1 \end{pmatrix} \begin{pmatrix} N_l - 1 \\ m_l N_l - 1 \end{pmatrix}}{\begin{pmatrix} N_l \\ m_l N_l \end{pmatrix}} \\
\leq\;& \sum_{k \in \mathcal{S}} \alpha_k^2 m_k^2 + \sum_{k \in \mathcal{S}} \alpha_k m_k \sum_{l \neq k} \alpha_l m_l \\
=\;& \left(\sum_{k \in \mathcal{S}} \alpha_k m_k \right)^2 .
\end{aligned}
$$

The probability that exactly one individual in $\boldsymbol{\alpha}$ migrates backward in time from subpopulation i to subpopulation $j (\neq i)$ while all other migrants are drawn from outside $\boldsymbol{\alpha}$, is

$$
\frac{\binom{\alpha_i}{1}\binom{N_i - \alpha_i}{m_{ij}N_i - 1}}{\binom{N_i}{m_{ij}N_i}} \cdot \frac{\binom{N_i - m_{ij}N_i - \alpha_i + 1}{m_i N_i - m_{ij}N_i}}{\binom{N_i - m_{ij}N_i}{m_i N_i - m_{ij}N_i}} \cdot \prod_{k \neq i} \frac{\binom{N_k - \alpha_k}{m_k N_k}}{\binom{N_k}{m_k N_k}}
$$

$$
= \alpha_i m_{ij} \cdot \frac{N_i}{N_i - m_i N_i - \alpha_i + 1} \prod_{k \in \mathcal{S}} \prod_{a=0}^{\alpha_k - 1} \frac{N_k - m_k N_k - a}{N_k - a},
$$

where $\prod_{a=0}^{-1} \equiv 1$. In that event, the backward migration step changes the value of the ancestral process to $\boldsymbol{\alpha} - \boldsymbol{\epsilon}^i + \boldsymbol{\epsilon}^j$. The ancestral process also takes this value when there are several migrations of individuals belonging to $\boldsymbol{\alpha}$ which, except for one migration from subpopulation i to subpopulation j, all compensate each other. Denoting by $R_N^{(m)}(\boldsymbol{\alpha}, \boldsymbol{\beta})$ the probability that the backward migration step changes the value of the ancestral process from $\boldsymbol{\alpha}$ to $\boldsymbol{\beta}$ and more than one individual in $\boldsymbol{\alpha}$ are migrants, the transition probability in one backward migration step is given by

$$
P_N^{(m)}(\boldsymbol{\beta}|\boldsymbol{\alpha}) =
$$

$$
\begin{cases}
1 - \sum_{i \in \mathcal{S}} \alpha_i m_i \frac{N_i}{N_i - m_i N_i - \alpha_i + 1} \prod_{k \in \mathcal{S}} \prod_{a=0}^{\alpha_k - 1} \frac{N_k - m_k N_k - a}{N_k - a} \\
\qquad - \sum_{\gamma \neq \alpha} R_N^{(m)}(\boldsymbol{\alpha}, \boldsymbol{\gamma}) & \text{if } \boldsymbol{\beta} = \boldsymbol{\alpha} \\[2ex]
\alpha_i m_{ij} \cdot \frac{N_i}{N_i - m_i N_i - \alpha_i + 1} \prod_{k \in \mathcal{S}} \prod_{a=0}^{\alpha_k - 1} \frac{N_k - m_k N_k - a}{N_k - a} \\
\qquad + R_N^{(m)}(\boldsymbol{\alpha}, \boldsymbol{\alpha} - \boldsymbol{\epsilon}^i + \boldsymbol{\epsilon}^j) & \text{if } \boldsymbol{\beta} = \boldsymbol{\alpha} - \boldsymbol{\epsilon}^i + \boldsymbol{\epsilon}^j (j \neq i) \\[2ex]
R_N^{(m)}(\boldsymbol{\alpha}, \boldsymbol{\beta}) & \text{otherwise,}
\end{cases}
$$

(9)

where

$$
\text{(10)} \qquad \sum_{\boldsymbol{\beta} \neq \boldsymbol{\alpha}} R_N^{(m)}(\boldsymbol{\alpha}, \boldsymbol{\beta}) \leq R_N^{(m)}(\boldsymbol{\alpha}) \leq \left(\sum_{k \in \mathcal{S}} \alpha_k m_k \right)^2.
$$

In the backward reproduction step, all individuals in subpopulation i choose their parent at random, independently and uniformly from among the N_i individuals which made up subpopulation i just before reproduction ($i \in \mathcal{S}$). Two individuals in subpopulation i choose the same parent with probability $1/N_i$. If $\boldsymbol{\alpha}$ is the present value of the ancestral process and exactly two of the α_i ancestors counted in subpopulation i share their parent, while all other ancestors in α have distinct parents, the value of the ancestral process changes to $\alpha - \epsilon^i$. The probability of this transition is

(11)
$$P_N^{(r)}(\alpha - \epsilon^i | \alpha) = \binom{\alpha_i}{2} \frac{1}{N_i} - R_N^{(r)}(\alpha, \alpha - \epsilon^i),$$

where $R_N^{(r)}(\alpha, \alpha - \epsilon^i)$ is a non-negative term arising from the possibility that two or more pairs of individuals belonging to α each share a parent. As

$$\sum_{i \in \mathcal{S}} P_N^{(r)}(\alpha - \epsilon^i | \alpha) =$$

$$P_N^{(r)}\{\text{exactly one pair of individuals in } \alpha \text{ share a parent}\}$$

and

$$\sum_{i \in \mathcal{S}} \binom{\alpha_i}{2} \frac{1}{N_i} =$$

$$\sum_{\nu=1}^{\infty} \nu P_N^{(r)}\{\text{exactly } \nu \text{ pairs of individuals in } \alpha \text{ each share a parent}\},$$

it follows that

$$\sum_{i \in \mathcal{S}} R_N^{(r)}(\alpha, \alpha - \epsilon^i) =$$

$$\sum_{\nu=2}^{\infty} \nu P_N^{(r)}\{\text{exactly } \nu \text{ pairs of individuals in } \alpha \text{ each share a parent}\}.$$

(12)

We denote by $V = \sum_{i \in \mathcal{S}} \binom{\alpha_i}{2}$ the number of pairs of individuals in α in which both individuals belong to the same subpopulation and hence can have a common parent. As there are at most n individuals in α (n is the size of the initial sample), we know that $V \leq \binom{n}{2}$. Hence, since $\forall i \in \mathcal{S} : N_i = 2c_i N \geq 2N$, the probability $R_N^{(r)}(\alpha)$ that two or more pairs of individuals belonging to α each share a parent is bounded by

(13)
$$R_N^{(r)}(\alpha) \leq \binom{V}{2} \left(\frac{1}{2N} \right)^2 \leq \frac{n^4}{32N^2}.$$

As

$$R_N^{(r)}(\alpha) = \sum_{\nu=2}^{\infty} P_N^{(r)}\{\text{exactly } \nu \text{ pairs of individuals in } \alpha \text{ each share a parent}\},$$

it follows from equation (12) that the quantities $R_N^{(r)}(\alpha, \alpha - \epsilon^i)$ in equation (11) are bounded in terms of the probability $R_N^{(r)}(\alpha)$ by

(14) $$2R_N^{(r)}(\alpha) \leq \sum_{i \in \mathcal{S}} R_N^{(r)}(\alpha, \alpha - \epsilon^i) \leq \left(\begin{array}{c} n \\ 2 \end{array} \right) R_N^{(r)}(\alpha).$$

Denoting by $R_N^{(r)}(\alpha, \beta)$ the probability that the backward reproduction step changes the value of the ancestral process from α to $\beta \notin \{\alpha\} \cup \{\alpha - \epsilon^i : i \in \mathcal{S}\}$, we have that

(15) $$R_N^{(r)}(\alpha) = \sum_{\beta \notin \{\alpha\} \cup \{\alpha - \epsilon^i : i \in \mathcal{S}\}} R_N^{(r)}(\alpha, \beta).$$

Combining the above, the transition probability of the ancestral process in one backward reproduction step is given by

$$
P_N^{(r)}(\beta | \alpha) =
\begin{cases}
1 - \sum_{i \in \mathcal{S}} \frac{1}{N_i} \left(\begin{array}{c} \alpha_i \\ 2 \end{array} \right) + \sum_{i \in \mathcal{S}} R_N^{(r)}(\alpha, \alpha - \epsilon^i) \\
\qquad - \sum_{\gamma \notin \{\alpha\} \cup \{\alpha - \epsilon^i : i \in \mathcal{S}\}} R_N^{(r)}(\alpha, \gamma) & \text{if } \beta = \alpha \\
\frac{1}{N_i} \left(\begin{array}{c} \alpha_i \\ 2 \end{array} \right) - R_N^{(r)}(\alpha, \alpha - \epsilon^i) & \text{if } \beta = \alpha - \epsilon^i \\
R_N^{(r)}(\alpha, \beta) & \text{otherwise,}
\end{cases}
$$

(16)

where

$$\sum_{\beta \neq \alpha} R_N^{(r)}(\alpha, \beta) \leq \left\{ \left(\begin{array}{c} n \\ 2 \end{array} \right) + 1 \right\} R_N^{(r)}(\alpha).$$

As migration and reproduction operate independently, the one-generation transition probabilities of the ancestral process, (8), are found from the transition probabilities in one backward migration step and one backward reproduction step as

$$P_N(\beta | \alpha) = \sum_{\gamma} P_N^{(m)}(\gamma | \alpha) P_N^{(r)}(\beta | \gamma).$$

In matrix notation:

(17) $$\mathbf{P}_N = \mathbf{P}_N^{(m)} \cdot \mathbf{P}_N^{(r)},$$

where $\mathbf{P}_N$, $\mathbf{P}_N^{(m)}$ and $\mathbf{P}_N^{(r)}$ are the transition matrices in one generation, one backward migration step and one backward reproduction step, respectively (for example, the entries of $\mathbf{P}_N$ are $(\mathbf{P}_N)_{\alpha,\beta} = P_N(\beta|\alpha)$). Denoting by $\mathbf{I}$ the identity matrix, with entries

$$I_{\alpha,\beta} = \delta_{\alpha,\beta} = \begin{cases} 1 & \text{if } \alpha = \beta \\ 0 & \text{otherwise,} \end{cases}$$

equation (9) for $P_N^{(m)}(\beta|\alpha)$ can be written as

$$(18) \qquad \mathbf{P}_N^{(m)} = \mathbf{I} + \frac{1}{2N}\mathbf{Q}_N^{(m)} + \mathbf{R}_N^{(m)},$$

where $\mathbf{Q}_N^{(m)}$ and $\mathbf{R}_N^{(m)}$ respectively denote the matrices with entries

$$\left(\mathbf{Q}_N^{(m)}\right)_{\alpha,\beta} =$$

$$\begin{cases} -\sum_{i \in \mathcal{S}} \alpha_i (2Nm_i) \dfrac{N_i}{N_i - m_i N_i - \alpha_i + 1} \displaystyle\prod_{k \in \mathcal{S}} \prod_{a=0}^{\alpha_k-1} \dfrac{N_k - m_k N_k - a}{N_k - a} \\ \qquad\qquad \text{if } \beta = \alpha \\[1em] \alpha_i(2Nm_{ij}) \cdot \dfrac{N_i}{N_i - m_i N_i - \alpha_i + 1} \displaystyle\prod_{k \in \mathcal{S}} \prod_{a=0}^{\alpha_k-1} \dfrac{N_k - m_k N_k - a}{N_k - a} \\ \qquad\qquad \text{if } \beta = \alpha - \epsilon^i + \epsilon^j \quad (j \neq i) \\[1em] 0 \qquad\qquad\qquad\qquad\qquad\qquad\qquad \text{otherwise} \end{cases}$$

and

$$\left(\mathbf{R}_N^{(m)}\right)_{\alpha,\beta} = \begin{cases} -\sum_{\gamma \neq \alpha} R_N^{(m)}(\alpha, \gamma) & \text{if } \beta = \alpha \\ R_N^{(m)}(\alpha, \beta) & \text{otherwise.} \end{cases}$$

From equation (16) it is seen that the matrix $\mathbf{P}_N^{(r)}$ can be partitioned similarly as

$$(19) \qquad \mathbf{P}_N^{(r)} = \mathbf{I} + \frac{1}{2N}\mathbf{Q}^{(r)} + \mathbf{R}_N^{(r)},$$

where $\mathbf{Q}^{(r)}$ is the matrix with entries

$$\left(\mathbf{Q}^{(r)}\right)_{\alpha,\beta} = \begin{cases} -\sum_{i \in \mathcal{S}} \dfrac{1}{c_i}\begin{pmatrix}\alpha_i \\ 2\end{pmatrix} & \text{if } \beta = \alpha \\[1em] \dfrac{1}{c_i}\begin{pmatrix}\alpha_i \\ 2\end{pmatrix} & \text{if } \beta = \alpha - \epsilon^i \\[1em] 0 & \text{otherwise} \end{cases}$$

(independent of N) and $\mathbf{R}_N^{(r)}$ is the matrix with entries

$$\left(\mathbf{R}_N^{(r)}\right)_{\alpha,\beta} = \begin{cases} \displaystyle\sum_{i \in \mathcal{S}} R_N^{(r)}(\alpha, \alpha - \epsilon^i) - \sum_{\gamma \notin \{\alpha\} \cup \{\alpha - \epsilon^i : i \in \mathcal{S}\}} R_N^{(r)}(\alpha, \gamma) & \text{if } \beta = \alpha \\[2mm] -R_N^{(r)}(\alpha, \alpha - \epsilon^i) & \text{if } \beta = \alpha - \epsilon^i \\[2mm] R_N^{(r)}(\alpha, \beta) & \text{otherwise.} \end{cases}$$

Note that when the number of subpopulations is infinite, the state space of the ancestral process, E, and hence the above matrices, are infinite. We consider the following norm on these matrices:

$$\|\mathbf{A}\| := \sup_{\alpha \in E} \sum_{\beta \in E} |A_{\alpha,\beta}|$$

where the matrix $\mathbf{A}$ has entries $A_{\alpha,\beta}$ ($\alpha, \beta \in E$). Using assumptions (i) and (ii), restricting to $N > \frac{1}{4}\sup_{i \in \mathcal{S}} M_i + \frac{1}{2}n$ and using the bounds in (10), (13), (14) and (15), all the matrices above have finite norm:

$$\|\mathbf{Q}_N^{(m)}\| \leq 4nN \frac{1}{1 - \frac{1}{4N}\sup_{i \in \mathcal{S}} M_i - \frac{n}{2N}} < \infty$$

$$\|\mathbf{Q}^{(r)}\| \leq 2\binom{n}{2}$$

$$\|\mathbf{R}_N^{(m)}\| \leq 2n^2$$

$$(20) \qquad \|\mathbf{R}_N^{(r)}\| \leq \frac{n^6}{32N^2},$$

while all three transition matrices have norm 1. Hence the product of any two of the above matrices exists (with finite entries) and has finite norm (bounded above by the product of the norms of these two matrices). Substituting equations (18) and (19) into equation (17), the one-generation transition matrix of the ancestral process α_N can be written as

$$(21) \qquad \mathbf{P}_N = \mathbf{I} + \frac{1}{2N}\left(\mathbf{Q}_N + \mathbf{\Delta}_N\right),$$

where

$$\mathbf{Q}_N = \mathbf{Q}_N^{(m)} + \mathbf{Q}^{(r)}$$
$$\mathbf{\Delta}_N = 2N\left(\mathbf{R}_N^{(m)} + \mathbf{R}_N^{(r)} + \mathbf{R}_N^{(m)}\mathbf{R}_N^{(r)} + \frac{1}{2N}\mathbf{Q}_N^{(m)}\mathbf{R}_N^{(r)} + \frac{1}{2N}\mathbf{R}_N^{(m)}\mathbf{Q}^{(r)} + \frac{1}{4N^2}\mathbf{Q}_N^{(m)}\mathbf{Q}^{(r)}\right).$$
$$(22)$$

As only a finite number of the α_i ($i \in \mathcal{S}$) are non-zero and with $\mathbf{Q}$ denoting the infinitesimal generator of the structured coalescent, given by equation (1), where M_{ij} and M_i are defined by the assumptions (i) and (ii), it is clear that

$$\lim_{N \to \infty} \mathbf{Q}_N = \mathbf{Q}, \text{ entry-wise,}$$

i.e.

(23) $$\forall \alpha, \beta \in E : \lim_{N \to \infty} (\mathbf{Q}_N)_{\alpha,\beta} = Q_{\alpha,\beta}.$$

We prove that for fixed $t \geq 0$, the transition matrix of the ancestral process α_N over $[2Nt]$ generations converges entry-wise to the transition matrix of the structured coalescent over time t:

(24) $$\lim_{N \to \infty} \mathbf{P}_N^{[2Nt]} = e^{t\mathbf{Q}}, \text{ entry-wise.}$$

First note that

$$\|\mathbf{Q}\| \leq n \sup_{i \in \mathcal{S}} M_i + 2 \binom{n}{2} < \infty,$$

so that $e^{t\mathbf{Q}} := \sum_{v=0}^{\infty} \frac{t^v \mathbf{Q}^v}{v!}$ exists, as

$$\forall \alpha, \beta \in E : |(e^{t\mathbf{Q}})_{\alpha,\beta}| \leq \sum_{v=0}^{\infty} \frac{t^v |(\mathbf{Q}^v)_{\alpha,\beta}|}{v!} \leq \sum_{v=0}^{\infty} \frac{t^v \|\mathbf{Q}\|^v}{v!} = e^{t\|\mathbf{Q}\|} < \infty.$$

Using equation (21), the matrix $\mathbf{P}_N^{[2Nt]}$ can be written as

$$
\begin{aligned}
\mathbf{P}_N^{[2Nt]} &= \left\{ \mathbf{I} + \frac{1}{2N} (\mathbf{Q}_N + \mathbf{\Delta}_N) \right\}^{[2Nt]} \\
&= \sum_{v=0}^{[2Nt]} \binom{[2Nt]}{v} \left(\frac{1}{2N} \right)^v (\mathbf{Q}_N + \mathbf{\Delta}_N)^v \\
&= \sum_{v=0}^{[2Nt]} \frac{[2Nt]([2Nt]-1) \cdot \ldots \cdot ([2Nt]-v+1)}{(2N)^v} \cdot \frac{(\mathbf{Q}_N + \mathbf{\Delta}_N)^v}{v!}.
\end{aligned}
$$

Hence for $\alpha, \beta \in E$:

(25) $$\left(\mathbf{P}_N^{[2Nt]} \right)_{\alpha,\beta} = \sum_{v=0}^{\infty} a_{v,N}$$

with

$$a_{v,N} = I_{\{v \leq [2Nt]\}} \cdot \frac{[2Nt]([2Nt]-1) \cdot \ldots \cdot ([2Nt]-v+1)}{(2N)^v} \cdot \frac{((\mathbf{Q}_N + \mathbf{\Delta}_N)^v)_{\alpha,\beta}}{v!}$$

(26)
where

$$I_{\{v \leq [2Nt]\}} = \begin{cases} 1 & \text{if } v \leq [2Nt] \\ 0 & \text{otherwise.} \end{cases}$$

As $N \to \infty$ and for fixed v, the first factor in the right-hand side of (26) converges to 1 and the second factor converges to t^v. We focus on the third factor. Assumptions (i) and (ii) imply that

$$(27) \qquad \qquad \sup_{i \in \mathcal{S}} m_i = O(1/N), \text{ as } N \to \infty.$$

Hence it follows from (10) that

$$\|\mathbf{R}_N^{(m)}\| \leq 2n^2 \left(\sup_{i \in \mathcal{S}} m_i \right)^2 = O(\frac{1}{N^2}),$$

while (20) yields that also

$$\|\mathbf{R}_N^{(r)}\| = O(\frac{1}{N^2}).$$

Restricting to

$$(28) \qquad \qquad N \geq \frac{1}{2} \sup_{i \in \mathcal{S}} M_i + n,$$

we have that

$$\|\mathbf{Q}_N^{(m)}\| \leq n \sup_{i \in \mathcal{S}} M_i \cdot \frac{1}{1 - \frac{1}{4N} \sup_{i \in \mathcal{S}} M_i - \frac{n}{2N}} \leq 2n \sup_{i \in \mathcal{S}} M_i,$$

which is independent of N. Substituting this information into equation (22), it follows that

$$(29) \qquad \qquad \|\mathbf{\Delta}_N\| = O(\frac{1}{N}), \text{ as } N \to \infty.$$

For N subject to (28), we have in addition that

$$(30) \qquad \qquad \|\mathbf{Q}_N\| \leq \|\mathbf{Q}_N^{(m)}\| + \|\mathbf{Q}^{(r)}\| \leq C,$$

where

$$(31) \qquad \qquad C := \left(2n \sup_{i \in \mathcal{S}} M_i + \|\mathbf{Q}^{(r)}\| \right)$$

is finite and independent of N. Expanding the matrix $(\mathbf{Q}_N + \mathbf{\Delta}_N)^v$, it takes the form

$$(32) \qquad \qquad (\mathbf{Q}_N + \mathbf{\Delta}_N)^v = \mathbf{Q}_N^v + \mathbf{A}_N$$

with

$$\|\mathbf{A}_N\| = O(\frac{1}{N}),$$

$\mathbf{A}_N$ being a finite sum of products of matrices $\mathbf{Q}_N$ and $\mathbf{\Delta}_N$, where each product contains at least one factor $\mathbf{\Delta}_N$. A fortiori, $\mathbf{A}_N = O(\frac{1}{N})$, entry-wise, i.e.

$$(33) \qquad \forall \alpha, \beta \in E : (\mathbf{A}_N)_{\alpha, \beta} = O(\frac{1}{N}).$$

Denoting by $\mathbf{V}$ the matrix with entries

$$V_{\alpha, \beta} = \begin{cases} \sum_{i \in \mathcal{S}} \left\{ \alpha_i M_i + \frac{1}{c_i} \binom{\alpha_i}{2} \right\} & \text{if } \beta = \alpha \\[2ex] \alpha_i M_{ij} & \text{if } \beta = \alpha - \epsilon^i + \epsilon^j \ (j \neq i) \\[2ex] \frac{1}{c_i} \binom{\alpha_i}{2} & \text{if } \beta = \alpha - \epsilon^i \\[2ex] 0 & \text{otherwise,} \end{cases}$$

we have for N subject to (28) that

$$\forall \alpha, \beta \in E : |(\mathbf{Q}_N)_{\alpha, \beta}| \leq V_{\alpha, \beta}.$$

As $\mathbf{V}$ does not depend on N and $\|\mathbf{V}\| \leq 2n \sup_{i \in \mathcal{S}} M_i + n(n-1) < \infty$, it follows from (23) by the dominated convergence theorem that

$$\begin{aligned} \lim_{N \to \infty} (\mathbf{Q}_N^v)_{\alpha, \beta} &= \lim_{N \to \infty} \sum_{\gamma_1, \dots, \gamma_{v-1}} (\mathbf{Q}_N)_{\alpha, \gamma_1} \cdot (\mathbf{Q}_N)_{\gamma_1, \gamma_2} \cdot \dots \cdot (\mathbf{Q}_N)_{\gamma_{v-1}, \beta} \\ &= \sum_{\gamma_1, \dots, \gamma_{v-1}} Q_{\alpha, \gamma_1} \cdot Q_{\gamma_1, \gamma_2} \cdot \dots \cdot Q_{\gamma_{v-1}, \beta} \\ &= (\mathbf{Q}^v)_{\alpha, \beta}. \end{aligned}$$

From (32) and (33) we obtain that

$$\lim_{N \to \infty} (\mathbf{Q}_N + \mathbf{\Delta}_N)^v = \mathbf{Q}^v, \text{entry-wise.}$$

Hence the quantities $a_{v,N}$ defined by equation (26), have as their respective limits

$$(34) \qquad \lim_{N \to \infty} a_{v,N} = \frac{t^v (\mathbf{Q}^v)_{\alpha, \beta}}{v!},$$

for all $v \in I\!\!N$. Using (29) and (30), we have for N sufficiently large that $\|\mathbf{Q}_N\| \leq C$ and $\|\mathbf{\Delta}_N\| \leq 1$, and hence that

$$|a_{v,N}| \leq \frac{t^v \|\mathbf{Q}_N + \mathbf{\Delta}_N\|^v}{v!} \leq \frac{t^v (C+1)^v}{v!}, \forall v \in I\!\!N.$$

As $\sum_{v=0}^{\infty} \frac{t^v(C+1)^v}{v!} = e^{t(C+1)} < \infty$, it follows from (25) and (34) by the dominated convergence theorem that

$$(35) \quad \forall \alpha, \beta \in E : \lim_{N \to \infty} \left(\mathbf{P}_N^{[2Nt]} \right)_{\alpha, \beta} = \sum_{v=0}^{\infty} \frac{t^v (\mathbf{Q}^v)_{\alpha, \beta}}{v!} = \left(e^{t\mathbf{Q}} \right)_{\alpha, \beta}.$$

Because of the Markov character of both the ancestral process α_N and the structured coalescent and because E is countable, this result implies that the finite-dimensional distributions of the process $\{\alpha_N([2Nt]); t \geq 0\}$ converge to those of the structured coalescent $\{\alpha(t); t \geq 0\}$ as $N \to \infty$.

Relative compactness. We regard E as a subspace of $\mathbb{R}^S$ (where $\mathbb{R}$ is the set of the real numbers), endowed with the norm

$$(36) \qquad \qquad \|\mathbf{x}\| = \sup_{i \in S} |x_i|,$$

for $\mathbf{x} = (x_i)_{i \in S} \in \mathbb{R}^S$. With this norm, E is complete, as every Cauchy-sequence in E is in the long run constant. According to Corollary 7.4 in Ethier and Kurtz [2], the relative compactness of $\{\alpha_N([2N\cdot])\}$ is guaranteed if we prove the following two conditions[2]:

(a) For every $\eta > 0$ and $t \geq 0$, there exists a compact set $\Gamma_{\eta,t} \subset E$ such that

$$\liminf_{N \to \infty} P\{\alpha_N([2Nt]) \in \Gamma_{\eta,t}\} \geq 1 - \eta.$$

(b) For every $\eta > 0$ and $T > 0$, there exists $\delta > 0$ such that

$$(37) \qquad \limsup_{N \to \infty} P\{w'(\alpha_N([2N\cdot]), \delta, T) \geq \eta\} \leq \eta,$$

where w' is the modulus of continuity:

$$w'(\alpha_N([2N\cdot]), \delta, T) = \inf_{\{t_i\}} \max_i \sup_{s,t \in [t_{i-1}, t_i)} \|\alpha_N([2Ns]) - \alpha_N([2Nt])\|,$$

where $\{t_i\}$ ranges over all partitions of the form $0 = t_0 < t_1 < \cdots < t_{k-1} < T \leq t_k$ with $\min_{1 \leq i \leq k}(t_i - t_{i-1}) > \delta$ and $k \geq 1$.

To verify condition (a), fix $t \geq 0$ and $0 < \eta < 1$. With the norm (36), E is not compact if the number of subpopulations is infinite. However, as E is countable, there exists a finite (and hence compact) set $\Gamma_{\eta,t} \subset E$ such that

$$(38) \qquad \qquad P\{\alpha(t) \in \Gamma_{\eta,t}\} \geq 1 - \eta.$$

---

[2] The condition (a) stated here is slightly stronger than that of Ethier and Kurtz. In particular, we have used that $\Gamma_{\eta,t}$ is a subset of the set $\Gamma_{\eta,t}^\eta$ in condition (a) of Ethier and Kurtz.

Since we have proved that the one-dimensional distributions of the ancestral process $\alpha_N([2N\cdot])$ converge to those of the structured coalescent $\alpha(\cdot)$ as $N \to \infty$,

$$P\{\alpha(t) \in \Gamma_{\eta,t}\} = \lim_{N \to \infty} P\{\alpha_N([2Nt]) \in \Gamma_{\eta,t}\}.$$

Combining this with (38), condition (a) is proved.

Condition (b) essentially requires that the ancestral process α_N, which is a pure jump process, doesn't jump too quickly, in the limit as $N \to \infty$, so that a partition $\{t_i\} = \{t_i^N\}$ can be found which satisfies the requirements set out in the definition of w' and which contains all the jump times of the ancestral process. To prove this, we couple the jumps of α_N to those of a process with a higher jump rate but whose inter-jump times are identically distributed and which is therefore easier to handle.

For $N \in I\!N_0$ we denote

$$p_N := \frac{1}{2N}(C+1),$$

where C is given by equation (31). In the remainder of the paper we restrict to N sufficiently large so that $p_N < 1$ and $\|Q_N + \Delta_N\| \leq (C+1)$ (see equations (29) and (30)). For each such N, we define the discrete-time Markov chain $(Z_N, \xi_N) \equiv \{(Z_N(\tau), \xi_N(\tau)); \tau = 0, 1, 2, \ldots\}$ to have state space $I\!N \times E$ and transition probabilities

$$P\{(Z_N(\tau+1), \xi_N(\tau+1)) = (j, \beta) | (Z_N(\tau), \xi_N(\tau)) = (i, \alpha)\}$$

$$= \begin{cases} 1 - p_N & \text{if } j = i \text{ and } \beta = \alpha \\[2mm] p_N - \sum_{\gamma \in E:\, \gamma \neq \alpha} P_N(\gamma|\alpha) & \text{if } j = i+1 \text{ and } \beta = \alpha \\[2mm] P_N(\beta|\alpha) & \text{if } j = i+1 \text{ and } \beta \neq \alpha \\[2mm] 0 & \text{otherwise,} \end{cases}$$

(39)

where $P_N(\beta|\alpha)$ is the transition probability of the ancestral process α_N from α to β in one generation. Using equation (21) and the restrictions made on N, we have $\forall \alpha \in E$ that

$$\sum_{\gamma \in E:\, \gamma \neq \alpha} P_N(\gamma|\alpha) = \frac{1}{2N} \sum_{\gamma \neq \alpha} (Q_N + \Delta_N)_{\alpha,\gamma}$$

$$\leq p_N,$$

so that (Z_N, ξ_N) is well defined, as the elements of its transition matrix, (39), are non-negative and, when summing over $(j, \beta) \in I\!N \times E$, add to one. From (39), it is clear that the marginal distribution of ξ_N is that of the ancestral process α_N, so that

$$(40) \qquad P\{w'(\alpha_N([2N\cdot]),\delta,T) \geq \eta\} = P\{w'(\xi_N([2N\cdot]),\delta,T) \geq \eta\}$$

for every $\eta > 0$ and $T > 0$. The process Z_N jumps with probability p_N, every generation; at each jump, its value increases by one. The construction is such that every time ξ_N jumps, Z_N jumps as well.

We denote by $0 = \rho_0^N < \rho_1^N < \ldots$ the jump times of the process (Z_N,ξ_N) and by $\tau_i^N := \rho_i^N - \rho_{i-1}^N$ ($i \in \mathbb{N}_0$) its inter-jump times. Because of the Markov character of (Z_N,ξ_N), the τ_i^N are mutually independent. As the probability of a jump of (Z_N,ξ_N) is p_N every generation, each τ_i^N is geometrically distributed with mean $1/p_N$.

Now fix $\eta > 0$ and $T > 0$. If for some $J \in \mathbb{N}_0$ and for some $\delta > 0$,

$$\rho_J^N \geq 2NT \text{ and } \tau_i^N > 2N\delta \text{ for } i = 1,\ldots,J,$$

then denoting $k_N := \min\{i : \rho_i^N \geq 2NT\}$ we have that $1 \leq k_N \leq J$ and the partition

$$t_i^N := \frac{\rho_i^N}{2N} \quad (i = 0,\ldots,k_N)$$

satisfies

$$0 = t_0^N < t_1^N < \cdots < t_{k_N-1}^N < T \leq t_{k_N}^N$$

and

$$t_i^N - t_{i-1}^N > \delta \quad (i = 1,\ldots,k_N);$$

as the process (Z_N,ξ_N) is constant between the jump times ρ_i^N, we have in that case that $\xi_N([2N\cdot])$ is constant on each interval $[t_{i-1}^N,t_i^N)$ ($i = 1,\ldots,k_N$), so that

$$w'(\xi_N([2N\cdot]),\delta,T) = 0.$$

Hence for every $J \in \mathbb{N}_0$ and $\delta > 0$:

$$P\{w'(\xi_N([2N\cdot]),\delta,T) < \eta\} \geq P\{\rho_J^N \geq 2NT \text{ and } \tau_i^N > 2N\delta \text{ for } i = 1,\ldots,J\}.$$
(41)

Thus in order to prove condition (b) it is sufficient to find $J \in \mathbb{N}_0$ and $\delta > 0$ such that

$$\liminf_{N\to\infty} P\{\rho_J^N \geq 2NT \text{ and } \tau_i^N > 2N\delta \text{ for } i = 1,\ldots,J\} \geq 1 - \eta.$$

Now

$$\begin{aligned}
&P\{\rho_J^N \geq 2NT \text{ and } \tau_i^N > 2N\delta \text{ for } i = 1,\ldots,J\} \\
&= P\{\rho_J^N \geq 2NT | \tau_i^N > 2N\delta \text{ for } i = 1,\ldots,J\} P\{\tau_i^N > 2N\delta \text{ for } i = 1,\ldots,J\} \\
&= P\{\rho_J^N \geq 2NT | \tau_i^N > 2N\delta \text{ for } i = 1,\ldots,J\} \left(P\{\tau_1^N > 2N\delta\}\right)^J,
\end{aligned}$$
(42)

because the τ_i^N $(i = 1, \ldots, J)$ are i.i.d. As $\rho_J^N = \sum_{i=1}^J \tau_i^N$, it is seen either by direct calculation or by a straightforward correlation argument that

$$(43) \quad P\{\rho_J^N \geq 2NT | \tau_i^N > 2N\delta \text{ for } i = 1, \ldots, J\} \geq P\{\rho_J^N \geq 2NT\}.$$

Since ρ_J^N is the time of the Jth jump of the process (Z_N, ξ_N) and Z_N counts the number of jumps of (Z_N, ξ_N),

$$(44) \quad P\{\rho_J^N > 2NT\} = P\{Z_N([2NT]) - Z_N(0) < J\}.$$

Combining (40), (41), (42), (43) and (44) we have that

$$P\{w'(\alpha_N([2N\cdot]), \delta, T) < \eta\} \geq P\{Z_N([2NT]) - Z_N(0) < J\} \left(P\left\{\frac{\tau_1^N}{2N} > \delta\right\}\right)^J,$$
(45)
for every $J \in \mathbb{N}_0$ and $\delta > 0$. Because the distribution of τ_1^N is geometric with mean $1/p_N$ and as $\lim_{N\to\infty}(2Np_N) = C + 1$, $\tau_1^N/(2N)$ converges in distribution, as $N \to \infty$, to an exponentially distributed random variable, X, with mean $1/(C + 1)$. As the probability of a jump of the process (Z_N, ξ_N) is p_N every generation and $Z_N([2NT]) - Z_N(0)$ is the number of jumps up to generation $[2NT]$, the distribution of $Z_N([2NT]) - Z_N(0)$ is binomial with parameters $[2NT]$ and p_N, so that as $N \to \infty$, $Z_N([2NT]) - Z_N(0)$ converges in distribution to a random variable, Z, which is Poisson distributed with mean equal to $\lim_{N\to\infty}([2NT]p_N) = T(C+1)$. Using (45), we obtain that

$$(46) \quad \liminf_{N\to\infty} P\{w'(\alpha_N([2N\cdot]), \delta, T) < \eta\} \geq P\{Z < J\}\left(P\{X > \delta\}\right)^J,$$

which can be made arbitrarily close to 1 by first choosing J sufficiently large and subsequently choosing δ small. Thus (37) follows and the proof is completed. □

Acknowledgements. I thank Peter Donnelly for valuable advice and suggestions on the work presented in this paper. This work was done while the author was a Research Assistant of the Belgian National Fund for Scientific Research. This research was in addition supported by the U.K. Science and Engineering Research Council (grant no. GR/G 11101). I thank the IMA for the excellent organization and facilities provided.

REFERENCES

[1] Cannings, C., 1974 The latent roots of certain Markov chains arising in genetics: A new approach. I. Haploid models. Adv. Appl. Prob. **6**: 260-290.

[2] Ethier, S. N., and Kurtz, T. G., 1986 *Markov Processes: Characterization and Convergence*. Wiley, New York.

[3] Griffiths, R. C., 1981 The number of heterozygous loci between two randomly chosen completely linked sequences of loci in two subdivided population models. J. Math. Biol. **12**: 251-261.

[4] Herbots, H. M., 1994 Stochastic Models in Population Genetics: Genealogy and Genetic Differentiation in Structured Populations. Ph.D. thesis, University of London. Submitted.

[5] Hey, J., 1991 A multi-dimensional coalescent process applied to multi-allelic selection models and migration models. Theoret. Popul. Biol. **39**: 30-48.

[6] Hudson, R. R., 1990 Gene genealogies and the coalescent process. In *Oxford Surveys in Evolutionary Biology* (D. J. Futuyma and J. Antonovics, eds.), Oxford University Press, Oxford, **7**: 1-44.

[7] Kingman, J. F. C., 1982 On the genealogy of large populations. Adv. Appl. Prob. **19A**: 27-43.

[8] Kingman, J. F. C., 1982 The coalescent. Stoch. Proc. Appl. **13**: 235-248.

[9] Kingman, J. F. C., 1982 Exchangeability and the evolution of large populations. In *Exchangeability in probability and statistics*, (G. Koch and F. Spizzichino, eds.), North-Holland, Amsterdam, pp. 97-112.

[10] Marjoram, P., and Donnelly, P., 1994 Pairwise comparisons of mitochondrial DNA sequences in subdivided populations and implications for early human evolution. Genetics **136**: 673-683.

[11] Marjoram, P., and Donnelly, P., 1994 Genealogical structure in populations of variable size, and the time since Eve. To appear in this volume.

[12] Nath, H. B., and Griffiths, R. C., 1993 The coalescent in two colonies with symmetric migration. J. Math. Biol. **31**: 841-852.

[13] Notohara, M., 1990 The coalescent and the genealogical process in geographically structured population. J. Math. Biol. **29**: 59-75.

[14] Slatkin, M., 1991 Inbreeding coefficients and coalescence times. Genet. Res., Camb., **58**: 167-175.

[15] Slatkin, M., 1993 Isolation by distance in equilibrium and non-equilibrium populations. Evolution **47**: 264-279.

[16] Strobeck, C., 1987 Average number of nucleotide differences in a sample from a single subpopulation: a test for population subdivision. Genetics **117**: 149-153.

[17] Tajima, F., 1989 DNA polymorphism in a subdivided population: the expected number of segregating sites in the two-subpopulation model. Genetics **123**: 229-240.

[18] Takahata, N., 1988 The coalescent in two partially isolated diffusion populations. Genet. Res., Camb., **52**: 213-222.

[19] Takahata, N., 1991 Genealogy of neutral genes and spreading of selected mutations in a geographically structured population. Genetics **129**: 585-595.

[20] Tavaré, S., 1984 Line-of-descent and genealogical processes, and their applications in population genetics models. Theor. Pop. Biol. **26**: 119-164.

[21] Wright, S., 1931 Evolution in Mendelian populations. Genetics **16**: 97-159.

[22] Wright, S., 1951 The genetical structure of populations. Annals of Eugenics **15**: 323-354.

AN ANCESTRAL RECOMBINATION GRAPH

ROBERT C. GRIFFITHS* AND PAUL MARJORAM*

Abstract. This paper describes a model of a gene as a continuous length of DNA represented by the interval $[0,1]$. The ancestry of a sample of genes is complicated by possible recombination events, where a gene can have two parent genes.

An analogue of Kingman's coalescent process, in which the ancestry of a sample of genes at a single locus is described by a stochastic binary tree, is a stochastic ancestral recombination graph, with vertices where coalescent or recombination events occur. All the information about ancestry is contained in this graph.

The sample DNA lengths have marginal ancestral trees at each point in $[0,1]$ which are imbedded in the graph. An upper bound is found for the expected number of distinct most recent common ancestors of these trees, and the expected maximum waiting time to these ancestors.

Key words. Coalescent process, Genealogical process, Population genetics, Recombination graph.

AMS(MOS) subject classifications. 60G35, 92A05, 92A10.

1. Introduction. An important ancestral process in population genetics is the coalescent process described in [10]. This represents the ancestry of a sample of n genes as a stochastic binary tree. A realization for 5 genes is illustrated in Figure 1.1. Measuring time backwards the number of ancestors $\{\xi_n(t), t \geq 0\}$ is a death process with $\xi_n(0) = n$ and rates $\mu_k = k(k-1)/2, k = n, \ldots, 2$. Vertices occur where two lines have a common ancestor. The rates are sufficiently fast to properly define a process $\{\xi_\infty(t), t \geq 0\}$ with an entrance boundary at infinity. The process has an absorbing state at unity, when the most recent common ancestor (MRCA) of the sample is found. The process arises as a limit from an ancestral process in a classical Wright-Fisher model with a fixed population size $2N$, and discrete generations when time is measured in units of $2N$ generations, and $N \to \infty$. The contents of a generation are formed by the $2N$ children choosing their parents at random from the previous generation. At a finer level a gene in this model might be thought of as a piece of DNA which does not break up along its ancestral lines (that is, there is no recombination).

This paper describes an analogue of the coalescent process when recombination is possible, the *ancestral recombination graph*. Such a graph for a two-locus model is described in [4].

It is convenient to represent a gene, thought of as a length of DNA, by the unit interval $[0,1]$. In a discrete Wright-Fisher model children in a generation choose one parent, with probability $1 - r$, or two parents, with probability r, when a recombination event, looking back in time, takes place. If recombination occurs a position for the break point, Z, is chosen

---

* Mathematics Department, Monash University, Clayton 3168, AUSTRALIA, Supported in part by Australian Research Grant A19131517.

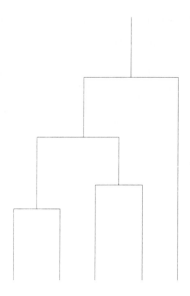

FIG. 1.1. *Coalescent tree.*

(independently from other break points) according to a given distribution, and the child gene is formed from the lengths $[0, Z]$ and $[Z, 1]$ from the first and second parents. Both of the parents are regarded as ancestors of any gene in a (forward) line of the child. Again time is measured in units of $2N$ generations and $N \to \infty$. The recombination rate per gene per generation r is scaled by holding $\rho = 2Nr$ fixed.

Particular cases for the break distribution are: Z is constant at 0.5, giving rise to a two-locus model; Z is discrete, taking values $\frac{1}{m}, \ldots, \frac{m-1}{m}$, giving rise to a m-locus model; and Z has a continuous distribution on $[0, 1]$, where breaks are possible at any point in $[0, 1]$.

A two-locus model is studied in [3], and finite-locus and continuous locus models in [5], [9].

Figure 1.2 illustrates a recombination graph for a sample of n genes. Looking back in time, coalescences occur when two edges join to a vertex, and recombination occurs when one edge joins to two. Positions $Z_1, Z_2, \ldots$ where breaks occur are labeled on the graph. The number of ancestors of the sample back in time $\{\xi(t), t \geq 0\}$ is a birth and death process with rates $\mu_k = k(k - 1)/2$ and $\lambda_k = k\rho/2$. Because of the quadratic death rate compared to the linear birth rate, with probability 1 there is a MRCA in the graph. As with the coalescent process, the the process can have an entrance boundary at infinity. It is implicit that the process is defined backward in time to negative infinity. Usually the graph is only of interest to the MRCA, since the whole ancestry of the sample is determined by then. However if the ancestry of a single individual is followed back in

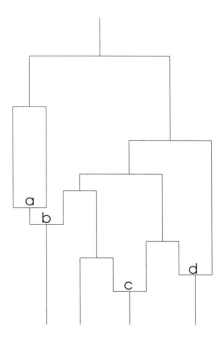

FIG. 1.2. *Two-locus recombination graph.*

time a graph is generated back to the next MRCA. Hitting a single edge (the MRCA) is a recurrent event, and the graphs between such hits are independent and identically distributed. Within a graph for n individuals a there is a subgraph for each individual which is either a single line to the grand MRCA, or is a graph which can be furthur decomposed into identically distributed subgraphs between hits of single edges. Subgraphs of the ancestry of n_0 of the n genes are consistent in the sense of being distributed as a recombination graph of a sample of n_0 genes.

It is shown in [4] that the expected waiting time to the grand MRCA from a sample of $n > 1$ genes is

$$(1.1) \qquad 2\rho^{-1} \int_0^1 \frac{1 - x^{n-1}}{1 - x} \left(e^{\rho(1-x)} - 1\right) dx.$$

The formula (1.1) also holds for the entrance boundary $n = \infty$. The expected time to a recombination event on a single line is $2/\rho$, therefore the expected time to generate a genealogy from a single gene to the next single edge is $2(e^\rho - 1)$.

The number of recombination vertices in the graph from a sample of n genes is distributed as the number of steps right in a random walk $\{\zeta_t, t = 0, 1, \ldots\}$ which starts at n, moves from k to $k + 1$ with probability $\rho/(\rho + k - 1)$ or to $k - 1$ with probability $(k - 1)/(\rho + k - 1)$, and has an absorbing state at 1. This is clearly true from the rates μ_k, λ_k. It is shown in [1] that

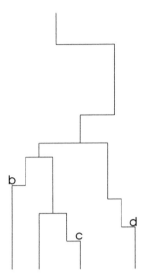

FIG. 1.3. *Marginal tree* $T(x)$, *when* $x > b$ *and* $x < c, d$.

the number of recombination vertices has a probability generating function $P_n(s) = Q_n(s)/Q_1(s)$, where

$$(1.2) \qquad Q_n(s) = \int_0^1 x^{\rho(1-s)-1}(1-x)^{n-1}e^{-\rho s(1-x)}dx.$$

Each point $x \in [0, 1]$ has a coalescent tree $T(x)$ associated with its ancestry. These trees are imbedded in the recombination graph. To obtain $T(x)$ trace from the leaves of the graph upward toward the MRCA in the graph. If there is a recombination vertex with label z, take the left path if $x \le z$, or right path if $x > z$. The MRCA in $T(x)$ may occur in the graph before the grand MRCA. Figure 1.3 shows an example of $T(x)$ when $x > b$ and $x < c, d$.

Since there are a finite number of recombination events in the graph, there are only a finite number of trees in $\{T(x); x \in [0, 1]\}$. There are potentially 2^R, if R recombination events occur, but some trees may be identical, or may not exist, depending on the ordering of the recombination break points. Recombination does not affect the marginal history of individual points, so for each $x \in [0, 1]$, $T(x)$ is distributed as a coalescent tree. Of course different trees share edges in the graph, and are not independently distributed.

Figure 1.4 contains all possible trees corresponding to the recombination graph in Figure 1.2. Trees in rows one and three in the same column are identical; the other trees are all distinct. If $b > a$ then all trees exist as marginal trees in the graph, otherwise if $b < a$ trees in Figure 1.4 with the right edge at vertex a do not exist as marginal trees.

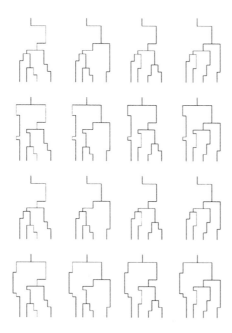

FIG. 1.4. *All possible marginal trees for the graph in Figure 1.2.*

Ancestor genes may only have part of their gametic material in common with the sample genes. It is even possible that some ancestor genes in the graph contain no material in common. A point x on an ancestor represented by an edge e in the graph has ancestral material in common with the sample if and only if e is included in $T(x)$. Thus the subset of $[0,1]$ over which that ancestor has ancestral material in common with the sample is $\mathcal{P}_e = \{x; T(x) \ni e, x \in [0,1]\}$. $\mathcal{P}_e$ is a union of a finite number of intervals, whose endpoints are a subset of the positions where recombination breaks have occured. If e and f are two edges, and $e \vee f$ denotes a coalesced edge from e and f, then $\mathcal{P}_{e\vee f} = \mathcal{P}_e \bigcup \mathcal{P}_f$. If a recombination break occurs at z, to an edge e, then the left and right hand edges from e in the graph are $\mathcal{P}_e \bigcap [0,z]$ and $\mathcal{P}_e \bigcap [z,1]$.

The embedded process describing the ancestral material in common with the sample is interesting. This Markov process has state $\{\mathcal{P}_{e_1}, \ldots, \mathcal{P}_{e_k}\}$ while there are k ancestors, and makes transitions according to how the random walk ζ_t moves, by either choosing two edges e, f at random to coalesce, or an edge chosen at random to split at Z, a recombination break point.

Edges $e_1, \ldots, e_k$ partition the interval $[0,1]$ of each of the n sample genes by their ancestral material in common. That is, $\mathcal{Q}_1, \ldots, \mathcal{Q}_n$ are partitions of $[0,1]$, $\mathcal{Q}_i = \{\mathcal{Q}_{ij}, j = 1, \ldots, k\}, i = 1, \ldots, n$ such that $\mathcal{Q}_{ij}$ is the material on edge e_j in common with sample gene i. Then also

$\mathcal{P}_{e_j} = \bigcup_{i=1}^{n} \mathcal{Q}_{ij}$. This way of thinking about ancestors is analogous to Kingman's scheme of labeling edges in a coalescent by an ancestral partition of the sample genes $1, \ldots, n$.

For example the imbedded ancestral partitions corresponding to the coalescent tree in Figure 1.1, with individuals labeled 1 to 5 from the left, are $\{\{1\}, [2], \{3\}, \{4\}, \{5\}\}, \{\{1, 2\}, \{3\}, \{4\}, \{5\}\}, \{\{1, 2\}, \{3, 4\}, \{5\}\},$ $\{\{1, 2, 3, 4\}, \{5\}\}, \{1, 2, 3, 4, 5\}$.

In the recombination graph each ancestor can be labeled by which sample genes, and subsets of material it is ancestral to. The sample is represented as $\bigotimes_{i=1}^{n}(i, [0, 1])$ and the ancestral partition is of this set. That is, while k ancestors, the jth ancestor would be labeled by $\bigotimes_{i=1}^{n}(i, \mathcal{Q}_{ij})$. The material on the n sample genes is partitioned by this labeling, as

$$\bigotimes_{i=1}^{n}(i, \mathcal{Q}_{ij}) \cap \bigotimes_{i=1}^{n}(i, \mathcal{Q}_{i\ell}) = \phi, j \neq \ell, \text{ and}$$
$$\bigcup_{j=1}^{k} \bigotimes_{i=1}^{n}(i, \mathcal{Q}_{ij}) = \bigotimes_{i=1}^{n}(i, [0, 1]).$$

2. Recombination events and MRCAs. In the following it is assumed that the recombination break distribution of Z is uniform in $[0, 1]$, unless otherwise mentioned.

THEOREM 2.1. *Let $R_{n,x,\delta}$ be the number of recombination events in $[x - \delta/2, x + \delta/2]$ before the MRCA at x. Define $h_n(x)$, the recombination density at $x \in [0, 1]$, by*

$$(2.1) \qquad h_n(x) = \lim_{\delta \to 0} \delta^{-1} P(R_{n,x,\delta} = 1) .$$

Then

$$(2.2) \qquad h_n(x) = \sum_{k=1}^{n-1} \frac{\rho}{k} ,$$

and

$$(2.3) \qquad \lim_{\delta \to 0} \delta^{-1} P(R_{n,x,\delta} > 1) = 0.$$

Proof. When there are k ancestors of the sample fragments $[x - \delta/2, x + \delta/2]$ then coalescence occurs at a rate $k(k-1)/2$, and recombination at rate $k\rho\delta/2$. The probability of no recombination events in $[x - \delta/2, x + \delta/2]$ in the graph, while k ancestors, is the probability that coalescence occurs before recombination,

$$\frac{k(k-1)/2}{k(k-1)/2 + k\rho\delta/2} = \frac{k-1}{k-1 + \rho\delta}.$$

The probability of no recombination events in $[x - \delta/2, x + \delta/2]$ before the MRCA at x is therefore

$$\prod_{k=2}^{n} \frac{k-1}{k-1+\rho\delta} \, ,$$

and

$$\lim_{\delta \to 0} \delta^{-1} P(R_{n,x,\delta} \geq 1) = \lim_{\delta \to 0} \delta^{-1} \left(1 - \prod_{k=2}^{n} \frac{k-1}{k-1+\rho\delta}\right) = \sum_{k=1}^{n-1} \frac{\rho}{k} \, .$$

Whatever the number of ancestors the recombination rate is proportional to δ, so $P(R_{n,x,\delta} > 1) = o(\delta^2)$ as $\delta \to 0$, and (2.1), (2.3) follow. □

COROLLARY 2.1. *The expected number of recombination events before the marginal MRCAs along the genes is* $\int_0^1 h_n(x)dx = \sum_{k=1}^{n-1} \frac{\rho}{k}$.

The result in this corollary is derived in [7].

Let $A_n(x,t)$ be the number of distinct ancestors of the sample of n, in $T(x)$, at time t back. $A_n(\cdot, t)$ is a random step function. Eventually $A_n(x, \tau) = 1, x \in [0,1]$ at the time τ when the MRCA of the graph is reached. If a cross-section of the graph at time t back has edges $e_1, \ldots, e_m$, then $A_n(x, t) = |\{e_i; x \in P_{e_i}, i = 1, \ldots, m\}|$. ¿From one locus coalescent theory, $E(A_n(x, t)) = E(\xi_n(t))$, not depending on x.

Of particular interest is the number of distinct MRCAs of a sample in the marginal trees at points along the genes,

$$|\{v(x); v(x) \text{ is the root of } T(x), x \in [0,1]\}| \, .$$

Clearly if there are R recombination events in the recombination graph then there can be at most $R + 1$ MRCAs of a sample along the genes.

For example suppose $a < b < c < d$ for the graph in Figure 1.1. Then for the sample $(0, b] \cup (d, 1]$ has the grand MRCA as the MRCA, and $(b, c]$, $(c, d]$ have other distinct MRCAs; a total of 3 distinct MRCAs.

The next theorem shows that multiple recombination events are rare in the ancestral lines of a sample of n genes as $n \to \infty$.

THEOREM 2.2. *Let* R_n *be the number of recombination events affecting the ancestory of a sample of* n *genes,* R_n^0 *the number of recombination events which occur to ancestors not having a previous recombination event in their lineage from the sample, and* R_n^1 *the number of recombination events to the grand first common ancestor. Then*

(2.4) $$R_n^0 \leq R_n \leq R_n^1,$$

(2.5) $$\sum_{j=2}^{n} \frac{\rho}{j+\rho-1} \leq E(R_n^0) \leq 1 + \sum_{j=2}^{n} \frac{\rho}{j+\rho-1},$$

(2.6) $$E(R_n^1) = \rho \int_0^1 \frac{1-(1-x)^{n-1}}{x} e^{\rho x} dx.$$

As $n \to \infty$, all three of $E(R_n^0)$, $E(R_n)$, $E(R_n^1)$ are asymptotic to $\rho \log(n)$, and $E(R_n^1 - R_n^0)$ is uniformly bounded above for $n = 2, 3, \ldots$.

Proof. For R_n^0 disregard the genealogy of any genes once they have been involved in recombination events, and consider the coalescence of ancestors where no recombination has taken place. Then ancestral lines in this modified coalescent are lost by coalescence or recombination at rates $k(k-1)/2$ and $k\rho/2$. The probability that the kth line is lost by recombination is $\rho/(k + \rho - 1)$, $k = n, \ldots, 1$. The last recombination event, if it occurs when one line, may, or may not, be before the MRCAs of the genes. This accounts for the inequality (2.5). The formula (2.6) for $E(R_n^1)$, and the result $E(R_n^1) \sim \rho \log(n)$ are derived in [1].

$E(R_n^1 - R_n^0)$ is shown to be uniformly bounded above by the following.

$$
\begin{aligned}
E(R_n^1 - R_n^0) &= \rho \int_0^1 \frac{1 - (1-x)^{n-1}}{x} e^{\rho x} dx - E(R_n^0) \\
&= \rho \int_0^1 \frac{1 - (1-x)^{n-1}}{x} (e^{\rho x} - 1) dx \\
&\quad + \rho \sum_{j=0}^{n-2} \int_0^1 (1-x)^j dx - E(R_n^0) \\
&\leq \rho \int_0^1 \frac{1 - (1-x)^{n-1}}{x} (e^{\rho x} - 1) dx \\
&\quad + \rho \sum_{j=2}^{n} \left(\frac{1}{j-1} - \frac{1}{j + \rho - 1} \right) \\
&\leq \rho \int_0^1 \frac{e^{\rho x} - 1}{x} dx + \rho^2 \sum_{j=1}^{\infty} \frac{1}{j(j + \rho)} \\
&< \infty.
\end{aligned}
$$

☐

This theorem provides an estimate for the expected logarithm of the number of distinct trees in the recombination graph, $\phi_n = E \log |\{T_n(x); x \in [0,1]\}|$. There are two possible paths to the grand MRCA in the graph along lines where there has been exactly one recombination event. Therefore

(2.7) $(\log 2) E(R_n^0) \leq \phi_n \leq (\log 2) E(R_n^1)$

and $\phi_n \sim \rho(\log 2)(\log n)$ as $n \to \infty$.

THEOREM 2.3. Let $p_n(x)$ be the probability that the MRCAs of $T(x-)$ and $T(x+)$ are identical in a recombination graph of n genes, given that a recombination event has occurred at x before the MRCA of $T(x)$. Then

$$
(2.8) \qquad
\begin{aligned}
p_n(x) &= 1 - \frac{n^2 + n - 2}{n(n+1) \sum_{k=1}^{n-1} \frac{1}{k}} \\
&\sim 1 - (\log(n))^{-1} \text{ as } n \to \infty.
\end{aligned}
$$

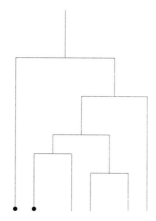

FIG. 2.1. *Ancestral tree of $\ell + 1$ genes after recombination.*

Proof. It is sufficient to consider just the one recombination event, since others do not affect $p_n(x)$. Suppose that this event occurs while there are ℓ ancestors of the sample.

The two MRCAs of the sample to the left and right of x are distinct if and only if one of the last two lines in the coalescence of the $\ell + 1$ genes after the recombination event is a single ancestor line of one of the genes involved in the recombination, and not an ancestor line of any of the other ℓ genes. Figure 2.1 illustrates this, with the two genes just after recombination shown by dots.

This occurs if and only if the subtree of the ℓ genes coalesces first, and so the probability of distinct ancestors is

$$2 \times \frac{\ell(\ell-1)}{(\ell+1)\ell} \cdot \frac{(\ell-1)(\ell-2)}{\ell(\ell-1)} \cdots \frac{2}{3} = \frac{4}{\ell(\ell+1)}.$$

The probability that there is one recombination event in $[x - \delta/2, x + \delta/2]$ while ℓ lines is

$$1 - \frac{\ell-1}{\ell-1+\rho\delta} + o(\delta^2) = \frac{\rho\delta}{\ell-1} + o(\delta^2),$$

so the conditional probability of a recombination event while ℓ lines is

$$\frac{1}{(\ell-1)\sum_{k=1}^{n-1}\frac{1}{k}}.$$

Finally, $p_n(x)$ is the sum of the probabilities, over ℓ, that the MRCAs of the sample of n genes immediately to the left and right of x are identical, and that the recombination event occurred while ℓ ancestors, given that a recombination event has occurred at x. That is

$$
\begin{aligned}
p_n(x) &= \sum_{\ell=2}^{n} \left(1 - \frac{4}{\ell(\ell+1)}\right) \frac{1}{(\ell-1)\sum_{k=1}^{n-1}\frac{1}{k}} \\
&= 1 - \frac{n^2+n-2}{n(n+1)\sum_{k=1}^{n-1}\frac{1}{k}}.
\end{aligned}
$$

□

THEOREM 2.4. *Let α_n be the expected number of distinct MRCAs of a sample of n genes. Then*

(2.9)
$$
\begin{aligned}
\alpha_n &\leq 1 + \left(1 - \frac{2}{n^2+n}\right)\rho \\
&\leq 1+\rho, \quad \text{for } n = 2, 3, \dots.
\end{aligned}
$$

(2.10)

Proof.

$$
\begin{aligned}
\alpha_n &\leq 1 + E(\text{Number of changes of ancestor along the genes}) \\
&= 1 + \int_0^1 P(\text{Change of ancestor at } x \mid \text{Recombination at } x) h_n(x) dx \\
&= 1 + \int_0^1 (1 - p_n(x)) h_n(x) dx \\
&= 1 + \left(1 - \frac{2}{n^2+n}\right)\rho.
\end{aligned}
$$

□

It is interesting that α_n is uniformly bounded in n, even though the expected number of recombination events before the MRCAs along the genes is asymptotic to $\rho \log(n)$.

3. Waiting times to MRCAs. Let $\{W_n(x), 0 \leq x \leq 1\}$ denote the collection of waiting times until the MRCAs at positions x for a sample of n genes. $W_n(x)$ is a random step function, depending on the number of recombination events in the ancestry.

Marginally

$$
W_n(x) = T_n(x) + \dots + T_2(x),
$$

where $\{T_k(x), k = 2, \dots n\}$ are the times while k ancestors, distributed as mutually independent exponential random variables with means

$\{\frac{2}{k(k-1)},\ k = 2,\ldots n\}$, and $E(W_n(x)) = 2\left(1 - \frac{1}{n}\right)$.

Although the marginal distribution of $W_n(x)$ does not depend on x or the recombination history of the sample, the distribution of $\{W_n(x),\ 0 \le x \le 1\}$ does.

THEOREM 3.1. Let $W = \max_{0 \le x \le 1} W_n(x)$, then

$$(3.1) \qquad E(W) \ \le \ 2 + \rho \frac{n^2 + n - 2}{2n(n+1)}$$

$$\le \ 2 + \frac{\rho}{2}, \quad \text{for } n = 2, 3, \ldots.$$

Proof. Let R_n be the number of recombination events before the MRCAs of the genes, occurring at positions $x_1, \ldots, x_{R_n}$. Define $x_0 = 0$, $x_{R_n+1} = 1$ for convenience. Let W_i be the time to the MRCA for the interval $[x_i, x_{i+1}]$. Then

$$W \ = \ \max_{i=0,\ldots,R_n} W_i$$

$$\le \ W_0 + \sum_{i=1}^{R_n}(W_i - W_{i-1}) \vee 0, \text{ and}$$

$$E(W) \ \le \ 2\left(1 - \frac{1}{n}\right) + E(R_n)E\Big((W_1 - W_0) \vee 0\Big).$$

$E(W_0) = 2\left(1 - \frac{1}{n}\right)$, since W_0 is the marginal waiting time to the MRCA at $x_0 = 0$. Now $P(W_1 > W_0) = \frac{1}{2}(1 - p_n(x_1))$, and $E(W_1 - W_0 | W_1 > W_0) = 1$, since $W_1 - W_0$ is the time taken for the last two lines of the genealogy at x_1 to coalesce, given $W_1 > W_0$. Thus

$$E(W) \ \le \ 2\left(1 - \frac{1}{n}\right) + \frac{1}{2}E(R_n)(1 - p_n(x_1))$$

$$= \ 2\left(1 - \frac{1}{n}\right) + \rho \frac{n^2 + n - 2}{2n(n+1)}$$

$$\le \ 2 + \frac{\rho}{2}.$$

$\square$

Bounds for α_n and $E(W)$, although derived for a uniform break distribution, carry through for any continuous distribution on $[0, 1]$.

4. Mutations on the graph. Let $\{V_n(x), x \in [0, 1]\}$ denote the collection of edge lengths of the marginal trees until the MRCAs at positions x for a sample of n genes. Using the notation in Section 3

$$V_n(x) = nT_n(x) + \cdots + 2T_2(x).$$

$V_n(x)$ is a random step function. A picture (essentially) of a simulated realization of $V_{10}(x)$ is shown in [6]. In a model with mutations occurring along the edges of the recombination graph according to a Poisson process of rate $\theta/2$, the total number of mutations occurring on lines with material in common to the sample is distributed as

(4.1)
$$N\left(\frac{\theta}{2}\int_0^1 V_n(dx)\right),$$

where $N(\cdot)$ is a Poisson process of unit rate.

The mean number of mutations is

(4.2)
$$\mu_m = \theta\sum_{j=1}^{n-1}\frac{1}{j},$$

and variance

$$
\begin{aligned}
\sigma_m^2 &= E\left((\theta/2)^2\int_0^1\int_0^1 V_n(dx)V_n(dy)\right) + \mu_m - \mu_m^2 \\
&= (\theta^2/2)\int_0^1(1-z)Q_n(z;\rho z)dz + \mu_m - \mu_m^2,
\end{aligned}
$$
(4.3)

where $Q_n(z;\rho z)$ is the product of edge lengths at two points distance z apart. This quantity is distributed as the product of edge lengths in a two locus model with recombination rate ρz between the loci.

Let M be the number of mutations and λ be the random variable $\frac{\theta}{2}\int_0^1 V_n(dx)$. Then (4.3) follows from

$$
\begin{aligned}
\sigma_m^2 &= E(M(M-1)) + \mu_m - \mu_m^2 \\
&= E(E(M(M-1))|\lambda) + \mu_m - \mu_m^2 \\
&= E(\lambda^2) + \mu_m - \mu_m^2.
\end{aligned}
$$

The formulae (4.2), (4.3) were derived by [5], (4.3) can be expressed as

(4.4)
$$\sigma_m^2 = \theta\sum_{i=1}^{n-1}\frac{1}{i} + \frac{1}{2}\frac{\theta^2}{\rho^2}\int_0^\rho(\rho - z)f_n(z)dz,$$

where $f_n(z)$ is the covariance of the edge lengths in a 2-locus model with recombination rate ρ and sample size n. In [5], time is measured in twice our units, accounting for a factor of four different in front of the integral expression in (4.4).

A particular case for $n = 2$, the only explicit formula for $f_n(z)$, is

(4.5)
$$f_2(z) = \frac{4(\rho + 18)}{\rho^2 + 13\rho + 18}.$$

A model with mutation is obtained by specifying the distribution of the position on a gene where mutation takes place. Mutations which occur in material ancestral to the sample will be represented in the sample genes. Suppose mutation occurs according to a continuous distribution in $[0, 1]$, and the label of a mutation is just the position where it occurs. An observed sample of n, then, is a collection of n sets of points where mutations have occurred. If there is a mutation at x_0, then some genes of the sample will contain the mutation, and others will not, being of the type of the MRCA at x_0.

There is an urn model representation of a two-locus sampling distribution in [1] which extends easily to the model in this paper where a gene is a length $[0, 1]$ of DNA. The idea in this representation is to produce the relative order of coalescent, mutation, and recombination events in the imbedded process in the graph. Then the shape of graph is filled in later.

Let $\{M(t), t = 0, 1, \ldots\}$, $M(0) = n$ be a random walk on positive integers with absorbing state 1, and transition probabilities for $m \geq 2$,

$$(4.6) \quad m = \begin{cases} m - 1, & \text{with probability } (m-1)/(m-1+\theta+\rho), \\ m, & \text{with probability } \theta/(m-1+\theta+\rho), \\ m + 1, & \text{with probability } \rho/(m-1+\theta+\rho). \end{cases}$$

Suppose τ is the absorption time. Keep a record of $M(0), \ldots, M(\tau)$. Begin with a sample of size 1 at τ, then construct samples at times $\tau - 1, \ldots, 1$. Let $D(t) = M(t) - M(t-1)$, $t = \tau, \ldots, 1$. If $D(t) = -1$, then choose a sample member at random to duplicate. If $D(t) = 0$, then choose a sample member at random to mutate at a random position in $[0, 1]$ according to a prescribed distribution. If $D(t) = 1$, then choose a pair at random to recombine at a position chosen according to a prescribed recombination break distribution. The sample at n at time 0 is distributed as a sample in the recombination graph.

REFERENCES

[1] ETHIER, S.N. AND GRIFFITHS, R.C. *On the two-locus sampling distribution*, **29**, 131-159, J. Math. Biol., (1990).

[2] ETHIER, S.N. AND GRIFFITHS, R.C., *The neutral two-locus model as a measure-valued diffusion*, **22**, 773-786, Adv. Appl. Prob., (1991)

[3] GRIFFITHS, R.C., *Neutral two-locus multiple allele models with recombination*, **19**, 169-186, Theoret. Popn. Biol., (1981).

[4] GRIFFITHS, R.C., *The two-locus ancestral graph*, **18**, 100-117, Selected proceedings of the symposium on applied probability, Sheffield, 1989., Institute of Mathematical Statistics, *ed*. I.V. Basawa and R.L. Taylor, IMS Lecture Notes–Monograph Series, (1991).

[5] HUDSON, R.R., *Properties of a neutral allele model with intragenic recombination*, **23**, 183-201,Theoret. Popn. Biol., (1983).

[6] HUDSON, R.R., *Gene genealogies and the coalescent process*, **7**, 1-44, Oxford Surveys in Evolutionary Biology, *ed*. Futuyma, D. and Antonovics, J. (1991).

[7] HUDSON, R.R. AND KAPLAN, N.L., *Statistical properties of the number of recombination events in the history of a sample of DNA sequences*, **111**, 147-164, Genetics, (1985).

[8] HUDSON, R.R. AND KAPLAN, N.L., *The coalescent process in models with selection and recombination*, **120**, 831-840, Genetics, (1988).

[9] KAPLAN, N.L. AND HUDSON, R.R., *The use of sample genealogies for studying a selectively neutral m-loci model with recombination*, **28**, 382-396, Theoret. Popn. Biol., (1985).

[10] KINGMAN, J.F.C., *The coalescent*, **13**, 235-248, Stoch. Proc. Applns., (1982).

THE EFFECT OF PURIFYING SELECTION ON GENEALOGIES

G. BRIAN GOLDING*

Abstract. Simulations are used to investigate the effect of purifying or directional selection on the branch lengths of a coalescent. It has been proposed by Fu & Li that differences among the branch lengths could be used to test for the presence of selection. The simulation results reported here suggest that directional selection does not have a large effect on branch lengths. An examination of populations with variable population sizes and of populations that exhibit sub-structure shows that both of these population processes can have a large effect on the patterns of branch lengths. Thus, the proposed test by Fu & Li does not appear to have sufficient power to detect the effects of simple directional, genic selection but it does appear to be a sophisticated and sensitive way to detect non- selective deviations from a simple neutral model of evolution.

1. Introduction. The neutral theory of molecular evolution proposed by Kimura (1968; see also Kimura 1983) has been one of the major driving forces of evolutionary studies. There have been many studies published that support the concept that most molecular variation is selectively neutral (Kimura 1983) and many other studies that question this conclusion (Gillespie 1991). But whether or not the neutral theory is ultimately proved to be correct or false, it has forced every suggested example of natural selection to be rigorously justified. This has been of great help to evolutionary studies and has shown that many examples of natural polymorphisms might have alternative explanations that have not been properly explored (e.g. Selander 1976).

Many of the techniques developed for the detection of natural selection rely on changes in the frequency of a morph either within the organism's lifetime or within a few generations. Absolute counts of individuals before and after selection are very simple and powerful ways to detect selection. But if all of the selective events are not accomplished within a single generation or if such counts can only be made between generations, then the effects of genetic mixing due to mating must also be considered. This is a large complication because the genotypes that were favoured in the previous generation may no longer exist in the current generation. Prout (1965) showed that counts from two consecutive generations are not sufficient to accurately measure selection. He later showed (Prout 1969) that at least four generations must be sampled in order to estimate fitness components and even then, the estimates have large variances. One way to get around this problem is to sample from more generations but again, even samples spread over 10 generations may have difficulty detecting selection (e.g. Watterson 1982).

* Department of Biology, McMaster University, 1280 Main Street West, Hamilton, Ontario, L8S 4K1. Tel, 905-525-9140 x24829, FAX, 905-522-6066, e-mail, Golding@McMaster.CA.

There are two large disadvantages to these studies. The first is the amount of work involved in sampling and analysing a single population over many generations. For some organisms this is simply not possible. The second unfortunate feature of this approach is that it is capable of detecting only very strong selection. For these methods to function, selection must be sufficiently strong to alter genotype frequencies over the short time span that can be observed. But in natural populations even very weak selection can be a strong evolutionary force if it is consistently present for a long period of time.

Other approaches for the detection of selection have thus attempted to look at patterns of genetic variability that might be caused by selection. This approach is capable of detecting small fitness differences so long as it has had an effect on the population. Some of these approaches examine geographical patterns (Lewontin and Krakauer 1975), some examine multi-locus associations (Hedrick et al. 1978), and others are based on the sampling theory of Ewens (1972). The most popular of the latter group is Watterson's (1978) test. Watterson showed that heterozygosity is a good statistic to test for the effects of overdominant selection. The application of these tests during the course of many studies has helped to fuel the fires of the neutralist controversy.

These tests are also not without problems. Watterson's test for example, assumes that an equilibrium state has been reached and it is known that even remote bottlenecks can affect allele distributions (Ewens 1977). Another problem with many of these studies has been the uncertainty of proper allelic identification. All these popular tests assume that alleles are accurately identified. But electrophoresis can hide many different alleles within a single allelic class (Singh et al. 1976).

Nucleotide sequence data permits accurate identification of every allelic difference. It also has the advantage that multiple changes can be observed and hence a study of sequences with just 100 nucleotides has an enormous potential number of different haplotypes (4^{100}; although this would never be realized in practice). This new information can be utilized to many purposes, including the search for selection.

Again, the neutral theory of molecular evolution forces us to search the nucleotide sequence data for rigorous evidence of the actions of natural selection. One suggested test for the presence of selection is to use Tajima's D test (Tajima 1989). This statistic looks at the relationship between two estimates of genetic variation - the number of segregating sites and the average pairwise nucleotide differences. Because deleterious mutations are maintained in low frequency and the number of segregating sites is not sensitive to frequencies but the pairwise differences are, these two estimates can be used to search for the influence of selection. One can also use maximum likelihood models of differences between the alleles fixed in different species and their phylogenetic histories to search for selection (Golding and Felsenstein 1990).

Another test for the presence of selection has been developed in a beautiful paper by Fu & Li (1993). In this paper they develop some remarkable results for the expected properties of the coalescent process. Fu & Li (1993) discovered that it is possible to build recursion relationships for the expected branch lengths of the coalescent based on how new branches or genes are added over time. Using these equations they are able to show the remarkable result that the expected sum of external branch lengths (those leading to "leaves" or "tips" of the tree; illustrated in Figure 1.1) is independent of sample size. Thus for a sample of n genes, they show that

$$E(L_n) = 4N_e,$$

where L_n is the sum of all external branch lengths and N_e is the effective population size. This amazing result implies that the sum of external branch lengths from 4 sequences should be equal to the sum obtained from 400 sequences. The same is not true for the expected sum of internal branch lengths (Figure 1.1). They show that in this case

$$E(I_n) = 4N_e(a_n - 1),$$

where

$$a_n = \sum_{k=1}^{n-1}(1/k),$$

and I_n is the sum of all internal branch lengths. Because mutations along external branches are effectively independent of mutations along internal branches, Fu & Li suggest using the difference between these expectations as a test for the presence of selection. They suggest that the normalized difference between the total number of mutations and the number of mutations along external branches should be used as a test statistic. Like Tajima's D, their statistic does not have a normal distribution and Fu & Li then determine the distribution via extensive simulations and finally supply an empirically determined bias correction.

They suggest that the action of selection will be reflected in the shape of the coalescent tree. Specifically, that there will be an excess of mutations in external branches if purifying or negative selection is present. While there will be a deficiency of mutations in external branches if balancing selection is present.

That selection may alter the shape of genealogies is a common theme throughout the literature and is even discussed in standard textbooks. In their discussion of selection on nucleotide sequences, Hartl and Clark (1989, pp.380-382) suggest that allelic genealogies will differ in the presence of selection and illustrate some example trees. However, in direct contradiction to Fu & Li's prediction, the inference from Hartl & Clark's example is that purifying selection would lead to shorter external branches and thus a deficiency of mutations in the external branches.

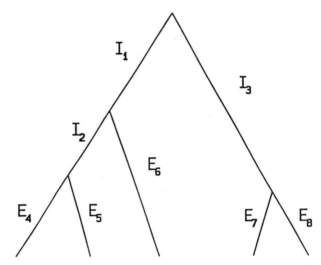

FIG. 1.1. *A simple tree or coalescent with a sample of 5 alleles. External branches are those leading to directly to the 5 alleles (they are labelled $E_4 - E_8$). Internal branches are those that do not lead directly to an allele (they are labelled $I_1 - I_3$). (from Fu & Li, 1993).*

In this note, the effects of directional selection on the shape of genealogies will be explored in an attempt to resolve this contradiction. Throughout, I will assume that the alleles influenced by selection cannot be identified *a priori* and instead that the search is simply for any indication of selection. Hence, it will be assumed that the allelic types are not labelled on the branches of the genealogical tree.

2. Simulations. The simulation keeps track of $2N = 100$ alleles over time. Generations are assumed to be distinct and non-overlapping. Each generation, a random sample of $2N$ alleles are sampled with replacement from the population to form the alleles for the next generation. This sampling process also defines an allelic genealogy that is stored for the entire population. After N generations, a sample of 10 alleles are randomly chosen. For these 10 alleles, their allelic genealogy is known and the lengths of all external and internal branches are stored. This population is used to begin another N generations, after which another sample of 10 alleles are chosen. This is continued until a thousand samples are obtained.

A two-allele model is assumed. Mutation may occur each generation and occurs equally to and from each allele. The mutation rate was set so that $4N_\mu = \theta = 0.1$. One allele is given a fitness of one while the other allele has a fitness of $1 - s$. Only simple, directional, genic selection is considered. The values of s are chosen so that $10^{-2} < 4Ns < 10^2$. These levels of $4Ns$ correspond to very weak selection through to very strong selection. Samples with increasing levels of selection were started from the previous populations simulated with lower levels of selection. This was done to ensure that an equilibrium had been reached.

The genealogy for the 10 sampled alleles is known in this simulation. This is not the case in real samples where the genealogy must be inferred. The inference of a genealogy from a sample would add further uncertainty to the results. Therefore, in this preliminary investigation only the actual effects of selection on the true genealogy will be examined and the added problems of trying to detect the effects of selection on inferred genealogies will be dealt with elsewhere.

3. A Simple Model. That selection is indeed functioning correctly in the simulation can be confirmed by examining the allele frequencies. Two examples of the deleterious allele frequencies are shown in Figure 3.1 for $10^{-2} < 4Ns < 10^2$. This shows that even $4Ns = 1$ is relatively effective selection and that deleterious allele frequencies have begun to decline with this level of selection. By the time that $4Ns = 10$, the deleterious allele is seldom found in the population. Figure 3.1 is present to remind us of the values of $4Ns$ where selection is effective. When $4Ns = 10^{-2}$ the alleles behave as if they are effectively neutral and this is reflected in the approximately 50% frequencies shown in the two examples in Figure 3.1. But when $4Ns = 10^2$, selection is so strong that the deleterious allele will seldom be seen. Indeed the deleterious allele is so rare that all but exceptional samples include only the non-deleterious alleles. In this case, the genealogy for the remaining allele(s) will reflect selective neutrality because only the relatively neutral alleles will still exist in the population. Thus, there is only a small window centred around $4Ns = 10^0$ where any selective effects on the genealogical branch lengths can be hoped to be observed.

The sum of the external branch lengths for 1000 samples are shown in Figure 3.2. As noted above, the expectation for the sum of external branch lengths is

$$E(L_n) = 4N_e = 200.$$

This expectation is confirmed by these simulations but they also demonstrate a very large variance around this mean. Figure 3.2 shows only the means and their standard errors but samples from individual trees can differ a great deal from the mean. The sum of the external branch lengths was as low as $41, 25, 41$ for $4Ns = 10^{-2}, 10^0, 10^2$ and as high as $858, 657, 728$ for $4Ns = 10^{-2}, 10^0, 10^2$.

There also does not appear to be any pattern to the external branch lengths that is induced by selection. When $4Ns = 10^1$ and $10^{1.25}$ there appears to be a slight decrease in the external branch lengths but this was not reproducible in simulations using other random number seeds. It seems that whatever deviations are caused by selection, they are completely masked by the natural variability of population based statistics.

A similar result is shown in Figure 3.3 for the sum of the internal

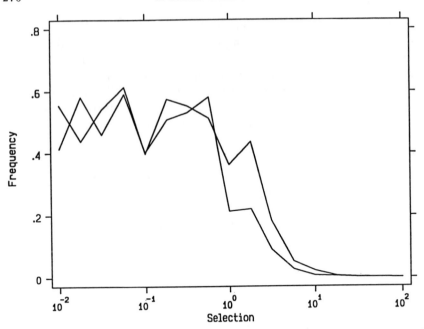

FIG. 3.1. *Two example allele frequencies generated by the simulation. The frequency of the deleterious allele is shown. Selection ranges from* $10^{-2} < 4Ns < 10^2$.

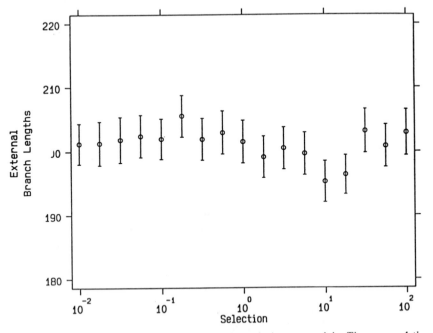

FIG. 3.2. *External branch lengths for a simple evolutionary model. The mean of the sum of external branch lengths for a sample of 10 alleles is shown as a circle. The bars illustrate one standard error above and below the mean.*

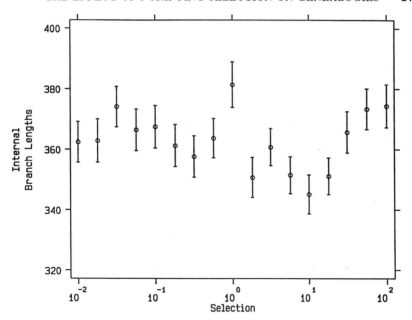

FIG. 3.3. *Internal branch lengths for a simple evolutionary model. The mean of the sum of internal branch lengths for a sample* 10 *alleles is shown as a circle. The bars illustrate one standard error above and below the mean.*

branch lengths. Here the expectation is

$$E(I_n) = 4N_e(a_n - 1) = 366.$$

Again this is confirmed by these simulations but the variation is large. The value when $4Ns = 10^0$ illustrates some of this variation. Figure 3.2 suggested that there might have been a minor decrease in the size of external branch lengths when $4Ns = 10^1$ and this would predict a corresponding increase in the internal branch lengths. But Figure 3.3 suggests that if anything, the internal branch lengths might also be decreased. However, the decrease is occurring at levels of $4Ns$ where selection should be sufficiently strong enough that only comparatively neutral alleles remain. In addition, where the effect of selection should be most apparent $(4Ns = 10^0)$ the internal branch lengths are actually increased. Again, different random number seeds suggest that there is little discernable effect of selection on internal branch lengths.

Because the expectations of the internal and external branch lengths are both a function of population size when evolution is selectively neutral, a comparison of their predictions can be examined. The statistic suggested by Fu & Li is to consider the difference

$$\eta - a_n \eta_e,$$

where η is the total number of mutations in the genealogy and η_e is the

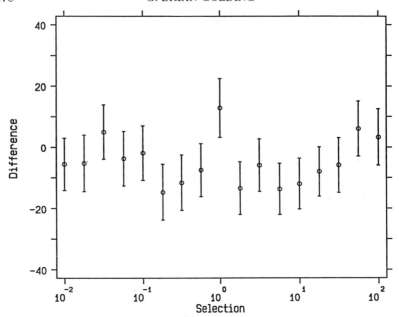

FIG. 3.4. *The mean value of D (as defined in the text) for a simlpe evolutionary model. The mean is based on a thousand samples and is shown as a circle. The bars illustrate one standard error above and below the mean.*

number of mutations along external branches (then normalized by a function of the expected variance). Here, since only a single two-allele locus is examined, a comparable test is to consider the difference in the total branch lengths versus the external branch lengths

$$D = J_n - a_n L_n,$$

where J_n is the total branch length. This difference is shown in Figure 3.4 and shows that throughout the range of $4Ns$, selection has a very minor effect on this statistic. On average, D is very close to zero whether or not selection is present. It appears from these results that variation in natural populations is large enough to mask any effect that selection might have on the branch lengths of the coalescent and that statistics based on these properties will have limited power.

4. Increasing Population Size. There are other forces that might act upon natural populations and some of these might be expected to alter the relative lengths of external/internal branches. One of these might be a recent increase in population size. This should cause the external branch lengths to be larger than otherwise expected. To test this idea, the simulation was modified such that it was run with $2N = 50$ for $2/3N$ generations and then $2N$ is increased back to 100 for $1/3N$ generations. At this point a sample of 10 alleles is taken and then again $2N$ is reduced to 50 and the simulation continues in this fashion until 1000 samples are obtained.

The results of this simulation are shown in Figures 4.1-4.3. Figure 4.1 shows the sum of the external branch lengths. For this situation there is not an established expectation. By analogy with the previous section, the expectation might be thought to be $4N_e$. The effective size for a population cycling between $2N = 50$ and $2N = 100$ is $2N_e = 60$ and hence $4N_e = 120$. But this is not a fair expectation since the population size has been just recently increased when the samples are taken. The means in Figure 4.1 are much larger than 120. But again, there is not any apparent effect of the selection acting on the sum of these branch lengths.

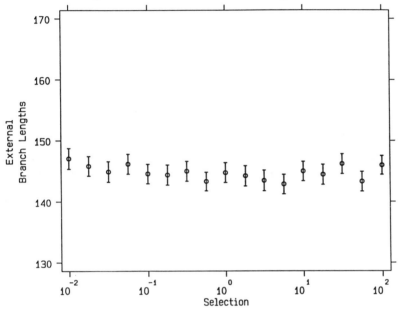

FIG. 4.1. *External branch lengths for samples from a variable sized population. The mean of the sum of external branch lengths for a sample of* 10 *alleles is shown as a circle. The bars illustrate one standard error above and below the mean.*

Figure 4.2 gives the sum of the internal branch lengths. Again an inappropriate expectation might be that

$$E(I_n) = 4N_e(a_n - 1) = 220.$$

The means of the internal branch lengths are much smaller than this number. Again there is a large variance but no apparent effect of selection on the branch lengths.

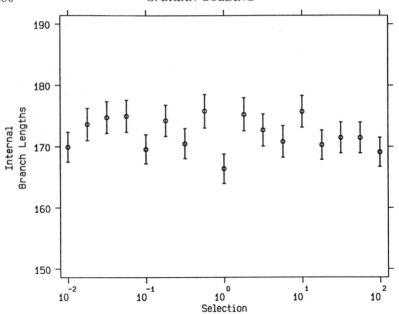

FIG. 4.2. *Internal branch lengths for samples from a variable sized population. The mean of the sum of internal branch lengths for a sample of 10 alleles is shown as a circle. The bars illustrate one standard error above and below the mean.*

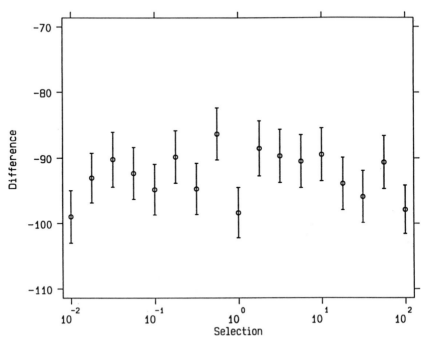

FIG. 4.3. *The value of D (defined in the text) for samples from a variable sized population. The mean is shown as a circle and the bars illustrate one standard error above and below the mean.*

It is the relative sizes of the external versus internal branch lengths that are important for the test of neutrality. The difference between these expectations is shown in Figure 4.3. The results are given for selection coefficients that range from $10^{-2} < 4Ns < 10^2$. The effect of this selection, if any, must be very slight since there is no consistent trend detectable as selection is increased. But the scale of this graph is quite different from that in Figure 3.4. There is a large departure from the "neutral" expectation that $D = 0$. The average D is -92 and can easily be statistically distinguished from zero.

5. Population Structure. Another situation that might lead to differences in the relative lengths of external and internal branch lengths is population structure. If the population is actually divided into two partially isolated populations with migration between them, then the internal branch lengths may be much longer than otherwise expected. To test this idea, a subdivided population was modelled. The population was arbitrarily divided in half. Then each generation, alleles were sampled for each half of the population but originating only from that half of the population. If this continued without change, then each half of the population would be completely isolated from the other half. Migration prevents this from happening. Migration between the two halves was modelled with an exchange rate such that $4Nm = 1.0$, where m is the probability that a pair of alleles are exchanged between the two halves.

The results of this simulation are shown in Figures 5.1-5.3. Figure 5.1 shows the sum of the external branch lengths. Again, there is not an established expectation but the external branch lengths are much larger than those observed in a population without substructure. Here the mean sum of external branch lengths is roughly 220 rather than 200. Similarly, the internal branch lengths shown in Figure 5.2 are longer than expected, roughly 560 rather than the 366 observed in Figure 3.3. Both Figures 5.1 and 5.2 also show even more variance than found in the previous simulations. This is due to the very different patterns of branch lengths that are observed depending on whether or not a migration event has occurred recently. In neither case is there a consistent effect of natural selection.

It is the relative sizes of the external/internal branch lengths that are important for the test. The difference between expectations is shown in Figure 5.3. Again, the results are given for selection coefficients that range from $10^{-2} < 4Ns < 10^2$ and again there is no discernable pattern but the variance has been increased. The scale of this graph is again quite different from Figures 3.4 and 4.3. There is a large departure from the "neutral" expectation that $D = 0$ but this time with D very large and positive. These values of D (whether selection is or is not acting) can easily be statistically distinguished from zero.

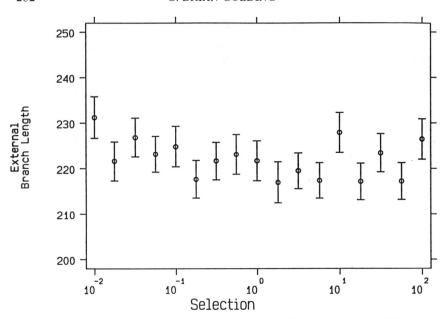

FIG. 5.1. *External branch lengths for samples from a subdivided population. The mean of the sum of external branch lengths for a sample of* 10 *alleles is shown as a circle. The bars illustrate one standard error above and below the mean.*

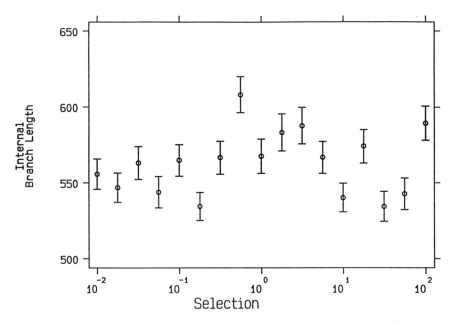

FIG. 5.2. *Internal branch lengths for samples from a subdivided population. The mean of the sum of internal branch lengths for a sample of* 10 *alleles is shown as a circle. The bars illustrate one standard error above and below the mean.*

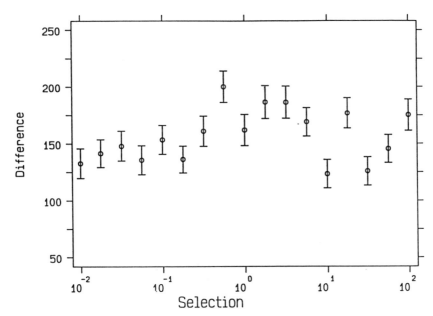

FIG. 5.3. *The value of D (defined in the text) for samples from a subdivided population. The mean is shown as a circle and the bars illustrate one standard error above and below the mean.*

6. Discussion . These results suggest that if selection has an effect on the "unlabelled" branch lengths of a coalescent, this effect must be largely hidden within the natural variability present in branch lengths. Hence the use of these branch lengths to search for the presence of selection is not likely to be successful. Given the level of variation present, it is unlikely that any test based on these properties would have great power.

However, such a test does seem to be a sensitive indicator of some other departures from a simplistic neutral model. When either the population size has recently expanded (within the last N generations in order to have an effect on the genealogy) or if there is nonrandom mating among the individuals composing the sample, then there will be significant departures from the "neutral" expectations. Both of these situations will affect the branch lengths within an unlabelled genealogy. Because the effects of these situations can cause large deviations they should be easily detectable.

These results must be considered tentative. More data are required to determine if selection actually does or does not influence the branch lengths of unlabelled coalescents. More parameter combinations must be examined and other forms of selection considered - most notably balancing selection. These studies are currently being pursued.

In general, it might be useful to develop a test of selective neutrality with a model that incorporates selection. In this way an alternative hypothesis can be specified and the data can be compared to an expectation both with and without the effects of selection. If the data fits neither model, then it becomes apparent that the models themselves are perhaps overly simplistic. Perhaps another way to make greater progress is to use more sequence information. When sequences are obtained from two species the levels of polymorphism can be compared with the levels of divergence. This is the idea behind McDonald & Kreitman's test (1991) and behind the popular HKA test (Hudson et al. 1987). The former has the appeal of a simple 2×2 contingency table while the latter compares levels of within species variation (measured by the number of segregating sites) and the between species divergence (measured by the number of differences) with their expectations based on a simple neutral model. This added information holds great promise that even more informative and more sensitive methods can be developed (Golding 1994).

7. Acknowledgments. The author gratefully thanks Dr. Simon Tavaré and a reviewer for their comments on this manuscript. GBG is a Fellow of the Evolutionary Biology Program of the Canadian Institute for Advanced Research and is supported by a Natural Sciences and Engineering Research Council of Canada grant.

REFERENCES

[1] Ewens, WJ, *The sampling theory of selectively neutral alleles*, Theor Pop Biol 3:87-112, 1972.

[2] Ewens, WJ, *Population genetics theory in relation to the neutralist-selectionist controversy*, Adv Hum Genet 8:67-134, 1977.

[3] Fu, Y-X and Li, W-H, *Statistical tests of neutrality of mutations*, Genetics 133:693-709, (1993).

[4] Gillespie, JH, *The causes of molecular evolution*, Oxford University Press, New York, N.Y., (1991).

[5] Golding, GB (ed) *Non-neutral evolution: Theories and molecular data*, Chapman and Hall, New York, NY,(1994).

[6] Golding, GB and Felsenstein, J, *A maximum likelihood approach to the detection of selection from a phylogeny*, J Mol Evol 31:511-523, (1990).

[7] Hartl, DL and Clark, AG, *Principles of Population Genetics*, 2nd edn. Sinauer Assoc. Inc, Sunderland, Mass., (1989).

[8] Hedrick, PW, Jain S, and Holden, L, *Multilocus systems in evolution*, Evol Biol 11:101-184, (1978).

[9] Hudson, RR, Kreitman, M, and Aguade, M, *A test of neutral molecular evolution based on nucleotide data*, Genetics 116:153-159, (1987).

[10] Kimura, M, *Evolutionary rate at the molecular level*, Nature 217:624-626 (1968).

[11] Kimura, M, *The neutral theory of molecular evolution*, Cambridge University Press, New York, N.Y., (1983).

[12] Lewontin, RC, and Krakauer, J., *Testing the heterogeneity of F values*, Genetics 80:397-398 (1975).

[13] McDonald J, and Kreitman, M. *Adaptive protein evolution at adh in Drosophila*, Nature 351:652-654, (1991).

[14] Prout, T, *The estimation of fitnesses from genotypic frequencies*, Evolution 19:546-551, (1965).

[15] Prout, T , *The estimation of fitnesses from population data*, Genetics 63:949-967, (1969).

[16] Selander, RK, (1976) *Genetic variation in natural populations* in:*Ayala F (ed) Molecular Evolution*, Sinauer, Sunderland, Massi (1976).

[17] Singh, RC, Lewontin, RC, Felton, A, *Genetic heterogeneity within electrophoretic "alleles" of xanthine dehydrogenase in Drosophila pseudoobscura*, Genetics 64:609-629 (1976).

[18] Tajima, F, *Statistical method for testing the neutral mutation hypothesis by DNA polymorphism*, Genetics 123:585–595, (1989).

[19] Takahata, N, *A simple genealogical structure of strongly balanced allelic lines and transspecies evolution of polymorphism*, Proc. Natl Acad Sci, USA 87:2419–2423, (1990).

[20] Watterson, GA, *The homozygosity test of neutrality*, Genetics 88:405–417 (1978).

[21] Watterson, GA, *Testing selection at a single locus*, biometrics 38:323–331, (1982).

HIERARCHICAL AND MEAN-FIELD STEPPING STONE MODELS*

D.A. DAWSON[†]

Abstract. The objective is to review some recent developments in the study of the long time behavior and spatial structure of a spatially distributed population with infinitely many types undergoing Fleming-Viot sampling, migration and selection. The methodology to be described includes the use of measure-valued diffusions and both mean-field and hierarchical mean-field limits.

Key words. Fleming-Viot process, stepping stone models, selection...

AMS(MOS) subject classifications. Primary 60H15; secondary 35R60

1. Introduction. Finite dimensional diffusion models have played an important role in the mathematics of population genetics involving finitely many alleles. Fleming and Viot (1979) introduced a class of measure-valued diffusion models of a population with infinitely many types which has subsequently developed into an important tool. One reason for this is that this setting is ideally suited to code more complex information such as genealogy, age of alleles, etc. (see Ethier and Kurtz (1992) for a recent survey of these develoments). The study of geographically structured populations was initiated by Wright (1931) and led to the development of stepping stone and island models. These models have been used to explore the applicability of Wright's three-phase shifting balance theory as well as for the analysis human mitochondria DNA ancestry (cf. Takahata (1991)). Until recently most of the work in this direction involved populations having finitely many types at each site. In this paper we review some recent work in which the population at each site is measure-valued. The motivation is that this provides a setting in which to study the spatial distribution of a given allele as well as to study the qualitative implications of selection in a spatially structured infinitely many alleles model.

2. Description of the infinitely many alleles stepping stone model . Consider a collection (finite or countable) of subpopulations (demes), indexed by S. The subpopulation at $\xi \in S$ at time t is described by a probability distribution $X_\xi(t)$ over a space $E = [0, 1]$ of posssible types (alleles). In other words, $X_\xi(t) \in M_1(E)$, the set of probability measures on E. Within each subpopulation there is mutation, selection and finite population sampling. Mutation is assumed to produce a new type chosen by sampling from a fixed source distribution $\theta \in M_1(E)$. Selection is prescribed by a fitness function $V(x)$ in the haploid case or by

---

* Supported by NSERC

† Department of Mathematics and Statistics, Carleton University, Ottawa, Canada K1S 5B6

$V(x, y) = V(y, x)$ in the diploid case. Migration from site ξ to site ξ' is assumed to occur at rate $q_{\xi,\xi'}$. Finally Fleming-Viot continuous sampling is assumed to take place within each subpopulation. It is a basic property of this model that for any $t > 0$, $X_\xi(t)$ is a purely atomic random measure (with countably many atoms) and therefore can be represented in the form

$$X_\xi(t) = \sum_{k \in I} m_{\xi,k}(t) \delta_{y_k}$$

where $m_{\xi,k}(t) \geq 0$ denotes the proportion of the population in subpopulation ξ of type $y_k \in E$. Note that in this model two individuals are related if and only if they are of the same type.

The questions which we wish to answer are:
  1. distribution in a given subpopulation, that is, what is the joint distribution of $\{m_{\xi,k}\}_{k \in I}$
  2. spatial distribution of relatives, that is, individuals of a given type, e.g.

$$\Psi_k(\{\xi_1, \ldots, \xi_r\}) = \frac{1}{r} \sum_{j=1}^{r} m_{\xi_j, k}$$

the proportion of type y_k in the combined subpopulations $\xi_1, \ldots, \xi_r$.
  3. how are these affected by the migration geometry.

In order to investigate the last two items we consider three different migration geometries:
  1. Euclidean stepping stone: $S = \mathbf{Z}^d$, $q_{\xi,\xi'} = \frac{1}{2d}$, $\xi \neq \xi'$
  2. Island model (mean-field): $S = \{0, \cdots, N-1\}$, $q_{\xi,\xi'} = 1/N$
  3. $S = \Omega_N$ = Hierarchical Group: (cf. Sawyer-Felsenstein)

$$\Omega_N = ((\xi_1, \xi_2, \ldots) : \xi_i \in \{0, \ldots, N-1\}, \xi_i = 0, \text{ a.a. } i)$$

$d(\xi, \xi') = \max\{i : \xi_i \neq \xi'_i\}$ *hierarchical distance*
$(d = 0, 1, 2, 3$ etc, individual, site, group, village, region, etc$)$

$$q_{\xi,\xi'} = \sum_{k=1}^{\infty} c_{k-1} N^{-2k+1} 1\{d(\xi', \xi) \leq k\}$$

Here c_k/N^k represents the rate of jumping k levels in the hierarchies.

3. Mathematical characterization of the process . At at given time t the process is described by the indexed family of random probability measures, $\{X_\xi(t)\}_{\xi \in S}$. Their joint distribution is completely determined by the joint moment measures, namely,

$$\int \cdots \int f(x_1, \cdots, x_n) M_{\xi_1, \cdots, \xi_n}(t; dx_1, \cdots, dx_n)$$

$$= E\left(\int_{[0,1]} \cdots \int_{[0,1]} f(x_1, \ldots, x_n) X_{\xi_1}(t, dx_1) \ldots X_{\xi_n}(t, dx_n)\right)$$

We also introduce an associated family of "test functions":

$$F((f, \pi), ((\mu_\xi)_{\xi \in S}) =$$
$$\int_{[0,1]} \cdots \int_{[0,1]} f(u_{\pi^n_k(1)}, \ldots, u_{\pi^n_k(n)}) \mu_{\tilde{\pi}_k(1)}(du_1) \ldots \mu_{\tilde{\pi}_k(k)}(du_k)$$

where $f \in C([0,1]^n)$, and $\pi = (\tilde{\pi}^n_k, \tilde{\pi}_k)$

$$\tilde{\pi}^n_k : \{1, \ldots, n\} \to \{1, \ldots, k\} \text{ and } \tilde{\pi}_k : \{1, \ldots, k\} \to S \cup \{\infty\}.$$

The mathematical characterization of the process is given as the unique solution of a martingale problem with generator L. In turn this martingale problem establishes a connection between the probability law of this process and that of a simpler process, called the dual process, which is closely related to the coalescent and directly gives the joint moment measures referred to above (cf. Dawson and Hochberg (1982), Vaillancourt (1990)).

We will begin with a description of the dual process $\pi(t)$ which has the form

$$\pi(t) = (\tilde{\pi}^n_k(t), \tilde{\pi}_k(t))$$

$$\tilde{\pi}^n_k : \{1, \ldots, n\} \to \{1, \ldots, k\} \text{ and } \tilde{\pi}_k : \{1, \ldots, k\} \to S \cup \{\infty\}$$

This means that $\tilde{\pi}^n_k$ is a partition of $\{1, \ldots, n\}$ and $\tilde{\pi}_k$ assigns locations to each element of the partition. The process evolves as follows

1. the partition elements perform continuous time symmetric random walks on S with rates $q_{\xi,\xi'}$ and in addition a partition element can jump to $\{\infty\}$ with rate c (once a partition element reaches ∞ it remains there without change or further coalescence).

2. each pair of partition elements during the period they reside at an element of S (but not $\{\infty\}$) coalesce to the partition element equal to the union of the two partition elements at rate d_0 Note that this is essentially equivalent to the coalescent geographically structured populations introduced by developed by Notohara (1990) and Takahata (1991).

3.1. Generator of the neutral stepping stone model. The generator of the neutral infinitely many types stepping stone model has the

usual form for a diffusion, namely a second order differential operator but
in this case involving the derivative

$$\frac{\delta F}{\delta \mu_\xi}(u) := \frac{d}{d\varepsilon} F(\mu + \varepsilon \cdot \delta_u)|_{\varepsilon=0}.$$

It is given by

$$L_\theta F((\mu_\xi)) = c \cdot \sum_{\xi \in S} \int_{[0,1]} \frac{\delta F}{\delta \mu_\xi}(u)(\theta(du) - \mu_\xi(du))$$

$$+b \cdot \sum_{\xi,\xi' \in S} q_{\xi,\xi'} \int_{[0,1]} \frac{\delta F}{\delta \mu_\xi}(u)(\mu_{\xi'}(du) - \mu_\xi(du))$$

$$+\frac{d}{2} \sum_{\xi \in S} \int_{[0,1]} \int_{[0,1]} \frac{\delta^2 F}{\delta \mu_\xi \delta \mu_\xi}(u,v)[\mu_\xi(du)\delta_u(dv) - \mu(du)\mu(dv)]$$

The first term corresponds to mutation with source distribution θ, the second to spatial migration and the last to continuous sampling. The sampling
rate coefficient d is inversely proportional to the effective population size
of a deme.

3.2. Dual process representation. For functions F which belong
to the special class of test functions defined above it can be shown that

$$LF(f,\pi)((\mu_\xi)) = KF(f,\pi)((\mu_\xi))$$

where K is the generator of a partition-valued dual $\pi(t)$ with $\pi(0) = (\tilde{\pi}_n^n, \tilde{\pi}_n)$. On the left hand side the operator L acts on the variables (μ_ξ)
and on the right hand side the operator K acts on the variable π.

It then follows (cf. Dawson (1993, section 5.5) that

$$E_{(\mu_\xi)_{\xi \in S}}(F((f,\pi(0)),X(t))) = E_{\pi(0)}(F_{f,\pi(t)}((\mu_\xi)_{\xi \in S})).$$

where $\pi(0) = (\tilde{\pi}_n^n(0), \tilde{\pi}_n(0))$ and $X_\infty(t) \equiv \theta$. Refer to Handa (1990) and
Vaillancourt (1990) for details on the existence and uniqueness of these
processes.

4. Long time behavior of island and stepping stone models.
In this section we review the known results on the long time behavior of
the neutral stepping stone and island models and then describe some preliminary results on the ergodic behavior of the island model with selection.

4.1. The neutral stepping stone model without mutation. In
this section we review the basic results relating the migration geometry to
the long-time behavior of the neutral stepping stone model.

THEOREM 4.1. *(a) If $c > 0$, then the process X has a unique invariant
measure ν_θ^c.*
*(b) If $c = 0$ and the migration random walk is transient, then the pro-
cess X has a one parameter family of nontrivial ergodic invariant measures
$(\{\nu_\mu\}_{\{\mu \in M_1(E)\}})$.*

(c) If $c = 0$ and the random walk is recurrent, the the extreme invariant measures are $\delta_a, a \in [0, 1]$, that is, there is - global fixation.

Refer to Sawyer (1976), Shiga (1980) for the case of finitely many alleles, Dawson, Greven and Vaillancourt (1994) for the measure-valued case. Recall that the nearest neighbour simple random walk in dimension d is transient if and only if $d \geq 3$. The hierarchical model with $c_k = r^k$ is transient if $r > 1$ and recurrent if $r \leq 1$.

QUESTION: Describe the structure of the invariant measure ν_μ. For example how many types are required to obtain "almost all" the mass in a region of radius R? i.e. How many different alleles do you expect to find when sampling in a region of radius R? In other words what is the distribution of the number of different alleles found when a sample is taken at random from a region of radius R. We will give some partial results in this direction in the hierarchical setting below.

4.2. Mean-field limit of the island model. We begin with island model with N "islands" denoted by X^N. Consider the population one particular island, ξ_0 as $N \to \infty$ and assume that at time $t = 0$ the $(x_\xi^N(0))_{\xi \in \{0, \ldots, N-1\}}$ are independent and identically distributed random measures with mean measure $\nu_0 \in M_1(E)$. Then it can be established that $X_{\xi_0}^N(t)$ converges weakly as $N \to \infty$ to the single site infinitely many alleles model with and mutation rate $c + b$ and time dependent mutation source $\tilde{\theta}(t) = \frac{c}{c+b}\theta + \frac{b}{c+b}E(X_{\xi_0}(t))$. This is a special case of the so-called mean-field or McKean-Vlasov limit of a system of exchangeable diffusions (see Gärtner (1988)) but extended to the infinite dimensional setting.

THEOREM 4.2. *Consider the infinitely many alleles model with mutation rate $c + b$, and time dependent mutation source $\tilde{\theta}(t) = \frac{c}{c+b}\theta + \frac{b}{c+b}E(X_{\xi_0}(t))$.*

1. *If $c \neq 0$, then there is a unique invariant probability measure $\Gamma_{\theta, \frac{d}{(c+b)}}$.*

2. *If $c = 0$, then there is a one parameter family of invariant probability measures $\{\Gamma_{\nu_0, \frac{d}{b}}\}_{\nu_0 \in M_1(E))}$.*

Here $\Gamma_{\nu_0, \gamma}$ is a probability measure on $M_1(E)$ given by the distribution of the random measure $\sum M_j \delta_{U_j}$, where

$$\{U_i\} \text{ are i.i.d. } \theta \text{ and } M_j = V_j \prod_{i=1}^{j-1}(1 - V_i)$$

where $\{V_j\}$ are i.i.d. and have densities

$$const \cdot (1 - x)^{(1-\gamma)/\gamma}, 0 \leq x \leq 1.$$

where $\gamma = d/c$ (recall that d and c are the sampling and mutation rates respectively). This is known as the GEM representation (cf. Ethier (1990)) of the random measure $\Gamma_{\nu_0, \gamma}$.

To verify this it suffices to note that the mean measure, ν_0 of an invariant probability measure must satisfy

$$\nu_0 = \frac{c}{c+b}\theta + \frac{b}{c+b}\nu_0$$

and then using the well-known results on the invariant measure for the infinitely many alleles model (e.g. Ethier (1990)).

4.3. The island model with selection. Now let us consider the island model with diploid selection with symmetric fitness function $V(x,y)$. The corresponding generator has the form

$$L_{\theta,V}^N F((\mu_j)\{j=0,...,N-1\}) = L_\theta^N F((\mu_j)\{j=0,...,N-1\}) + \sum_{j=0}^{N-1} < \frac{\delta V(\mu)}{\delta \mu_j}, \frac{\delta F}{\delta \mu_j} >_{\mu_j}$$

where $V(\mu) = \int\int V(x,y)\mu(dx)\mu(dy)$ and

$$< \frac{\delta V(\mu)}{\delta \mu_j}, \frac{\delta F}{\delta \mu_j} >_{\mu_j}$$
$$= \int\int\int [V(x,y)\tfrac{\delta F}{\delta \mu_j}(x) - V(y,z)\tfrac{\delta F}{\delta \mu_j}(x)]\mu_j(dx)\mu_j(dy)\mu_j(dz).$$

See Dawson (1993, Section 10.1.1) for the existence and uniqueness of the process with generator $L_{\theta,V}^N$.

The term $<,>_{\mu_j}$ can be interpreted as the "Riemannian inner product" at μ_j associated to the infinitely many alleles Fleming-Viot model (cf. Overbeck-Röckner-Schmuland (1993)). In other words the selection term involves a gradient with respect to a Riemannian structure. If we consider the special case of one site ($N = 1$) and E finite, this gradient, denoted by ∇_μ, coincides with the Shashahani gradient (cf. Hofbauer and Sigmund (1991)) which is used in the deterministic theory to prove Kimura's maximum principle, namely, that the system converges to a fixed point of maximal mean fitness. Therefore the gradient which appears here can be viewed as the infinite dimensional generalization of the Shashahani gradient. The corresponding infinite dimensional deterministic dynamics has the form

$$\frac{\partial \mu(t)}{\partial t} = \nabla_{\mu(t)} V(\mu(t)).$$

Moreover for the one site model with selection there is a reversible invariant probability measure

$$\Gamma_{\theta,V,\gamma}(d\mu) := Z^{-1} \exp(2V(\mu)/d)\Gamma_{\theta,\gamma}(d\mu).$$

The "Gibbs" form of this measure is also a consequence of the fact that the selection term is of gradient form (see Ethier and Kurtz (1992), Theorem 8.6).

We next consider the mean-field limit of the N colony system with mutation source θ and fitness V as $N \to \infty$.

THEOREM 4.3. *Let X^N denote the island model with N islands and fitness function V. Consider the population of one particular island, ξ_0, as $N \to \infty$ and assume that at time $t = 0$ the $(x_\xi^N(0))_{\xi \in \{0,...,N-1\}}$ are independent and identically distributed random measures with mean measure $\mu_0 \in M_1(E)$. Then $(X_{\xi_0}^N(t))_{t \geq 0}$ converges in distribution as $N \to \infty$ to the solution (of the weak form) of the following McKean-Vlasov equation for the probability law of a typical island, $P(t) \in M_1(M_1(E))$:*

$$(4.1) \qquad \frac{\partial P(t)}{\partial t} = (L^1_{\alpha\nu(t)+(1-\alpha)\theta,V})^* P(t), \quad P(0) = \delta_{\mu_0},$$

where $*$ denotes the adjoint, $\alpha := \frac{b}{b+c}$ denotes the relative rate of migration compared to mutation and $\nu(t) = \int \mu P(t, d\mu)$.

Equation 4.1 can be rigorously reformulated in a weak form and questions of existence and uniqueness studied (cf. Gärtner (1988) for the finite dimensional case and Dawson and Gärtner (1994) for the measure-valued case). In addition large deviation methods can be used to study the probability that the N-island models for large N will deviate from the mean-field limit dynamics (cf. Dawson and Gärtner (1994)).

We obtain the following self-consistent nonlinear equation for the fixed points of this equation, that is, the corresponding equilibrium distributions:

$$\nu_0 = Z^{-1} \int \mu \exp(2V(\mu)/d) \Gamma_{(\alpha\nu_0+(1-\alpha)\theta),\gamma}(d\mu)$$

where Z is a normalization factor.

Remarks 1. *Consider the case in which selection is additive, that is $V(x,y) = V(x) + V(y)$, $c = 1$, there is no mutation and the initial measure μ_0 is supported on finitely many alleles $\{x_j : j = 1 \cdots, J\}$ (with $\mu_0(x_j) > 0$). Then every solution of the mean field equilibrium is supported on $\{x_j : V(x_j) = \max V(x_i)\}$ and the mean fitness, $E(\int \int V(x)X_t(dx))$ monotonically increases to $\max V(x_i)$ (Dawson and Greven (1994)).*

2. Consider the case in which $V \in \{0,1\}$ and V is symmetric thus defining a graph with S as vertices. Then (Sigmund and Hofbauer (1991) section 23.5) showed that the stable fixed points in the deterministic case (i.e. infinite population) are in 1-1 correspondence with the cliques of the graph. Consider the mean field limit of the corresponding island model with small mutation rate. It is an interesting problem to describe the fixed points of the McKean-Vlasov equation and to determine whether these change as the sampling rate is varied, that is, whether or not "phase transitions" can occur.

3. Shiga and Uchiyama (1986) obtained results on the characterization of invariant measures for stepping stone models with selection in the case of two alleles. In particular their results imply that a type, $e \in E$ of sufficiently *high fitness can be maintained in the population even if the mutation rate is positive and $\theta(\{e\}) = 0$.*

4. *Kimura (1983) also introduced the notion of "intergroup selection" in the island model. In particular this includes the possibility of the death of the entire population of an island and its replacement by a colony of another island thus duplicating the allelic distribution of the latter. The relative probabilities determining which island produces the successsful colony depends on its fitness which is a function of the allelic distribution of the colony.*

5. Hierarchical mean field methodology. In this section we return to the neutral case and address the question of the spatial distribution of relatives in the neutral case. This turns out to be a rather diffficult question to answer for the general stepping stone model. An important advantage of the hierarchical model is that the methodology developed in Dawson and Greven (1993) can be applied leading to rigorous results in this setting which can also serve as a prototype for the more general case. In fact the qualitative results described in this section are expected to have analogues for the euclidean lattice case.

The results of section 4 on the island model imply that the empirical average distribution

$$N^{-1} \sum_{j=0}^{N-1} x_j^N(t)$$

is almost constant for times of order $o(N)$. We can view the island model as the first level in the hierarchical group model. The second level would then consist of N collections of N islands and in which effects of spatial migration between these collections occurs at a rate of order $\frac{1}{N}$. Thus we can view the hierarchical system as one in which macroscopic changes at level k take place over times of order N^k. In fact this heuristic picture can be made rigorous in the *hierarchical mean field limit (HMF)*, that is the limit of the hierarchical system as the parameter $N \to \infty$. This turns out to be a powerful tool and allows us give rigorous formulations of multiple time scale effects which in turn give insight into the behavior of the hierarchical systems even for N fixed and suggests conjectures for the behavior of lattice systems as well. We will now outline the main steps in this program which is carried out in detail in Dawson, Greven and Vaillancourt (1994).

The neutral hierarchical stepping stone model has generator

$$L_N F((\mu_\xi))$$
$$= \sum_{k=1}^{\infty} \frac{c_{k-1}}{N^{2k-1}} \sum_{d(\xi',\xi)\leq k} \int_{[0,1]} \frac{\delta F}{\delta \mu_\xi}(u)(\mu_{\xi'}(du) - \mu_\xi(du))$$
$$+ d_0/2 \sum_{\xi \in \Omega_N} \int_{[0,1]} \int_{[0,1]} \frac{\delta^2 F(\mu)}{\delta \mu_\xi \delta \mu_\xi}(u,v)[\mu_\xi(du)\delta_u(dv) - \mu_\xi(du)\mu_\xi(dv)]$$

The subset of Ω_N given by $\{\xi' : d(\xi, \xi') \leq k\}$ is called the *k-block* containing ξ. The first step involves the consideration of the *k-block averages* in the *natural time scale* $\{N^k t : t > 0\}$.

$$X^N_{\xi,k}(N^k t) = N^{-k} \sum_{\xi':d(\xi,\xi')\leq k} x^N_{\xi'}(N^k t) \text{ and } X^N_{\xi,0} := x^N_\xi \quad \forall \xi \in \Omega_N.$$

In the limit $N \rightarrow \infty$ we will show that
(i) the various time scales $t, tN, tN^2, \ldots$ *completely separate* and
(ii) there exists a sequence of *quasistationary states*, one for each level $k, k = 1, 2, 3, \ldots$

5.1. The mean-field limit in a fast time scale. Consider the one level process $x^N_{\xi,1}$.

THEOREM 5.1. *Let $c_k = 0$ for $k > 1$. Then as $N \longrightarrow \infty$*

$$\mathcal{L}(x^N_{\xi,1}(Ns))_{s\geq 0} \rightarrow \mathcal{L}(Z^{\theta,1}_s)_{s\geq 0}$$

where $Z^{\theta,1}$ is also a Fleming-Viot process (this is called the fixed point property of Fleming-Viot) with state space $M_1[0,1]$ and generator

$$\mathcal{G}^{\theta,1} F(\mu)$$
$$= c_1 \int_{[0,1]} \int_{[0,1]} \frac{\delta F}{\delta \mu} [\theta(dy) - \mu(dy)]$$
$$+ \int_{[0,1]} \int_{[0,1]} \frac{\delta^2 F(\mu)}{\delta \mu \delta \mu}(u, v) Q^1(\mu, du, dv)$$

$Q^1(\mu, dx, dy) = d_0 c_0/(d_0 + c_0)[\mu(du)\delta_u(dv) - \mu(du)\mu(dv)].$
This process can now be carried out inductively at each successive level.

5.2. HMF level k-diffusions and their equilibrium measures. $(Z^{\theta,k}_t)_{t>0}$ is a Fleming-Viot process with state space $M_1[0,1]$ and generator

$$\mathcal{G}^{\theta,k} F(\mu) = c_k \int_{[0,1]} \frac{\delta F}{\delta \mu}(u)(\theta(du) - \mu(du))$$
$$+ \int_{[0,1]} \int_{[0,1]} \frac{\delta^2 F}{\delta \mu \delta \mu}(u, v) Q^{\theta,k}(\mu; du, dv)$$
with $Q^{\theta,k}(\mu; du, dv) = d_k[\mu(du)\delta_u(dv) - \mu(du)\mu(dv)].$

$\Gamma^k_\theta(.) \in M_1(M_1([0,1]))$ is the unique equilibrium state of the process $Z^{\theta,k}_t$ called the *level k quasiequilibrium*.
The diffusion constants d_k are defined via

$$d_{k+1} = \frac{d_k c_k}{c_k + d_k}, \quad d_0 = 1.$$

THEOREM 5.2. **Multiple time scale behavior**

Consider $X^N(t) = ((x_\xi^N(t)_{\xi \in \Omega_N})$ *and let* $(x_\xi^N(0))_{\xi \in \Omega_N}$ *be i.i.d. random measures with mean* $E(x_\xi^N(0)) = \theta \in M_1[0,1]$.
For $k, j \in \mathbf{N}$ *we have the following behavior*
Case 1: $k > j$. *Law of large numbers*

$$\mathcal{L}((x_{\xi,k}^N(N^j t))_{t \in \mathbf{R}^+}) \Longrightarrow_{N \to \infty} \delta_\theta$$

Case 2: $k = j$. *Level* k *diffusion*

$$\mathcal{L}((x_{\xi,k}^N(N^k t)_{t \in \mathbf{R}^+}) \Longrightarrow_{N \to \infty} \mathcal{L}((Z_t^{k,\theta})_{t \in \mathbf{R}^+})$$

Case 3: $k < j$ *Level* j *quasiequilibrium*

$$\mathcal{L}((x_{\xi,k}^N((N^j)^{1+\delta} s + N^k t)_{t \in \mathbf{R}^+}) \Longrightarrow_{s,N \to \infty} \mathcal{L}((Z_t^{k,\theta^*})_{t \in \mathbf{R}^+})$$

where $0 < \delta < 1$ *and* $\mathcal{L}(\theta^*) = \mu_\theta^{j,k}(\cdot)$ *which is defined below.*

5.3. HMF interaction markov chain and global equilibrium .
Given $\{c_k\}$ we have a reverse time inhomogeneous $M_1[0,1]$-valued Markov chain with transition function $\Gamma_\theta^k(d\theta')$ at time, k.
For $k > j$, we have the $(k - j)$ step transition on $M_1[0,1]$

$$\mu_\theta^{k,j}(.) = \int_{M_1[0,1]} \cdots \int_{M_1[0,1]} \Gamma_\theta^k(d\theta_1)\Gamma_{\theta_1}^{k-1}(d\theta_2)\ldots\Gamma_{\theta_{k-j+1}}^{j+1}(.),$$

This describes the distribution of the j-block average in time scale $N^k t$ with constant initial condition given by $\theta \in M_1(E)$.

Definition: A hierarchical mean field *global equilibrium* is an entrance law for this Markov chain which describes the macroscopic behavior in exponential (or superpolynomial) time scales.

Global Equilibrium and Spatial Ergodic Theorem
1. *There is a one parameter family* $\{\mu_\theta^{\infty,j}\}$ *with* $\theta \in M_1[0,1]$ *of nontrivial extremal entrance laws for the interaction chain (global equilibria) if and only if the transience condition for the migration random walk is satisfied. Otherwise the only extremal entrance laws are* $\{\delta_{\delta_a}, a \in [0,1]\}$.
2. *Under the global equilibrium with mean measure* θ, $\lim_{k \to \infty} x_{\xi,k} = \theta$ *a.s.*

5.4. Spatial distribution of alleles.
Each global equilibrium defines the distribution $\mu_\theta^{\infty,j}$ which is a probability measure on $M_1[0,1]$ supported by the set of purely atomic measures with countably many non-zero

atoms (i.e. countably many coexisting *alleles*). We now consider the spatial distribution of these alleles.

Note that countably many distinct alleles survive and even every block contains these countable many alleles due to the long range jumps. Recall that the rate of jumping k levels in the hierarchy is $\frac{c_k}{N^k}$.

Let $(M_j^{(k)})_{j=1,2,\ldots}$ denote the *size ordered* masses of the atoms in a k-block. In other words, M_1^k is the mass of the most frequent allele in a k-block, M_2^k is the mass of the second most frequent allele, etc.

THEOREM 5.3. **Number of Alleles occupying a Spatial Area.** *Let* $S_\ell^{(k)} = \sum_{j=0}^\ell M_j^{(k)}$. *Then*

$$S_{n(k)}^{(k)} \longrightarrow 1 \quad \text{if} \quad n(k)/c_k \to \infty$$

$$S_{n(k)}^{(k)} \longrightarrow 0 \quad \text{if} \quad n(k)\sum_{\ell=k}^\infty 1/c_\ell \to 0$$

For example, if $c_k = c^k$ with $c > 1$, then $S_{c^{(1+\varepsilon)k}}^{(k)} \to 1$ if and only if $\varepsilon > 0$. This means that the types of almost all the subpopulation in a given k-block belong to a subset of at most $c^{(1+\varepsilon)k}$ distinct alleles if $\varepsilon > 0$.

The interaction Markov chain can also be used to compute a number of other quantities of interest, for example, to obtain approximations to the distribution of the population mass of the most frequent allele in a k-block.

REFERENCES

[1] D.A. DAWSON, *Measure-valued Markov processes* in Ecole d'Été de Probabilités de Saint Flour XXI - 1991, ed. P.L. Hennequin, Lecture Notes in Mathematics 1541, pp. 1–260, Springer, Berlin, Heidelberg, New York (1993).

[2] D.A. DAWSON AND J. GÄRTNER, *Multilevel large deviations and interacting diffusions II*, in preparation (1994).

[3] D.A. DAWSON AND A. GREVEN, *Hierarchical models of interacting diffusions: multiple time scale phenomena, phase transition and cluster of pattern formation*, Prob. Th. Rel. Fields, 96, pp. 435–473, (1993).

[4] D.A. DAWSON, A. GREVEN AND J. VAILLANCOURT, *Equilibria and quasiequilibria for infinite collections of interacting Fleming-Viot processes*, preprint, (1994).

[5] D.A. DAWSON AND A. GREVEN, *Hierarchically interacting Fleming-Viot processes with selection: Multiple space-time scale analysis and quasi equilibria*, in preparation, (1994).

[6] D.A. DAWSON AND K.J. HOCHBERG, *Wandering random measures in the Fleming-Viot model*, 10, pp. 693–703 (1982).

[7] P. DONNELLY AND T.G. KURTZ, *A countable representation of the Fleming-Viot measure-valued diffusion*, preprint (1994).

[8] S.N. ETHIER, *On the stationary distribution of the neutral one-locus diffusion model in population genetics*, Ann. Appl. Prob., 2, pp. 24–35, (1992).

[9] S.N. ETHIER AND T.G. KURTZ, *On the statioary distribution of the neutral diffusion model in population genetics*, Ann. Appl.Probab., 2, pp. 24–35, (1992).

[10] S.N. ETHIER AND T.G. KURTZ, *Fleming-Viot processes in population genetics*, SIAM J. Control Opt., 31, pp. 345–386, (1992).

[11] W.H. FLEMING AND M. VIOT, *Some measure-valued Markov processes in population genetics theory*, Indiana Univ. Math. J., 28, pp. 817–843 (1979).

[12] J. GÄRTNER, *On the McKean-Vlasov limit for interacting diffusions*, Math. Nachr., 137, pp. 197–248, (1988).

[13] K. HANDA, *A measure-valued diffusion process describing the stepping stone model with infinitely many alleles* , Stoch. Proc. Appl., 36, pp. 269–296, (1990).

[14] J. HOFBAUER AND K. SIGMUND, *The Theory of Evolution and Dynamical Systems*, Cambridge Univ. Press, Cambridge, (1991).

[15] M. KIMURA AND G.H. WEISS, *The stepping stone model of population structure and the decrease of genetics correlation with distance*, Genetics, 49, pp. 561–576 (1964).

[16] M. KIMURA, *Diffusion model of intergroup selection, with special reference to evolution of an altruistic character*, Proc. Nat. Acad. Sci. USA, 80, pp. 6317–6321, (1983).

[17] M. KIMURA, *Diffusion model of population genetics incorporating group selection, with special reference to an altruistic trait*, in Stochastic Processes and their Applications, Eds. K. Itô and T. Hida, Lecture Notes in Math. 1203, Springer-Verlag, pp. 101–118, (1986).

[18] M. NOTOHARA, *The coalescent and the genealogical process in geographically structured population*, J. Math. Biol., 29, pp. 59–75 (1990).

[19] Y. OGURA AND N. SHIMAKURA, *Stationary solutions and their stability of Kimura's diffusion model with intergroup selection*, J. Math Kyoto Univ.

[20] L. OVERBECK, M. RÖCKNER AND B. SCHMULAND, *An analytic approach to Fleming-Viot processes*, Ann. Probab, (1993).

[21] S. SAWYER, *Results for the stepping stone model for migration in population genetics*, Ann. Probab. , 4, pp. 699–726, (1976).

[22] S. SAWYER AND FELSENSTEIN, *Isolation by distance in a hierarchically clustered population*, J. Appl. Prob., 20 , pp. 1–10, 1983).

[23] T. SHIGA, *An interacting system in population genetics*, J. Math. Kyoto Univ., 20, pp. 213–242 (1980).

[24] T. SHIGA, *Mathematical results on the stepping stone model in population genetics*, in Population Genetics and Molecular Evolution T. Ohta and K. Aoki, eds., pp. 267–279, Japan Sci. Soc. Press and Springer-Verlag, Tokyo and Berlin, (1985).

[25] T. SHIGA, *Existence and uniqueness of solutions for a class of non-linear diffusion equations*, J. Math. Kyoto Univ., 27, pp. 195–215, (1987).

[26] T. SHIGA AND K. UCHIYAMA, *Stationary states and their stability of the stepping stone model involving mutation and selection*, Probab. Th. Rel.Fields, 73, pp. 87–117, (1986).

[27] N. TAKAHATA, *Genealogy of neutral genes and spreading of selected mutations in a geographically structured population*, Genetics, 129, pp. 585–595.

[28] J. VAILLANCOURT, *Interacting Fleming-Viot processes*, Stoch. Proc. Appl., 36, pp. 45–57, (1990).

[29] S. WRIGHT, *Evolution in Mendelian populations*, Genetics, 16, pp. 97–159, (1931).

A NOTE ON THE STEPPING STONE MODEL WITH EXTINCTION AND RECOLONIZATION

CLAUDIA NEUHAUSER[*], STEPHEN M. KRONE[†], AND HYUN-CHUNG KANG[‡]

1. Introduction. An important issue in population genetics is to explain the amount of genetic diversity in nature. Simple models can frequently assist in obtaining a better understanding of observed phenomena. In population genetics, the Wright-Fisher model serves such a function. It describes a population in which gene frequencies change due to random mating, mutation and selection. The Wright-Fisher model ignores the fact that natural populations are frequently subdivided into local populations. A first step to include this important feature is to allow for subdivision but ignore the geographical structure. Two popular models which incorporate this are Wright's (1931) "island" model and the "metapopulation" model of Levins (1970). In each of these, the population is usually divided into a finite number of colonies. In the first model, migrants come from an external source and choose islands uniformly from among all local populations. In the second model, migrants also choose their locations randomly, but this time the migrants themselves come from the overall population. One step further is to give a geographical structure to the local populations. This was done by Malécot (1969) and Kimura and Weiss (1964) in the stepping stone model, the topic of our discussion.

The stepping stone model is now widely used in population genetics to describe the evolution of a population with mating and geographical structure. It is typically formulated on a countable set, where each element of the set corresponds to a colony. Each colony consists of a population of N haploid individuals. For each such individual, we are interested in a single locus on a single chromosome at which two alleles (or genes), A_1 and A_2, can occur. Changes in the gene frequencies are caused by migration between colonies, in addition to mutation, selection and random mating within each colony.

Models for the evolution of the genetic composition of populations in subdivided habitats, such as island models, metapopulation models, or stepping stone models, frequently assume that local populations persist indefinitely. This is a reasonable assumption only if the average time until

---

[*] Department of Mathematics, 480 Lincoln Drive, University of Wisconsin, Madison, WI 53706.

Partially supported by the National Science Foundation under grant DMS-9206139. The first author also wishes to acknowledge support from the Institute for Mathematics and its Applications during the winter and spring quarter 1994.

[†] Center for the Mathematical Sciences, 1308 W. Dayton, University of Wisconsin, Madison, WI 53715-1149.

[‡] Department of Forestry, Russel Laboratories, University of Wisconsin, Madison, WI 53706.

extinction of a local population is much larger than the time interval over which the whole population is studied. Several studies [see, e.g., Simberloff and Wilson (1969) or Crowell (1973)] found that local extinctions may in fact be quite frequent in nature. Since extinction and recolonization may affect the population structure, it is important to incorporate these features into theoretical models, as was pointed out by McCauley (1991).

It is not obvious how extinction and recolonization affect the spatial distribution of gene frequencies. Opposing views can be found in the literature. Based on the loss of genetic diversity that a population experiences when going through small numbers of breeding adults ("bottlenecks" or "founder events"), S. Wright (1940) suggested that extinction and recolonization would enhance genetic differentiation. The universality of this claim was contradicted by Slatkin (1977) who investigated the effect of extinction and recolonization on Wright's island model with different modes of migration. He found situations in which extinction and recolonization acted in a way similar to gene flow and thus diminished genetic differentiation. Wade and McCauley (1988) and Whitlock and McCauley (1990) continued to study variations of Slatkin's models and confirmed Slatkin's observations: Depending on the mode of migration and the number of "founders" of new colonies, extinction can either enhance or diminish genetic differentiation. A clear and short summary on this controversy can be found in McCauley (1991). We note that none of the above models have a geographical structure.

Maruyama and Kimura (1980) investigated the effect of extinction and recolonization in both the island model and the stepping stone model. They studied an infinite-alleles model with mutation in which the population consists of a finite number of demes (sites) each with a finite carrying capacity. In both of their models, recolonization of a site occurs immediately after extinction of a colony at that site. In the island model, a certain number of colonizing individuals are chosen from among all existing colonies with equal probabilities; in their stepping stone model, colonizing individuals are only chosen from neighboring colonies. The new colony then grows to carrying capacity at a finite rate. Both their theoretical results and computer simulations showed that extinction increases the probability that two randomly chosen genes, each one from a different colony, are identical by descent.

Mathematical results on both the discrete-time and the continuous-time stepping stone model are quite numerous. We will only mention a few here and point out later how these results are related to ours. Sawyer (1976, 1977, 1979) investigated the discrete-time stepping stone model on $\mathbf{Z}^d$. He was particularly concerned with the rate at which clusters of individuals of common origin grow in the selectively neutral case with or without mutation. Clustering rates were also the focus of two articles by Cox and Griffeath (1987, 1990) who studied a simplified version of the continuous-time stepping stone model (the so-called multitype voter model). Their main result says that the density of the different alleles in appropriately scaled boxes converges to a time change of the Wright-Fisher diffusion. In a series of papers, Shiga obtained results on the stepping stone model via diffusion approximations [see Shiga (1985) for a summary of these and related results]. Itatsu (1985, 1989) investigated equilibrium measures of the discrete-time stepping stone model on a countable set with mutation and selection. There are, however, no rigorous results on the stepping stone model with extinction and recolonization.

In this note, we formulate a stepping stone model with extinction and recolonization as an interacting particle system and point out some of the differences that occur when one allows for extinction. The proofs will be presented elsewhere [cf. Kang, Krone, and Neuhauser (1994)].

2. The model. We now carefully define the model we wish to investigate. Let the d-dimensional integer lattice, $\mathbf{Z}^d$, be the set of possible locations of colonies. Each existing colony has a population of N genes evolving according to a continuous-time Moran model. (The Moran model is a continuous-time analogue of the usual Wright–Fisher model.) In addition, we allow for extinction of colonies and migration from neighboring colonies.

More explicitly, we consider a continuous-time Markov process $\xi_t = (\eta_t, \zeta_t)$, $t \geq 0$, where $\eta_t : \mathbf{Z}^d \to \{0, 1\}$ is the occupancy process and $\zeta_t : \mathbf{Z}^d \to \{1, 2\}^N \cup \Upsilon$ gives the allele compositions of the colonies. If $\eta_t(x) = 0$ (resp., 1), then we say that the colony at $x \in \mathbf{Z}^d$ is vacant (resp., occupied) at time t. If $\eta_t(x) = 1$, then $\zeta_t(x) = (\zeta_t(x, 1), \ldots, \zeta_t(x, N))$ gives the alleles of the N individuals in colony x at time t, 1 for allele A_1 and 2 for allele A_2; if $\eta_t(x) = 0$, then set $\zeta_t(x) = \Upsilon$. The symbol Υ is just used to denote a "cemetery state" for ζ_t; think of it as a vector of 0's if you prefer. The A_1-allele frequency at time t in an occupied colony x is denoted $\rho_t(x)$. Note that $\frac{1}{N} \sum_{k=1}^N (2 - \zeta_t(x, k)) = \rho_t(x)$. The dynamics of the process are as follows:

(i) For each pair (x, y) with $\|x - y\|_1 = 1$ (i.e., nearest neighbors), if x is occupied at time t, then at rate 1 the colony at x sends an exact copy of its gene composition, i.e., $\zeta_t(x)$, to y. If y is vacant when this happens, then y becomes occupied and $\zeta_t(y) = \zeta_{t-}(x)$. If y is already occupied, then for each k, $k = 1, 2, \ldots, N$, an independent coin is tossed with probability $\theta \in [0, 1]$ of heads. If heads comes up, set $\zeta_t(y, k) = \zeta_{t-}(x, k)$; if tails comes up, $\zeta_t(y, k) = \zeta_{t-}(y, k)$.

(ii) If x is occupied, then it becomes vacant at rate $\gamma(x)$. In general, this extinction rate may depend on the location and the gene composition of the colony.

(iii) If x is occupied, then a selection/mating event at that location occurs at rate $N\lambda$. Selection occurs first, immediately followed by mating in which a single individual in the colony is chosen to give birth to a single offspring of the same type, and a compensating death occurs. The relative fitnesses of A_1 and A_2 in colony x may depend on x in general, say $1 + s_1(x)$ and $1 + s_2(x)$, respectively, with $s_i(x) \in [0, \infty)$. If the A_1-allele frequency at x is $\rho(x)$, then, after taking selection into account, its weighted frequency is

$$(2.1) \qquad \sigma(\rho(x)) = \frac{(1 + s_1(x))\rho(x)}{(1 + s_1(x))\rho(x) + (1 + s_2(x))(1 - \rho(x))}.$$

Mating is defined as follows: At the time t of a selection/mating event, a single individual $k \in \{1, \ldots, N\}$ is randomly chosen and a coin is tossed, with probability $\sigma(\rho_{t-}(x))$ of heads. If heads comes up, the value of $\zeta_t(x, k)$ is replaced by 1 at time t; if tails comes up, it is replaced by 2 at time t. All other individuals are unaffected.

Step (i) describes migration to neighboring colonies. If a colony is vacant, it can be recolonized from a neighboring "parent" colony and this new colony has the same gene composition as the parent colony. Already occupied colonies can receive genes from neighboring colonies. The parameter θ is a measure of how successfully the invading population can replace members of the original population. Note that migration here is somewhat different from the usual stepping stone model in that colonies do not actually migrate (i.e., swap positions with neighboring demes) but send offspring to neighboring sites. This leaves the parent colony unchanged. Also note that by rescaling time, the migration rate can be set equal to one. Step (ii) describes the extinction mechanism. In this note, we will treat two different cases. In the case of a homogeneous environment, we will set $\gamma(x) = \Delta$, for all $x \in \mathbf{Z}^d$, regardless of the gene composition of the colony at x. We will also treat the case of a periodic environment. We call site x a δ-colony if the colony at x dies at rate δ if $\rho(x) \geq 1/N$ and at rate Δ if $\rho(x) = 0$. All other sites are called Δ-colonies. The death rate on a Δ-colony is Δ, regardless of the gene composition of the colony. In the periodic environment, δ-colonies are placed D units apart and the remaining lattice sites are filled with Δ-colonies. Step (iii) describes the selection and mating mechanism of a continuous-time Moran model. Mating here is simply resampling: The single individual chosen at the time of selection/mating is replaced by another individual. The replaced individual can be thought of as being killed. In this way, the population size of a colony is always N. (In discrete-time models, this could be replaced by the Wright–Fisher model in which all N individuals "mate" at the same time.) In the homogeneous environment, the selection coefficients do not depend on the location. In the periodic environment, the selection coefficients may depend on the location: If x is a δ-colony, then we set $s_1(x) = s_1$ and $s_2(x) = 0$; if x is a Δ-colony, we set $s_1(x) = 0$ and $s_2(x) = s_2$, where s_1 and s_2 are nonnegative constants. In other words, A_1 has a selective advantage (or, at least no disadvantage) on δ-colonies and A_2 has a selective advantage on Δ-colonies. Note that when $\rho_t(x) = 1$ (resp., 0), fixation of allele A_1 (resp., A_2) will have occurred. Since there is no mutation in our model, this fixation can only be interrupted by gene flow (migration) from neighboring sites. We call the above process a *stepping stone model with extinction and recolonization*.

Note that (i) and (ii) imply that η_t is a basic contact process in a homogeneous or periodic environment [cf. Liggett (1985) or Durrett (1988) for results on the basic contact process]. Later, we will need the notion of the *critical death rate* for a basic contact process in a homogeneous environment. Consider a contact process in which particles die at rate δ and give birth on neighboring vacant sites at rate 1. Since this process is attractive, starting from a configuration of all 1's, there is a unique stationary measure ν_1. The all 0 configuration is a trap. It is known that there is a unique number, $\delta_c > 0$, depending on the dimension, for which ν_1 is the point mass on the configuration of all 0's when $\delta \geq \delta_c$, and is a nontrivial invariant measure when $\delta < \delta_c$. It is clear that, to get any interesting behavior in our model, we will need both extinction rates Δ and δ to be less than δ_c.

Using the standard tool of graphical representation [Harris (1974)], this process can be constructed starting from any initial configuration.

3. Results. Having defined the model, we will now present the results. Let $\mathcal{P}(\mathcal{X})$ (resp., $\mathcal{S}$) denote the set of probability measures (resp., translation invariant probability measures) on the space of configurations,

$$\mathcal{X} \equiv (\{1,2\}^N \cup \Upsilon)^{\mathbf{Z}^d},$$

and let μ_1 (resp., μ_2) $\in \mathcal{P}(\mathcal{X})$ be the limit law corresponding to initial distribution δ_1 (resp., δ_2) $\in \mathcal{P}(\mathcal{X})$, which puts 1's (resp., 2's) at each $(x, k) \in \mathbf{Z}^d \times \{1, \dots, N\}$. These invariant measures exist because the particle system starting with all 1's (resp., all 2's) is attractive. Note that, due to the absence of mutation, once the process is void of 1's or 2's, it will remain so. In addition, let $\pi_\beta \in \mathcal{P}(\mathcal{X})$ denote the law which assigns values $\{1, 2\}$ in an i.i.d. manner to the coordinates (x, k), with probability β for 1 and $1 - \beta$ for 2. We write ζ_t^μ for the process ζ_t with initial distribution $\mu \in \mathcal{P}(\mathcal{X})$; when $\mu = \pi_\beta$, we simply write ζ_t^β. If $\mu, \nu \in \mathcal{P}(\mathcal{X})$, then we use the notation $\zeta_t^\mu \Rightarrow \nu$ to denote the weak convergence $\mu S_t \Rightarrow \nu$, where μS_t is the law of ζ_t^μ. (Note that weak convergence here is the same as convergence of finite dimensional distributions.)

We remark that, throughout the paper, the colony size, N, and the mating rate, λ, are fixed, strictly positive numbers.

The first theorem considers the case of a homogeneous environment in which one of the alleles has a selective advantage. Not surprisingly, the allele with the selective advantage takes over.

THEOREM 3.1. *Suppose* $\gamma(x) = \Delta$ $(0 \leq \Delta < \delta_c)$ *for all* $x \in \mathbf{Z}^d$. *Further assume that, for each* x, $s_1(x) \equiv s_1$ *and* $s_2(x) \equiv s_2$, *with* $s_1 < s_2$. *Then the "1's die out". That is, if* ζ_0 *is translation invariant and* $P(\zeta_0(j, k) = 2) > 0$, *then* $\zeta_t \Rightarrow \mu_2$.

The key to proving Theorem 3.1 is duality. We will give more details after the next theorem. Our next result deals with the case of selectively neutral alleles in a homogeneous environment.

THEOREM 3.2. *Suppose* $\gamma(x) = \Delta$ $(0 \leq \Delta < \delta_c)$ *and* $s_1(x) = s_2(x) = 0$, *for all* $x \in \mathbf{Z}^d$.
(a) *If* $d \leq 2$, *then clustering occurs: For any initial configuration* ζ_0,

$$P(\zeta_t(x, j) = 1, \zeta_t(y, k) = 2) \to 0 \qquad \text{as } t \to \infty,$$

holds for all $(x, j) \neq (y, k) \in \mathbf{Z}^d \times \{1, \ldots, N\}$. *Furthermore, if* $\mu \in \mathcal{S}$, *then there is a constant* $q \in [0, 1]$ *such that*

$$\zeta_t^\mu \Rightarrow q\mu_1 + (1 - q)\mu_2$$

as $t \to \infty$.

(b) *If* $d \geq 3$, *then coexistence is possible: For any* $\beta \in (0, 1)$, *there is a unique translation invariant stationary distribution* ν_β, *such*

$$\zeta_t^\beta \Rightarrow \nu_\beta$$

as $t \to \infty$; ν_β *gives* 0 *probability to the set of configurations which do not have positive proportions of both 1's and 2's.*

Comparing this theorem with Sawyer (1976), we see that the dichotomy between clustering and coexistence is the same, with and without extinction. Without extinction, the model closely resembles the voter model [see, e.g., Liggett (1985) or Durrett (1988)]; with extinction, it can be viewed as a mixture of the voter model and the multitype contact process [see Neuhauser (1992) for results on the multitype contact process]. In both models, this dichotomy between clustering in $d \leq 2$ and coexistence in $d \geq 3$ can be observed. Duality is the key to the proof. The dual process provides ancestral information by looking at the process in reverse time. Roughly speaking, the duals exhibit the same behavior as coalescing random walks. Two individuals are of the same type if their duals coalesce and may be of different type if they do not coalesce. The dichotomy can thus be explained by the recurrence of random walks in $d \leq 2$ and transience in $d \geq 3$. The proofs of both theorems rely heavily on Neuhauser (1992) in which similar results for the multitype contact process were proved.

Even though the model with extinction does not exhibit qualitatively different behavior compared to the one without extinction, the next result will show that extinction causes a change in the rate of clustering. We show that extinction speeds up clustering when $d = 1$. To do this, we set

(3.1)
$$g_t(x, y; \Delta) = \max_{1 \leq j, k \leq N} P_\Delta(\zeta_t(\lfloor x\sqrt{t} \rfloor, j) = 1,$$

$$\zeta_t(\lfloor y\sqrt{t} \rfloor, k) = 2 \mid \zeta_t(\lfloor x\sqrt{t} \rfloor, j) > 0, \zeta_t(\lfloor y\sqrt{t} \rfloor, k) > 0)$$

for $x, y \in \mathbf{Z}$ and $\Delta \geq 0$, where the subscript Δ refers to the extinction probability in each colony and $\lfloor \cdot \rfloor$ denotes the integer part.

THEOREM 3.3. *Assume the neutral setting of Theorem 3.1, with $d = 1$ and initial distribution π_β, $\beta \in (0, 1)$. Then, for any $\Delta \in (0, \delta_c)$ and $x \neq y$ in $\mathbf{Z}$, there exists $\theta_0 > 0$ such that $\theta \in (0, \theta_0)$ implies*

$$g_t(x, y; 0) > g_t(x, y; \Delta)$$

for all sufficiently large t.

The proof of this result, again, uses duality. This time, one makes use of the fact that increasing the death rate Δ makes it easier for colonies to invade neighboring sites, since the density of occupied sites is decreasing in Δ and it is much easier to colonize a vacant site than an occupied site for small θ. (Recall that a coin with probability θ of heads is tossed to decide whether an allele can successfully invade a neighboring site.) Theorem 3.3 and, to a larger extent, its proof show that the probability of two sampled individuals (from different colonies) being identical by descent is higher with extinction. So, in our setting, extinction enhances genetic differentiation.

The main point of this note is to illustrate some of the differences between stepping stone models with and without extinction. In a homogeneous environment, the differences appear only on a quantitative level; qualitatively, the systems behave quite similarly. In an inhomogeneous environment, the behavior can change drastically as illustrated in the following theorem. It concerns the behavior of the above stepping stone model in the one dimensional periodic environment defined in Section 2, in which the A_1 alleles have a strong selective advantage on δ-colonies and the A_2 alleles have a strong selective advantage on Δ-colonies. That is, we place δ-colonies at $x = 0, \pm D, \pm 2D, \ldots$, where D is some positive integer, and Δ-colonies at the remaining lattice points. On δ-colonies, put $s_1(x) = s_1$ and $s_2(x) = 0$; on Δ-colonies, put $s_1(x) = 0$ and $s_2(x) = s_2$. The death rate of colonies with at least one A_1-allele (i.e., $\rho > 0$) is δ on a δ-colony and Δ on a Δ-colony. All other death rates are Δ.

THEOREM 3.4. *Assume the periodic environment defined above with $\delta = 0$ and suppose ζ_0 is translation invariant.*

(i) Let $\Delta = 0$ and fix $N \geq 1$. Assume initially that there are infinitely many Δ-colonies with $\rho(x) < 1$. Then, for any $D \geq 3$, $s_1 \geq 0$, and

$\theta \in (0,1)$, *the A_1 alleles die out with probability one for sufficiently large s_2 and λ.*

(ii) Let $\Delta > 0$ and $D > 0$ be fixed. Assume initially that there are infinitely many δ-colonies with $\rho(x) > 0$. For any $s_2 \geq 0$ and $\theta \in (0,1)$, there exists $N_0 = N_0(\theta, D, \Delta)$ so that for all $N \geq N_0$, the A_2 alleles die out with probability one for sufficiently large s_1 and λ.

The first part of the theorem is not surprising. Here, there is no extinction on any site, and the heterogeneous environment gives strong selective advantage to different alleles on different sites. The A_2 alleles, however, have a "spatial advantage" since there are more sites where A_2 alleles are favored over A_1 alleles than vice versa. The second part says that this "spatial advantage" can be outweighed by a "survival disadvantage" due to extinction.

REFERENCES

[1] Cox, J.T. and D. Griffeath, *Recent results for the stepping stone model*, pp. 73-83. In *Percolation Theory and Ergodic Theory of Infinite Particle Systems*, Springer-Verlag, New York (1987).

[2] Cox, J.T. and D. Griffeath, (1990) *Mean field asymptotics for the planar stepping stone model*, Proc. London Math. Soc., **61**, pp. 189-208 (1990).

[3] Crowell, K.L. , *Experimental zoogeography: introductions of mice onto small islands*, Amer. Nat. **107**, pp. 535-558 (1973).

[4] Durrett, R., *Lecture Notes on Particle Systems and Percolation*, Wadsworth (1988).

[5] Itatsu, S., *Equilibrium measures of the stepping stone model with selection in population genetics*, pp. 257-266, In *Population Genetics and Molecular Evolution*, Ohta, T. and Aoki, K., eds. Japan Sci. Soc. Press, Tokyo/Springer, Berlin (1985).

[6] Itatsu, S., *Ergodic properties of the stepping stone model*, Nagoya Math. J., **114**, pp. 143-163 (1989).

[7] Kang, H., S. Krone, and C. Neuhauser, *Stepping stone models with extinction and recolonization*, Preprint (1994).

[8] Levins, R., *Extinction In Some Mathematical Problems in Biology* (M. Gerstenraber, ed.), Amer. Math. Soc. **2**, 75-108 (1970).

[9] Liggett, T. *Interacting Particle Systems*, Springer-Verlag (1985).

[10] Kimura, M. and G. H. Weiss, *The stepping stone model of population structure and the decrease of genetic correlation with distance*, Genetics, **49**, pp. 561-576 (1964).

[11] Malécot, G. *The Mathematics of Heredity*, English translation, W.H. Freeman, San Francisco, (1969).

[12] Maruyama, T. and M. Kimura, *Genetic variability and effective population size when local extinction and recolonization of subpopulations are frequent*, Proc. Natl. Acad. Sci. USA, **77**, pp. 6710-6714 (1980).

[13] McCauley, D.E., *Genetic consequences of local population extinction and recolonization*, TREE, **6**, pp. 5-8 (1991).

[14] Neuhauser, C., *Ergodic theorems for the multitype contact process*, Probab. Theory Rel. Fields, **91**, pp. 467-506 (1992).

[15] Sawyer, S., *Results for the stepping stone model for migration in population genetics*, Ann. Probab., **4**, pp. 699-728 (1976).

[16] Sawyer, S. *Rates of consolidation in a selectively neutral migration model*, Ann. Probab., **5**, pp. 486-493 (1977).

[17] Sawyer, S., *A limit theorem for patch sizes in a selectively-neutral migration model*, J. Appl. Probab., **16**, pp. 482-495 (1979).

[18] Shiga, T., *Mathematical results on the stepping stone model in population genetics*, pp. 267-279. In *Population Genetics and Molecular Evolution*, Ohta, T. and Aoki, K., eds. Japan Sci. Soc. Press, Tokyo/Springer, Berlin (1985).

[19] Simberloff, D.S. and E.O. Wilson, *Experimental zoogeography of islands: the colonization of empty islands*, Ecology, **50**, 278-296 (1969).

[20] Slatkin, M., *Gene flow and genetic drift in a species subject to frequent local extinctions*, Theor. Population Biology, **12**, 253-262 (1977).

[21] Wade, M.J. and D.E. McCauley, *Extinction and recolonization: their effects on the genetic differentiation of local populations*, Evolution, **42**, pp. 995-1005 (1988).

[22] Whitlock, M.C. and D.E. McCauley *Some population genetic consequences of colony formation and extinction: genetic correlations within founding groups*, Evolution, **44**, pp. 1717-1724 (1990).

[23] Wright, S., *Evolution in Mendelian populations*, Genetics, **16**, pp. 97-159 (1931).

[24] Wright, S., *Breeding structure of populations*, Amer. Natur., **74**, pp. 232-248 (1940).

ON THE NORMAL-SELECTION MODEL[*]

S.N. ETHIER[†]

Abstract. In the normal-selection model, every new mutant has a selection intensity that is normally distributed with mean 0 and variance σ_0^2. The resulting diffusion model is a Fleming–Viot process with an unbounded selection intensity function. A proper characterization of this process remains an open problem, but one can derive a number of results concerning its stationary distribution.

1. The normal-selection model. Tachida (1991) described what he referred to as a "nearly neutral mutation model" as follows:

> Consider a random mating population of N diploid individuals. The standard Wright-Fisher model in population genetics is assumed .... A mutation occurs with a rate u per generation per gene. If mutation occurs, the selection coefficient of the mutated gene is a random number s drawn from a fixed distribution $f(s)$ regardless of the original state of the gene. This mutation model is the house-of-cards model .... The fitness of a genotype $A_i A_j$ is $1 + s_i + s_j$ where s_i is the selection coefficient of the allele A_i. In the present study, we assume that $f(s)$ is normally distributed with mean zero and variance σ^2. Because we are interested in mutations with very small effects, the magnitude of σ is assumed to be $O(1/N)$ ....

Actually, Tachida used the term "nearly neutral" to refer to the case in which $0.2 < 4N\sigma < 5$. Since it is useful to have a name for the model that does not depend on the sizes of the parameters, we refer to it hereafter as the normal-selection model.

We begin by formulating a Wright–Fisher model that is general enough to include the normal-selection model. It depends on several parameters:

- E (a locally compact separable metric space) is the set of all possible alleles; it is often referred to as the "type space."
- N (a positive integer) is the diploid population size.
- u (in $[0,1]$) is the mutation rate per generation per gene.
- ν_0 (in $\mathcal{P}(E)$, the set of Borel probability measures on E) is the distribution of the type of a new mutant; this is the "house-of-cards" assumption.
- $s(x)$ (a Borel function defined for each $x \in E$ and bounded below by $-\frac{1}{2}$) is the selection coefficient of allele x.

[*] Research supported in part by National Science Foundation grant DMS-9311984.
[†] Department of Mathematics, University of Utah, Salt Lake City, UT 84112.

The Wright–Fisher model is a Markov chain describing the evolution of the composition of the population of gametes $(x_1, \ldots, x_{2N}) \in E^{2N}$ or, since the order of the gametes is unimportant, $(2N)^{-1} \sum_{i=1}^{2N} \delta_{x_i} \in \mathcal{P}(E)$. (Here $\delta_x \in \mathcal{P}(E)$ is the unit mass at $x \in E$.) Time is discrete and measured in generations. The transition mechanism is specified by

$$(1.1) \qquad \mu = \frac{1}{2N} \sum_{i=1}^{2N} \delta_{x_i} \longmapsto \frac{1}{2N} \sum_{i=1}^{2N} \delta_{Y_i},$$

where

$$(1.2) \qquad Y_1, \ldots, Y_{2N} \text{ are i.i.d. } \mu^{**} \qquad [\text{random sampling}],$$

$$(1.3) \qquad \mu^{**}(\Gamma) = (1-u)\mu^*(\Gamma) + u\nu_0(\Gamma) \qquad [\text{mutation}],$$

$$(1.4) \qquad \mu^*(\Gamma) = \frac{\int_E \int_\Gamma (1 + s(x) + s(y))\, \mu(dx)\, \mu(dy)}{\int_E \int_E (1 + s(x) + s(y))\, \mu(dx)\, \mu(dy)} \qquad [\text{selection}].$$

We now consider the normal-selection model in this framework. The type space E is unspecified. However, ν_0 and the function s must jointly satisfy the following condition: If X is a random variable with distribution ν_0, then $s(X)$ has the normal distribution with mean 0 and variance σ^2. Furthermore, $\sigma = \sigma_0/(2N)$ for an appropriate constant σ_0. There are therefore a number of possible choices for E, ν_0, and s, including:

$$(1.5) \quad E = (0,1), \qquad \nu_0 = \text{uniform distribution}, \qquad s(x) = \sigma\Phi^{-1}(x),$$

where Φ is the standard normal cumulative distribution function,

$$(1.6) \qquad E = \boldsymbol{R}, \qquad \nu_0 = N(0, \sigma^2), \qquad s(x) = x,$$

and

$$(1.7) \qquad E = \boldsymbol{R}, \qquad \nu_0 = N(0, \sigma_0^2), \qquad s(x) = x/(2N).$$

(Strictly speaking, we should replace $s(x)$ above by $s(x) \vee (-\frac{1}{2})$, so that a genotype's fitness will always be nonnegative.) Notice that each of these choices yields an equivalent model; they are simply reparametrizations of each other.

In what follows we will consider a diffusion approximation to the normal-selection model. Since such an approximation ordinarily requires that the selection coefficients be $O(1/N)$, (1.5) and (1.7) are preferable to (1.6), and we adopt (1.7) as the choice of parameters because it seems relatively simple.

2. The Fleming–Viot process. In the Wright–Fisher model (1.1)–(1.4), suppose we assume that

$$(2.1) \qquad u = \frac{\theta}{4N} \quad \text{and} \quad s(x) = \frac{h(x)}{2N},$$

where θ is a positive constant and h is a bounded continuous function on E. Let us denote by $\{\mu_\tau^{(N)}, \ \tau = 0, 1, 2, \ldots\}$ the resulting Markov chain. Then it is well known (see, e.g., Ethier and Kurtz (1987)) that, assuming convergence of initial distributions, $\{\mu_{[2Nt]}^{(N)}, \ t \geq 0\}$ converges in distribution in $D_{\mathcal{P}(E)}[0, \infty)$ (with $\mathcal{P}(E)$ given the topology of weak convergence) to a diffusion process in $\mathcal{P}(E)$ with generator $\mathcal{L}_h$ defined for

$$(2.2) \qquad \varphi(\mu) = F(\langle f_1, \mu \rangle, \ldots, \langle f_k, \mu \rangle) = F(\langle \boldsymbol{f}, \mu \rangle),$$

where $k \geq 1$, $f_1, \ldots, f_k \in B(E)$, $F \in C^2(\boldsymbol{R}^k)$, and $\langle f, \mu \rangle = \int_E f \, d\mu$, by

$$(2.3) \quad (\mathcal{L}_h \varphi)(\mu) = \tfrac{1}{2} \sum_{i,j=1}^k (\langle f_i f_j, \mu \rangle - \langle f_i, \mu \rangle \langle f_j, \mu \rangle) \, F_{z_i z_j}(\langle \boldsymbol{f}, \mu \rangle)$$

$$+ \tfrac{1}{2}\theta \sum_{i=1}^k (\langle f_i, \nu_0 \rangle - \langle f_i, \mu \rangle) \, F_{z_i}(\langle \boldsymbol{f}, \mu \rangle)$$

$$+ \sum_{i=1}^k (\langle f_i h, \mu \rangle - \langle f_i, \mu \rangle \langle h, \mu \rangle) \, F_{z_i}(\langle \boldsymbol{f}, \mu \rangle).$$

The diffusion with generator $\mathcal{L}_h$ is known as a Fleming–Viot process. (See Ethier and Kurtz (1993) for a survey article concerning such processes.)

Before specializing to (1.7), we record several facts about this process that will be useful in what follows.

LEMMA 2.1. *Under the above assumptions, the Fleming–Viot process with generator $\mathcal{L}_h$ has a unique stationary distribution Π_h, is strongly ergodic, and is reversible. In fact,*

$$(2.4) \qquad \Pi_0(\cdot) = \boldsymbol{P}\Big\{ \sum_{i=1}^\infty \rho_i \delta_{\xi_i} \in \cdot \Big\},$$

where $\xi_1, \xi_2, \ldots$ are i.i.d. ν_0 and $(\rho_1, \rho_2, \ldots)$ is Poisson–Dirichlet with parameter θ and independent of $\xi_1, \xi_2, \ldots$. Furthermore,

$$(2.5) \qquad \Pi_h(d\mu) = e^{2\langle h, \mu \rangle} \, \Pi_0(d\mu) \Big/ \int_{\mathcal{P}(E)} e^{2\langle h, \nu \rangle} \, \Pi_0(d\nu).$$

These results can be found in Ethier and Kurtz (1994a,b).

3. Diffusion approximation of the normal-selection model. We assume (1.7) and (2.1). This should lead to a Fleming–Viot process with generator $\mathcal{L}_h$ as in (2.2)–(2.3), where

$$(3.1) \qquad E = \boldsymbol{R}, \qquad \nu_0 = N(0, \sigma_0^2), \qquad h(x) = x.$$

However, the function h in (3.1) is unbounded, and this turns out to be a serious complication. Indeed, note in particular that $\mathcal{L}_h$ is not well defined on all of $\mathcal{P}(\boldsymbol{R})$ because of terms such as $\langle h, \mu \rangle$ in (2.3), which require that μ have finite mean. More importantly, the theorem that characterizes the Fleming–Viot process is not applicable here, and so the limit theorem that justifies the diffusion approximation does not apply either. The solution, we believe, is to replace $\mathcal{P}(\boldsymbol{R})$ by a smaller state space. This problem remains open, but some partial results are given in Section 5.

Nevertheless, we can be quite explicit about the stationary distribution for the yet-to-be-characterized limiting diffusion.

THEOREM 3.1. *Assume (3.1). Then* Π_h, *defined by (2.5), is such that* $\mathcal{L}_h$ *is a symmetric linear operator on* $L^2(\Pi_h)$.

Remark. Once the diffusion associated with $\mathcal{L}_h$ is characterized, this result will tell us that Π_h is a reversible stationary distribution. A separate argument will be needed to show that it is the unique stationary distribution.

Proof. For each positive integer K, let $h_K : \boldsymbol{R} \mapsto \boldsymbol{R}$ be given by $h_K(x) = (-K) \vee (x \wedge K)$. Then, by Lemma 2.1 and the fact that h_K is bounded and continuous,

$$\int_{\mathcal{P}(\boldsymbol{R})} \varphi(\mu) \mathcal{L}_{h_K} \psi(\mu) e^{2\langle h_K, \mu \rangle} \, \Pi_0(d\mu) = \int_{\mathcal{P}(\boldsymbol{R})} \psi(\mu) \mathcal{L}_{h_K} \varphi(\mu) e^{2\langle h_K, \mu \rangle} \, \Pi_0(d\mu)$$

(3.2)

for all φ and ψ of the form (2.2). Now $|h_K| \leq |h|$ for all K (recall that $h(x) = x$), so by the dominated convergence theorem it is enough to verify that

$$(3.3) \qquad \int_{\mathcal{P}(\boldsymbol{R})} e^{\alpha \langle |h|, \mu \rangle} \, \Pi_0(d\mu) < \infty$$

for some $\alpha > 2$. Again by Lemma 2.1, Π_0 is the distribution of $\sum_{i=1}^{\infty} \rho_i \delta_{\sigma_0 Z_i}$, where $Z_1, Z_2, \ldots$ are i.i.d. standard normal and $(\rho_1, \rho_2, \ldots)$ is Poisson–Dirichlet with parameter θ and independent of $Z_1, Z_2, \ldots$. Consequently, the integral in (3.3) is equal to

$$(3.4) \qquad \boldsymbol{E}\left[e^{\alpha \langle |h|, \sum_{i=1}^{\infty} \rho_i \delta_{\sigma_0 Z_i} \rangle} \right] = \boldsymbol{E}\left[e^{\alpha \sigma_0 \sum_{i=1}^{\infty} \rho_i |Z_i|} \right]$$

$$\leq \boldsymbol{E}\left[\prod_{i=1}^{\infty} (1 + C\alpha \sigma_0 \rho_i) \right] \leq e^{C\alpha \sigma_0} < \infty,$$

where the first inequality uses the following estimate on the moment generating function of the absolute value of a standard normal random variable

Z: For each $T > 0$ there exists $C > 0$ such that

$$(3.5) \qquad E[e^{t|Z|}] = \sum_{n=0}^{\infty} \frac{t^n E[|Z|^n]}{n!} \leq 1 + Ct, \qquad 0 \leq t \leq T.$$

□

The next section gives some applications of this result.

4. Applications. Tachida (1991) used computer simulation to study the homozygosity, the average selection coefficient, and the substitution rate in the normal-selection model. In certain cases, we can provide exact results for the diffusion model at equilibrium.

We denote by F the homozygosity and by S the average selection intensity. These are of course functions on $\mathcal{P}(\mathbf{R})$, given by

$$(4.1) \qquad F(\mu) = \mu^2(D) = \sum_{x \in \mathbf{R}} \mu(\{x\})^2, \qquad S(\mu) = \langle h, \mu \rangle,$$

where $\mu^2 = \mu \times \mu$, D denotes the diagonal of $\mathbf{R}^2$, and as usual $h(x) = x$. For simplicity of notation, we write $E_{\text{neu}}[\varphi]$ for $\int \varphi \, d\Pi_0$ and $E_{\text{sel}}[\varphi]$ for $\int \varphi \, d\Pi_h$. A similar convention applies to variances.

THEOREM 4.1. *For each $g \in B([0,1])$,*

$$(4.2) \qquad E_{\text{sel}}[g(F)] = E_{\text{neu}}[g(F)e^{2\sigma_0^2 F}] / E_{\text{neu}}[e^{2\sigma_0^2 F}].$$

In addition,

$$(4.3) \quad E_{\text{sel}}[S] = 2\sigma_0^2 E_{\text{sel}}[F], \quad Var_{\text{sel}}(S) = \sigma_0^2 E_{\text{sel}}[F] + 4\sigma_0^4 Var_{\text{sel}}(F).$$

Proof. As in the proof of Theorem 3.1, we use the fact that Π_0 is the distribution of $\sum_{i=1}^{\infty} \rho_i \delta_{\sigma_0 Z_i}$ to show that

(4.4)

$$E_{\text{sel}}[g(F)] = \int_{\mathcal{P}(\mathbf{R})} g(\mu^2(D)) \, \Pi_h(d\mu)$$

$$= \int_{\mathcal{P}(\mathbf{R})} g(\mu^2(D)) e^{2\langle h, \mu \rangle} \, \Pi_0(d\mu) \Big/ \int_{\mathcal{P}(\mathbf{R})} e^{2\langle h, \mu \rangle} \, \Pi_0(d\mu)$$

$$= E\left[g\left(\sum_{i=1}^{\infty} \rho_i^2\right) e^{2\sigma_0 \sum_{i=1}^{\infty} \rho_i Z_i}\right] \Big/ E\left[e^{2\sigma_0 \sum_{i=1}^{\infty} \rho_i Z_i}\right]$$

$$= E\left[g\left(\sum_{i=1}^{\infty} \rho_i^2\right) e^{2\sigma_0^2 \sum_{i=1}^{\infty} \rho_i^2}\right] \Big/ E\left[e^{2\sigma_0^2 \sum_{i=1}^{\infty} \rho_i^2}\right]$$

$$= \int_{\mathcal{P}(\mathbf{R})} g(\mu^2(D)) e^{2\sigma_0^2 \mu^2(D)} \, \Pi_0(d\mu) \Big/ \int_{\mathcal{P}(\mathbf{R})} e^{2\sigma_0^2 \mu^2(D)} \, \Pi_0(d\mu)$$

$$= E_{\text{neu}}[g(F)e^{2\sigma_0^2 F}] / E_{\text{neu}}[e^{2\sigma_0^2 F}],$$

proving (4.2). As for (4.3),

(4.5)

$$
\begin{aligned}
E_{\mathrm{sel}}[S] &= \int_{\mathcal{P}(\mathbf{R})} \langle h, \mu \rangle \, \Pi_h(d\mu) \\
&= \int_{\mathcal{P}(\mathbf{R})} \langle h, \mu \rangle e^{2\langle h, \mu \rangle} \, \Pi_0(d\mu) \Big/ \int_{\mathcal{P}(\mathbf{R})} e^{2\langle h, \mu \rangle} \, \Pi_0(d\mu) \\
&= E\Big[\sigma_0 \sum_{i=1}^{\infty} \rho_i Z_i e^{2\sigma_0 \sum_{j=1}^{\infty} \rho_j Z_j}\Big] \Big/ E\Big[e^{2\sigma_0 \sum_{i=1}^{\infty} \rho_i Z_i}\Big] \\
&= \sigma_0 \sum_{i=1}^{\infty} E\Big[\rho_i Z_i e^{2\sigma_0 \rho_i Z_i} \prod_{j: j \neq i} e^{2\sigma_0 \rho_j Z_j}\Big] \Big/ E\Big[e^{2\sigma_0^2 \sum_{i=1}^{\infty} \rho_i^2}\Big] \\
&= \sigma_0 \sum_{i=1}^{\infty} E\Big[\rho_i 2\sigma_0 \rho_i e^{2\sigma_0^2 \rho_i^2} \prod_{j: j \neq i} e^{2\sigma_0^2 \rho_j^2}\Big] \Big/ E\Big[e^{2\sigma_0^2 \sum_{i=1}^{\infty} \rho_i^2}\Big] \\
&= 2\sigma_0^2 E\Big[\sum_{i=1}^{\infty} \rho_i^2 e^{2\sigma_0^2 \sum_{j=1}^{\infty} \rho_j^2}\Big] \Big/ E\Big[e^{2\sigma_0^2 \sum_{i=1}^{\infty} \rho_i^2}\Big] \\
&= 2\sigma_0^2 \int_{\mathcal{P}(\mathbf{R})} \mu^2(D) e^{2\sigma_0^2 \mu^2(D)} \, \Pi_0(d\mu) \Big/ \int_{\mathcal{P}(\mathbf{R})} e^{2\sigma_0^2 \mu^2(D)} \, \Pi_0(d\mu) \\
&= 2\sigma_0^2 E_{\mathrm{neu}}[F e^{2\sigma_0^2 F}] / E_{\mathrm{neu}}[e^{2\sigma_0^2 F}] \\
&= 2\sigma_0^2 E_{\mathrm{sel}}[F],
\end{aligned}
$$

where the last equality uses (4.2). The derivation of the second formula in (4.3) is similar and is omitted. □

These quantities can be evaluated numerically using Griffiths' (1988) recursion for the moments of neutral homozygosity. For example,

(4.6)
$$
\begin{aligned}
E_{\mathrm{sel}}[F] &= E_{\mathrm{neu}}[F e^{2\sigma_0^2 F}] / E_{\mathrm{neu}}[e^{2\sigma_0^2 F}] \\
&= \sum_{n=0}^{\infty} \frac{(2\sigma_0^2)^n}{n!} E_{\mathrm{neu}}[F^{n+1}] \Big/ \sum_{n=0}^{\infty} \frac{(2\sigma_0^2)^n}{n!} E_{\mathrm{neu}}[F^n].
\end{aligned}
$$

We have done this for the various choices of the parameters in Tachida (1991), and have found rather good agreement between his simulation results for the Wright–Fisher model with $2N = 100$ and our exact results for the diffusion model.

5. The characterization problem. Here we consider the problem of proving existence and uniqueness of solutions of the martingale problem for $\mathcal{L}_h$. Although other choices are possible, we take as our state space the set of Borel probability measures on $\mathbf{R}$ with finite moment generating functions.

Given $\mu \in \mathcal{P}(\mathbf{R})$, we denote its moment generating function by

(5.1)
$$
M_\mu(\rho) = \int_{\mathbf{R}} e^{\rho x} \, \mu(dx), \qquad \rho \in \mathbf{R}.
$$

Let us define

$$(5.2) \qquad \mathcal{P}_{\mathrm{mgf}}(\boldsymbol{R}) = \{\mu \in \mathcal{P}(\boldsymbol{R}) : M_\mu(\rho) < \infty \text{ for all } \rho \in \boldsymbol{R}\}$$

and

$$(5.3) \quad d(\mu, \nu) = \int_0^\infty \left(1 \wedge \sup_{|\rho| \le r} |M_\mu(\rho) - M_\nu(\rho)|\right) e^{-r} \, dr, \quad \mu, \nu \in \mathcal{P}_{\mathrm{mgf}}(\boldsymbol{R}).$$

Then $(\mathcal{P}_{\mathrm{mgf}}(\boldsymbol{R}), d)$ is a complete separable metric space and $d(\mu_n, \mu) \to 0$ if and only if $\mu_n \Rightarrow \mu$ and the function $x \mapsto e^{\rho x}$ is $\{\mu_n\}$-uniformly integrable for each $\rho \in \boldsymbol{R}$. Thus, the topology on $\mathcal{P}_{\mathrm{mgf}}(\boldsymbol{R})$ is slightly stronger than the topology of weak convergence.

5.1. Existence and uniqueness in the case of bounded h. Let us begin by trying to understand the case in which h is bounded. (Here we do not assume (3.1).) We will need the following extension of the first part of Lemma 4.3.2 of Ethier and Kurtz (1986).

LEMMA 5.1. *Let X be a progressively measurable process with values in a complete separable metric space, and let f, g, c be bounded Borel functions on that space. If*

$$(5.4) \qquad f(X(t)) - \int_0^t g(X(s)) \, ds$$

is an $\{\mathcal{F}_t^X\}$-martingale, then so is

$$(5.5) \quad f(X(t)) e^{-\int_0^t c(X(s)) \, ds}$$
$$- \int_0^t \{g(X(s)) - c(X(s))f(X(s))\} e^{-\int_0^s c(X(r)) \, dr} \, ds.$$

Let $\mathcal{P}(E)$ have the topology of weak convergence, let $\Omega \equiv C_{\mathcal{P}(E)}[0, \infty)$ have the topology of uniform convergence on compact sets, let $\mathcal{F}$ be the Borel σ-field, let $\{\mu_t, \, t \ge 0\}$ be the canonical coordinate process, and let $\{\mathcal{F}_t\}$ be the corresponding filtration.

PROPOSITION 5.2. *Let h_0 and h_1 be bounded Borel functions on E. If $P \in \mathcal{P}(\Omega)$ is a solution of the martingale problem for $\mathcal{L}_{h_0}$, then*

$$(5.6)$$
$$R_t = \exp\left\{\langle h_1, \mu_t\rangle - \langle h_1, \mu_0\rangle - \int_0^t e^{-\langle h_1, \mu_s\rangle} \mathcal{L}_{h_0} e^{\langle h_1, \mu_s\rangle} \, ds\right\}$$
$$= \exp\left\{\langle h_1, \mu_t\rangle - \langle h_1, \mu_0\rangle - \int_0^t \left[\tfrac{1}{2}(\langle h_1^2, \mu_s\rangle - \langle h_1, \mu_s\rangle^2)\right.\right.$$
$$\left.\left. + \tfrac{1}{2}\theta(\langle h_1, \nu_0\rangle - \langle h_1, \mu_s\rangle) + \langle h_0 h_1, \mu_s\rangle - \langle h_0, \mu_s\rangle\langle h_1, \mu_s\rangle\right] ds\right\}$$

is a mean-one $\{\mathcal{F}_t\}$-martingale on $(\Omega, \mathcal{F}, P)$. Furthermore, the measure $Q \in \mathcal{P}(\Omega)$ defined by

$$(5.7) \qquad dQ = R_t\, dP \quad \text{on} \quad \mathcal{F}_t, \qquad t \geq 0,$$

is a solution of the martingale problem for $\mathcal{L}_{h_0+h_1}$.

Proof. By Lemma 4.3.2 of Ethier and Kurtz (1986),

$$(5.8) \qquad \frac{\varphi(\mu_t)}{\varphi(\mu_0)} \exp\left\{ -\int_0^t \frac{(\mathcal{L}_{h_0}\varphi)(\mu_s)}{\varphi(\mu_s)}\, ds \right\}$$

is a mean-one $\{\mathcal{F}_t\}$-martingale on $(\Omega, \mathcal{F}, P)$ for each $\varphi \in \mathcal{D}(\mathcal{L}_{h_0})$ with $\inf \varphi > 0$. In particular, if $\varphi(\mu) \equiv \exp\{\langle h_1, \mu \rangle\}$, then (5.8) reduces to (5.6).

Fix $\varphi \in \mathcal{D}(\mathcal{L}_{h_0+h_1})$, $t > s \geq 0$, and a bounded $\mathcal{F}_s$-measurable function H_s. To complete the proof, it is enough to show that

$$(5.9) \qquad E^Q\left[\left(\varphi(\mu_t) - \varphi(\mu_s) - \int_s^t (\mathcal{L}_{h_0+h_1}\varphi)(\mu_r)\, dr \right) H_s \right] = 0.$$

But the left side of (5.9) is

(5.10)

$$E^P[\varphi(\mu_t)R_t H_s] - E^P[\varphi(\mu_s)R_s H_s] - \int_s^t E^P[(\mathcal{L}_{h_0+h_1}\varphi)(\mu_r)R_r H_s]\, dr$$

$$= E^P\left[e^{-\langle h_1, \mu_0 \rangle} \left\{ \varphi(\mu_t)e^{\langle h_1, \mu_t \rangle} \exp\left(-\int_0^t e^{-\langle h_1, \mu_r \rangle} \mathcal{L}_{h_0} e^{\langle h_1, \mu_r \rangle}\, dr \right) \right.\right.$$

$$- \varphi(\mu_s)e^{\langle h_1, \mu_s \rangle} \exp\left(-\int_0^s e^{-\langle h_1, \mu_r \rangle} \mathcal{L}_{h_0} e^{\langle h_1, \mu_r \rangle}\, dr \right)$$

$$- \int_s^t (\mathcal{L}_{h_0+h_1}\varphi)(\mu_r)e^{\langle h_1, \mu_r \rangle}$$

$$\left.\left. \exp\left(-\int_0^r e^{-\langle h_1, \mu_q \rangle} \mathcal{L}_{h_0} e^{\langle h_1, \mu_q \rangle}\, dq \right) dr \right\} H_s \right]$$

$$= E^P\left[e^{-\langle h_1, \mu_0 \rangle} \left\{ \varphi(\mu_t)e^{\langle h_1, \mu_t \rangle} \exp\left(-\int_0^t e^{-\langle h_1, \mu_r \rangle} \mathcal{L}_{h_0} e^{\langle h_1, \mu_r \rangle}\, dr \right) \right.\right.$$

$$- \varphi(\mu_s)e^{\langle h_1, \mu_s \rangle} \exp\left(-\int_0^s e^{-\langle h_1, \mu_r \rangle} \mathcal{L}_{h_0} e^{\langle h_1, \mu_r \rangle}\, dr \right)$$

$$- \int_s^t (\mathcal{L}_{h_0}[\varphi(\mu_r)e^{\langle h_1, \mu_r \rangle}] - \varphi(\mu_r)\mathcal{L}_{h_0} e^{\langle h_1, \mu_r \rangle})$$

$$\left.\left. \exp\left(-\int_0^r e^{-\langle h_1, \mu_q \rangle} \mathcal{L}_{h_0} e^{\langle h_1, \mu_q \rangle}\, dq \right) dr \right\} H_s \right]$$

$$= 0,$$

where the last equality uses Lemma 5.1, and the next-to-last equality uses (2.2) and

(5.11)

$$\mathcal{L}_{h_0}[\varphi(\mu)e^{\langle h_1,\mu\rangle}] - \varphi(\mu)\mathcal{L}_{h_0}e^{\langle h_1,\mu\rangle}$$

$$= (\mathcal{L}_{h_0}\varphi)(\mu)e^{\langle h_1,\mu\rangle} + \sum_{i=1}^{k}(\langle f_i h_1,\mu\rangle - \langle f_i,\mu\rangle\langle h_1,\mu\rangle)F_{z_i}(\langle \boldsymbol{f},\mu\rangle)e^{\langle h_1,\mu\rangle}$$

$$= (\mathcal{L}_{h_0+h_1}\varphi)(\mu)e^{\langle h_1,\mu\rangle}.$$

This completes the proof. □

COROLLARY 5.3. *Let h be a bounded Borel function on E. Then the martingale problem for $\mathcal{L}_h$ is well posed.*

Proof. For existence, apply Proposition 5.2 with $h_0 = 0$ and $h_1 = h$. For uniqueness, apply the same result with $h_0 = h$ and $h_1 = -h$. We are using the fact that existence and uniqueness are known for $\mathcal{L}_0$. □

We remark that Corollary 5.3 also follows by a duality argument, but the present approach gives additional information, such as the form of the Radon–Nikodym derivative.

5.2. The normal-selection model: existence. We assume (3.1) hereafter and define

(5.12) $$\Omega^\circ = C_{(\mathcal{P}_{\mathrm{mgf}}(\boldsymbol{R}),d)}[0,\infty) \subset \Omega = C_{\mathcal{P}(\boldsymbol{R})}[0,\infty).$$

For $\mu \in \mathcal{P}(\boldsymbol{R})$ we denote by $P_\mu \in \mathcal{P}(\Omega)$ the unique solution of the martingale problem for $\mathcal{L}_0$ (i.e., the distribution of the neutral model) starting at μ.

LEMMA 5.4. *For each $\mu \in \mathcal{P}_{\mathrm{mgf}}(\boldsymbol{R})$,*

(5.13) $$\boldsymbol{E}^{P_\mu}\left[\sup_{0 \le t \le T} M_{\mu_t}(\rho)^2\right] \le 8T M_\mu(2\rho) + (8T + \tfrac{1}{2}\theta^2 T^2)e^{2\sigma^2\rho^2}$$

for all $T > 0$ and $\rho \in \boldsymbol{R}$. In particular, $x \mapsto e^{\rho x}$ is $\{\mu_t, \ 0 \le t \le T\}$-uniformly integrable P_μ-a.s. for each $\rho \in \boldsymbol{R}$ and $T > 0$, and therefore $P_\mu(\Omega^\circ) = 1$.

Proof. For each $g \in \overline{C}(\boldsymbol{R})$,

(5.14) $$Z^g(t) \equiv \langle g,\mu_t\rangle - \tfrac{1}{2}\theta \int_0^t (\langle g,\nu_0\rangle - \langle g,\mu_s\rangle)\,ds$$

is a continuous $\{\mathcal{F}_t\}$-martingale on $(\Omega,\mathcal{F},P_\mu)$ with quadratic variation process

(5.15) $$\langle Z^g\rangle_t = \int_0^t (\langle g^2,\mu_s\rangle - \langle g,\mu_s\rangle^2)\,ds.$$

If, in addition, g is nonnegative, then $\langle g, \mu_t \rangle \leq Z^g(t) + \frac{1}{2}\theta t \langle g, \nu_0 \rangle$ for all $t \geq 0$, so

(5.16)

$$
\begin{aligned}
\boldsymbol{E}^{P_\mu} \left[\sup_{0 \leq t \leq T} \langle g, \mu_t \rangle^2 \right] &\leq 2\boldsymbol{E}^{P_\mu} \left[\sup_{0 \leq t \leq T} Z^g(t)^2 \right] + \frac{1}{2}\theta^2 T^2 \langle g, \nu_0 \rangle^2 \\
&\leq 8\boldsymbol{E}^{P_\mu} [Z^g(T)^2] + \frac{1}{2}\theta^2 T^2 \langle g, \nu_0 \rangle^2 \\
&= 8 \int_0^T \boldsymbol{E}^{P_\mu} [\langle g^2, \mu_s \rangle - \langle g, \mu_s \rangle^2] \, ds + \frac{1}{2}\theta^2 T^2 \langle g, \nu_0 \rangle^2 \\
&\leq 8 \int_0^T \langle U(s)g^2, \mu \rangle \, ds + \frac{1}{2}\theta^2 T^2 \langle g, \nu_0 \rangle^2 \\
&\leq 8T\langle g^2, \mu \rangle + (8T + \frac{1}{2}\theta^2 T^2)\langle g^2, \nu_0 \rangle
\end{aligned}
$$

for all $T > 0$, where $\{U(t)\}$ is the semigroup on $\overline{C}(\boldsymbol{R})$ with generator $Af = \frac{1}{2}\theta(\langle f, \nu_0 \rangle - f)$; it is given by

(5.17) $$U(t)f = e^{-\theta t/2}f + (1 - e^{-\theta t/2})\langle f, \nu_0 \rangle.$$

Now define $g : \boldsymbol{R} \mapsto (0, \infty)$ by $g(x) = e^{\rho x}$, where $\rho \in \boldsymbol{R}$. Applying (5.16) with $g \wedge K$ in place of g, and noting that

(5.18) $$\sup_{0 \leq t \leq T} \langle g, \mu_t \rangle^2 = \lim_{K \to \infty} \sup_{0 \leq t \leq T} \langle g \wedge K, \mu_t \rangle^2$$

and that $M_{\nu_0}(\rho) = e^{\sigma^2 \rho^2 / 2}$ for all $\rho \in \boldsymbol{R}$, we have (5.13). $\square$

Let Ω° have the topology of uniform convergence on compact sets, let $\mathcal{F}^\circ$ be the Borel σ-field, let $\{\mu_t, \ t \geq 0\}$ be the canonical coordinate process on Ω°, and let $\{\mathcal{F}_t^\circ\}$ be the corresponding filtration. We do not distinguish notationally between the canonical coordinate process on Ω and that on Ω°, between $P_\mu \in \mathcal{P}(\Omega)$ and its restriction to $\mathcal{F}^\circ$ (note that $\mathcal{F}^\circ \subset \mathcal{F}$), or between R_t of (5.6) and its restriction to Ω°. We denote R_t by $R_t^{h_0, h_0 + h_1}$ to indicate its dependence on h_0 and h_1. Recall h of (3.1).

LEMMA 5.5. *For each* $\mu \in \mathcal{P}_{\mathrm{mgf}}(\boldsymbol{R})$, $\{R_t^{0,h}, \ t \geq 0\}$ *is a mean-one* $\{\mathcal{F}_t^\circ\}$-*martingale on* $(\Omega^\circ, \mathcal{F}^\circ, P_\mu)$.

Proof. As in the proof of Theorem 3.1, define $h_K : \boldsymbol{R} \mapsto \boldsymbol{R}$ by $h_K(x) = (-K) \vee (x \wedge K)$, and note that $\{R_t^{0,h_K}, \ t \geq 0\}$ is a mean-one $\{\mathcal{F}_t\}$-martingale on $(\Omega, \mathcal{F}, P_\mu)$ by Proposition 5.2. Moreover, $R_t^{0,h_K} \to R_t^{0,h}$ pointwise on Ω° for all $t \geq 0$, and

(5.19)

$$
\begin{aligned}
e^{\langle h_K, \mu \rangle} R_t^{0,h_K} &\leq \exp\left\{ \langle |h|, \mu_t \rangle + \frac{1}{2}\theta \int_0^t \langle |h|, \mu_s \rangle \, ds \right\} \\
&\leq \exp\left\{ (1 + \frac{1}{2}\theta t) \sup_{0 \leq s \leq t} \langle |h|, \mu_s \rangle \right\}
\end{aligned}
$$

$$\leq \sup_{0 \leq s \leq t} \langle e^{(1 + \frac{1}{2}\theta t)|h|}, \mu_s \rangle$$

$$\leq \sup_{0 \leq s \leq t} \{ M_{\mu_s}(1 + \tfrac{1}{2}\theta t) + M_{\mu_s}(-1 - \tfrac{1}{2}\theta t) \}, \qquad t \geq 0,$$

P_μ-a.s., so by Lemma 5.4 we can apply the dominated convergence theorem.
□

For each $\mu \in \mathcal{P}_{\mathrm{mgf}}(\boldsymbol{R})$, Lemma 5.5 allows us to define $Q_\mu \in \mathcal{P}(\Omega^\circ)$ by

$$(5.20) \qquad dQ_\mu = R_t^{0,h} \, dP_\mu \quad \text{on} \quad \mathcal{F}_t^\circ, \qquad t \geq 0.$$

We now show that Q_μ solves the Ω° martingale problem for $\mathcal{L}_h$ starting at μ.

LEMMA 5.6. *Let $\mu \in \mathcal{P}_{\mathrm{mgf}}(\boldsymbol{R})$. Then*

$$(5.21) \qquad \varphi(\mu_t) - \int_0^t (\mathcal{L}_h \varphi)(\mu_s) \, ds$$

is an $\{\mathcal{F}_t^\circ\}$-martingale on $(\Omega^\circ, \mathcal{F}^\circ, Q_\mu)$ for each $\varphi \in \mathcal{D}(\mathcal{L}_h)$.

Proof. Let h_K be as in the preceding proof, and define $Q_\mu^K \in \mathcal{P}(\Omega^\circ)$ as in (5.20) but with h_K in place of h. Then, by Proposition 5.2, Q_μ^K solves the Ω° martingale problem for $\mathcal{L}_{h_K}$ starting at μ. Consequently, if $\varphi \in \mathcal{D}(\mathcal{L}_h)$, $t > s \geq 0$, and H_s is bounded and $\mathcal{F}_s^\circ$-measurable on Ω°, then

$$(5.22)$$

$$\begin{aligned}
0 &= E^{Q_\mu^K}\left[\left(\varphi(\mu_t) - \varphi(\mu_s) - \int_s^t (\mathcal{L}_{h_K}\varphi)(\mu_r) \, dr \right) H_s \right] \\
&= E^{P_\mu}\left[\left(\varphi(\mu_t) - \varphi(\mu_s) - \int_s^t (\mathcal{L}_{h_K}\varphi)(\mu_r) \, dr \right) R_t^{0,h_K} H_s \right] \\
&\to E^{P_\mu}\left[\left(\varphi(\mu_t) - \varphi(\mu_s) - \int_s^t (\mathcal{L}_h\varphi)(\mu_r) \, dr \right) R_t^{0,h} H_s \right] \\
&= E^{Q_\mu}\left[\left(\varphi(\mu_t) - \varphi(\mu_s) - \int_s^t (\mathcal{L}_h\varphi)(\mu_r) \, dr \right) H_s \right],
\end{aligned}$$

where we are using the bound (5.19), and the result follows. □

5.3. The normal-selection model: uniqueness. Given $\mu \in \mathcal{P}_{\mathrm{mgf}}$ $(\boldsymbol{R})$, let $Q_\mu \in \mathcal{P}(\Omega^\circ)$ be a solution of the Ω° martingale problem for $\mathcal{L}_h$ starting at μ. We would like to establish the following two conjectures: (a) $\{ R_t^{h,0}, \, t \geq 0 \}$ is a mean-one $\{\mathcal{F}_t^\circ\}$-martingale on $(\Omega^\circ, \mathcal{F}^\circ, Q_\mu)$, and (b) the measure $P_\mu \in \mathcal{P}(\Omega^\circ)$ defined by

$$(5.23) \qquad dP_\mu = R_t^{h,0} \, dQ_\mu \quad \text{on} \quad \mathcal{F}_t^\circ, \qquad t \geq 0,$$

is a solution of the Ω° martingale problem for $\mathcal{L}_0$ starting at μ. Since this property uniquely determines P_μ, and since $(R_t^{h,0})^{-1} = R_t^{0,h}$ for all $t \geq 0$,

(5.23) would imply that Q_μ is given by (5.20), and the required uniqueness would follow.

We have been unable to verify these conjectures, so this will have to remain an open problem for now.

It might help to work in an even smaller state space. For example, it should be possible to prove existence in the set of Borel probability measures μ on $\mathbf{R}$ with the property that $\int_{\mathbf{R}} e^{\alpha x^2} \mu(dx) < \infty$ for each $\alpha < (2\sigma_0^2)^{-1}$.

There is a well-known dual process for the Fleming–Viot process, and it can be used to establish uniqueness under certain conditions. The only complication in the present situation is that the corresponding dual process takes values in a space of *unbounded* functions, and so verification of the duality identity is not straightforward.

Once uniqueness is established, there are a host of additional problems that should be addressed, involving diffusion approximations, ergodic theorems, rates of convergence to equilibrium, and so on. As noted by a referee, the diffusion approximation of the stationary distribution of the Wright–Fisher model requires justification, and indeed, even the uniqueness of this stationary distribution is open, as far as we know.

REFERENCES

ETHIER, S. N. AND KURTZ, T. G., *Markov Processes: Characterization and Convergence.* Wiley, New York (1986).

ETHIER, S. N. AND KURTZ, T. G., The infinitely-many-alleles model with selection as a measure-valued diffusion. In *Stochastic Models in Biology.* M. Kimura, G. Kallianpur, and T. Hida, eds. Lecture Notes in Biomathematics **70** (1987), pp. 72–86. Springer–Verlag, Berlin.

ETHIER, S. N. AND KURTZ, T. G., Fleming–Viot processes in population genetics. *SIAM J. Control Opt.* **31** (1993), 345–386.

ETHIER, S. N. AND KURTZ, T. G., Convergence to Fleming–Viot processes in the weak atomic topology. *Stochastic Processes Appl.* **54** (1994a), 1–27.

ETHIER, S. N. AND KURTZ, T. G., Coupling and ergodic theorems for Fleming–Viot processes, (1994b), preprint.

GRIFFITHS, R. C., Distribution of F in the infinitely-many-alleles model. Statistics Research Report 183, Dept. of Mathematics, Monash University (1988).

TACHIDA, H., A study on a nearly neutral mutation model in finite populations. *Genetics* **128** (1991), 183–192.

BRANCHING PROCESSES AND EVOLUTION[*]

ZIAD TAIB[†]

1. Introduction. One of the aims of mathematical population genetics is to provide mathematical models able to explain (at least certain aspects of) the genetic diversity which is observed in nature. Such models have two main ingredients: Evolutionary forces and population dynamics. The latter ingredient is often of a very simple nature, like the Wright-Fisher model, the Moran model and, to some extent, birth and death processes. In Section 2 we describe a quite general branching process model where individuals are subject to neutral mutations according to the infinite alleles hypothesis. Some of the results concerning forward aspects of this process are reviewed in Section 3 while backward aspects are shortly discussed in Section 4. In Section 5 the model is enlarged to allow individuals to have different fitnesses. The resulting multitype branching process is then used to model situations involving both selection and mutation.

2. Preliminaries. A single-type, general branching process is to be thought of as a population model where an individual, x, chooses, at birth, a life history $\omega \in \Omega$ using a probability measure P, the life law. The set of all possible individuals is denoted by I and is formally defined by $I = \{0\} \bigcup_{n=0}^{\infty} N^n$. 0 stands for the single ancestor of the population and an individual in the nth generation is described using the birth orders of her predecessors back to the ancestor. The life history space Ω is equipped with some σ-algebra $\mathcal{F}$. Interesting aspects of the life of an individual can be defined as random elements on Ω. $\lambda : \Omega \to R^+$ gives thus the life span of an individual and $\tau : N \times \Omega \to R^+ \cup \{+\infty\}$ gives the successive ages at childbearing. $\rho : N \times \Omega \to \{0,1\}$ describes the genetic status of the offspring (0 for mutant and 1 for non-mutant). The sequence $(\tau(k,\omega))_k$ defines a point process ξ according to $\xi(A) = \#\{k; \tau(k) \in A\}$, where A is a (Borel) set in R_+. The expectation of this point process is a measure $\mu = E[\xi]$ that we call the reproduction measure, and that turns out to be the most important parameter of the model. In the sequel we write $\mu(t)$ for $\mu([0,t])$. We also make the usual assumption that the life histories of different individuals are independent so the whole process is defined on $(\Omega^I, \mathcal{F}^I, P^I)$. Furthermore, it will be assumed that the ancestor is born at time zero. Then the ages at childbearing can be used to define the birth moments, σ_x, of different individuals $x \in I$ iteratively. In what follows the following conditions will be assumed: (1) $\mu(\infty) \geq 1$ (supercriticality) (2) There exists an $\alpha \in R$ such that $\hat{\mu}(\alpha) = 1$. α is called the Malthusian

[*] Supported by the Swedish Natural Science Research Council.

[†] Department of Mathematics, Chalmers University of Technology and the University of Göteborg, S-412 96 Göteborg, Sweden, e-mail ziad@math.chalmers.se

parameter. (3) $\beta = \int_0^\infty ue^{-\alpha u}\mu du < \infty$ (β can be interpreted as the mean generation length). (4) μ is non-lattice, i.e. it is not concentrated on a set of the form $\{dk; k \in N\}$ for some d. (5) Some moment condition on ξ (for example $Var[\hat{\xi}(\alpha)] < \infty$ or $E[\hat{\xi}(\alpha)\log^+(\hat{\xi}(\alpha))] < \infty$).

Obviously, ρ can be used to split ξ into two point processes ξ_n and ξ_m, the point processes according to which individuals beget non-mutant and mutant children respectively. The corresponding reproduction measures are denoted μ_n and μ_m ($\xi = \xi_n + \xi_m$ and $\mu = \mu_n + \mu_m$). Finally, we assume that whenever a mutation occurs, it leads to a new allelic form never seen before.

3. Forward aspects. It turns out that there are three different interesting processes: (i) The underlying process, which uses ξ and μ. This process is an ordinary general branching process consisting of all the individuals regardless their genetic status. (ii) The ancestral process which uses ξ_n and μ_n and consists of those individuals who carry the same allele as the ancestor. Any mutant initiates a new process which behaves like the ancestral process. (iii) An embedded process where alleles are seen as (macro-) individuals in a population of alleles. This processes uses a complicated point process ξ' having $\mu'(t) = \sum_{k=0}^\infty \mu_n^{*k} \star \mu_m(t)$ as its reproduction measure. It turns out that $\hat{\mu}'(\alpha') = 1$ is solved for $\alpha' = \alpha$, and that $\beta' = \frac{\beta}{\hat{\mu}(\alpha)}$.

Let $Z_n(t)$ be the number of individuals in the ancestral process at time t. Let further $q_n(t) = P(Z_n(t) = 0)$. Under suitable conditions (cf. [10]), it can be shown that as $t \to \infty$, the extinction probability, q_n, of the ancestral process is the solution to the equation $s = E[e^{s\xi_n(\infty)}]$ and that with α_n and β_n denoting the Malthusian parameter and the mean generation age in the ancestral process and $\sigma_n^2 = Var[\xi_n(\infty)]$

$$\mu_n(\infty) < 1 \Rightarrow q_n = 1 \quad and \quad 1 - q_n(t) \sim ce^{\alpha_n t}$$

$$\mu_n(\infty) = 1 \Rightarrow q_n = 1 \quad and \quad 1 - q_n(t) \sim \frac{2\beta_n}{\sigma_n^2 t}$$

$$\mu_n(\infty) > 1 \Rightarrow q_n < 1.$$

Let M_t stand for the number of alleles at time t, and $M_t(j)$ for the number of alleles with j representatives at time t. Using the convergence results for branching processes counted with random characteristics (cf. [3] or the Appendix in [10]), it is straightforward to see that

RESULT 1.

$$\frac{M_t(j)}{M_t} \longrightarrow P_j = \frac{\alpha}{1 - \hat{q}_n(\alpha)} \int_0^\infty e^{-\alpha u} P(Z_n(u) = j)du$$

almost surely on the set of non-extinction. P_j is usually called the frequency spectrum and can obviously not be further specified at this level of generality (i.e. without specifyng the probability measure P). Its tail however can be described quite generally. With $Q(k) = \sum_{j>k} P_j$ we can show that $Q(c_1 e^{\alpha_n t}) \sim c_2 e^{\alpha t}$ as $t \to \infty$ for some (complicated) constants c_1 and c_2.

RESULT 2. Let $M_t(a) =$ the number of alleles older than a at time t. Then

$$\frac{M_t(a)}{M_t} \longrightarrow \frac{\alpha}{1 - \hat{q}_n(\alpha)} \int\limits_a^\infty e^{-\alpha u}(1 - q_n(u))du$$

in the same sense as above.

Much more than what is done in Result 2. can be achieved; in [10] it is shown that the whole point process of the ages of the oldest alleles converges, as time passes, towards a mixed Poisson point process.

4. Backward aspects. As time evolves the branching population (being supercritical) will either die out or grow (exponentially) forever. In the latter case it also stabilizes its composition over age, kinship structure, types (if any) and all other aspects. Such a stable population follows stable laws, which can be understood in terms of a randomly sampled individual (cf. [3] or Chapter 5 in [10]). To see this heuristically, let $Z_t^\chi = \sum_{x \in I} \chi_x(t - \sigma_x)$ stand for the branching process counted by the random characteristic χ, where $(t - \sigma_x)$ stands for x's age at time t. The idea here is to associate a weight, χ, with every individual and instead of counting the individuals themselves, we sum all their χ-values. Let further Y_t be the total population count at time t. The ratio Z_t^χ / Y_t can now be interpreted as an empirical mean; every term in the sum Z_t^χ is chosen with probability $1/Y_t$. In other words the ratio can be understood as the mean χ-value of a randomly sampled individual chosen from the population at time t. Using some version of the law of large numbers, such a ratio can be shown to converge, as t tends to infinity, to the expected χ-value $\tilde{E}[\chi]$, of a randomly chosen individual from a very old population. If χ is the indicator function of some event A, we can interpret $\tilde{E}[\chi]$ as the probability, $\tilde{P}(A)$, of A. Since this can be done for any relevant event A, the above defines a probability measure $\tilde{P}$, the stable pedigree measure (for the details of how $\tilde{P}$ is formally constructed as well as for proofs, cf. [3]). $\tilde{P}$ can be used to describe the probabilities of all kinds of events related to a randomly sampled individual, including her ancestry. Just to mention some examples of such events in the context of our model, we mention that the backward process along the ancestry line of the randomly sampled individual is a renewal process with interarrival time distribution $\mu_\alpha(t) = \int_0^t e^{-\alpha u} \mu(du)$. Also the successive intervals between mutant ancestors turn out to follow a delayed renewal process where the delay time distribution follows $\sum_{k=0}^\infty (\mu_\alpha^n)^{*k}(t) \hat{\mu}_m(\alpha)$ and the interarrival time distribution is $\sum_{k=0}^\infty (\mu_\alpha^n)^{*k} \star \mu_m(t)$, where $\mu_\alpha^n(t) = \int_0^t e^{-\alpha u} \mu_n(du)$.

It turns also out that the rate of this renewal process is $\frac{\hat{\mu}_m(\alpha)}{\beta}$ and that $\hat{\mu}_m(\alpha)$ stands for the probablity that any ancestor is a mutant. Using the central limit theorem for renewal processes it is possible to make postdictions and confidence intervals for the time moment in the past when the kth (k assumed to be large) mutant ancestor of the randomly sampled individual appeared into the population (cf. [4] for details).

We now, shortly, discuss the molecular clock hypothesis. In the context of our model this can be formulated in the following way: "The rate of evolution is independent of the population size and the generation length". In our case the first statement is true while the second one implies that $\hat{\mu}_m(\alpha) = \beta c$ for some constant c. If we for example consider age-dependent mutations $p(t)$, we get

$$\int e^{-\alpha u} p(u)\mu(u)du = c \int u e^{-\alpha u}\mu(u)du$$

so roughly speaking $p(u) = cu$, i.e. the rate of evolution is the same as the mutation rate, a result which is due to Kimura (more about this topic in [4]). For a discussion of how this interpretation of the molecular clock relates to the usual one, cf. [5] (in particular, the comment by Peter Donnelly).

5. The probability of identity by descent.

Let ϕ denote the number of individuals carrying the same allele as a randomly sampled one. As will be seen below, ϕ is the key ingredient to the probability of identity by descent, i.e. the probability that two randomly sampled individuals carry the same allele. To understand what is going on, consider the following simple balls in urns problem. Assume an urn containing a total of n balls of which n_i are of colour i, $i = 1, 2..., k$, and $\sum_{i=1}^k n_i = n$. Taking two balls at random and checking if they are of the same colour, is equivalent to taking one first ball, b_1, at random and then choosing a second ball, b_2, and checking if it is of the same colour as b_1. Let F_x stand for the set of all balls carrying the same colour as some particular ball x. With this notation, we obtain for large values of n

$$
\begin{aligned}
Prob(b_2 \in F_{b_1}) &= \sum_x Prob(b_2 \in F_{b_1}|b_1 = x)Prob(b_1 = x) \\
&= \frac{1}{n}\sum_x Prob(b_2 \in F_x)
\end{aligned}
$$

$$= \frac{1}{n} \sum_x \frac{n(x)-1}{n-1}$$

$$\simeq \frac{1}{n} E[\phi]$$

where ϕ stands for the cardinality of the colour of a randomly sampled ball. Instead of sampling balls, we can of course sample colours and use "biased sampling". This gives

$$E[\phi] = \sum_{l=1}^{k} \frac{n_l}{n} n_l = \sum_{l=1}^{k} \frac{1}{n} n_l^2$$

so we see that the 2nd moment of the biased sampling distribution is the main ingredient in ϕ's mean, which is at its turn all we need to calculate the probability of identity by descent. This heuristic argument is meant as a motivation (and not a proof) to the following result

RESULT 3.

1. Let ϕ be as above. Then

$$\tilde{E}[\phi] = \frac{\alpha \hat{\mu}_m(\alpha)}{1 - \hat{q}_n(\alpha)} \int_0^\infty e^{-\alpha u} E[Z_n(u)^2] du$$

and is finite if there exists a constant γ such that (i) $\gamma < \frac{\alpha}{2}$ (ii) $E[\hat{\xi}_n(\gamma)] < 1$ (iii) $E[\hat{\xi}_n(\gamma)^2] < \infty$.

2. Let P_t stand for the probability of identity by descent at time t. Then for large values of t we have the approximation $P_t \sim \tilde{E}[\phi]/Y_t$.

EXAMPLE

To see how Result 3. can be used, we show how $E[Z_n(u)^2]$ can be computed in the case of a so called Markovian branching process. In this case, the life span is exponentially (δ) distributed and an individual reproduces by splitting into a random number of offspring which is independent of the life span and which we assume to follow the Poisson distribution with parameter m. Since we are in the supercritical case, we take $m > 1$. We will also assume a fix mutation probability, θ. Then the reproduction of non-mutant offspring will also be Poissonian with parameter $m(1-\theta)$. As to the Malthusian parameter of the process, this turns out to be $\alpha = \delta(m-1)$ and it is easy to see that $\alpha_n = \delta(m-1) - \delta m\theta$. We can now compute $E[Z_n(u)^2]$ which we denote $m_2(u)$. A standard argument shows that $m_2(u)$ satisfies the backward equation

$$\frac{d}{dt} m_2(u) = \delta m^2 (1-\theta)^2 e^{2(\delta(m-1)-\delta m\theta)u} + \delta m_2(u)$$

under the boundary condition $m_2(0) = 1$. Modulo some work, the solution, $E[Z_n(u)^2]$, to this equation can be determined as

$$m_2(u) = \begin{cases} \delta u & \text{if } m = \frac{1}{1-\theta}; \\ \frac{m^2(1-\theta)^2}{m(1-\theta)-1} e^{-(m-1-m\theta)\delta u} \left(e^{-(m-1-m\theta)\delta u} - 1 \right) & \text{otherwise} \end{cases}$$

and this can of course be used in Result 3.

6. The balance between mutation and selection. The essence of selection is that some individuals do better than others. This can be described in the context of our model by extending our single type process into a multitype branching process. We denote by Γ the (geno-) type space, with elements γ. An individual of type γ will now choose her life according to a probability measure P_γ. The fitness of such an individual will be denoted by $\omega(\gamma)$. As before we also allow for mutations splitting the reproduction point process ξ into ξ_n and ξ_m. But since the alleles are now well defined (as elements of Γ) we can be more specific about the effect of a mutation. Given that a mutation has taken place, the type of the offspring is assumed to change from γ to γ' (say) according to some probability density function $u(\gamma, \gamma')d\gamma'$. Our basic parameter μ is replaced by the kernel $\mu(\gamma, d\gamma' \times du)$ to be understood as the mean number of children with types in the infinitesimal set $d\gamma'$ pertaining to a mother of type γ in the infinitesimal age interval da. More precisely we have

$$\mu(\gamma, d\gamma' \times du) = \omega(\gamma)\mu_n(du)1(\gamma \in d\gamma') + \omega(\gamma)\mu_m(du)u(\gamma, \gamma')d\gamma'.$$

The most interesting case is the Malthusian supercritical case when it is possible to find a constant α and an eigen-function h satisfying

$$h(\gamma) = \int_\Gamma \int_{R_+} e^{-\alpha u} h(\gamma')\mu(\gamma, d\gamma' \times du).$$

In this case, $e^{-\alpha u}h(\gamma')\mu(\gamma, d\gamma' \times du)/h(\gamma)$ can be seen as the transition kernel of an (embedded) Markov renewal process . The Markov chain part of that process is governed by the transition probabilities

$$\mu_\alpha(\gamma, d\gamma') = \int_{R_+} e^{-\alpha u} h(\gamma')\mu(\gamma, d\gamma' \times du)/h(\gamma).$$

Let φ be the invariant probability measure of this chain and define $\pi(\gamma) = \varphi(\gamma)/h(\gamma)$ such that $\int_\Gamma \pi(d\gamma) = 1$. π describes the stable birth type distribution and is the crucial ingredient of all kinds of stable composition laws. For a detailed description of branching processes with general type spaces as well as for exact conditions, cf. [2] (cf. also [11]).

The above model can be used in a variety of situations involving models for the balance between selection and mutations of metric traits (cf. [1]). Also models of the kind described in [8] for the evolution of transposable elements in haploid populations can be seen as special cases of our model.

In order to illustrate how the above model can be used, we apply it to Kingman's house-of-cards model (cf. [7]) which has the advantage of admitting a "fairly simple and explicit analysis" (cf. also the contribution by S. Ethier to this volume). The idea of that model is to consider mutations having the effect of bringing down the "biochemical house-of-cards painfully build up by past evolution". Here Γ is taken to be the positive integers and $u(i,j) = u(j)$, $\sum_j u(j) = 1$. For this model we obtain

$$h(i) = \int_{R_+} \sum_j e^{-\alpha u} h(j)\omega(i)1(i=j)\mu_n(du) + \int_{R_+} \sum_j e^{-\alpha u} h(j)\omega(i)u(j)\mu_m(du)$$

or, equivalently

$$h(i) = \omega(i)\hat{\mu}_n(\alpha)h(i) + \omega(i)\hat{\mu}_m(\alpha)\sum_j h(j)u(j)$$

so we take $h(i) = \frac{\omega(i)\hat{\mu}_m(\alpha)}{1-\omega(i)\hat{\mu}_n(\alpha)}c$ for any constant c. It is now not very difficult to see that

$$\mu_\alpha(i,j) = \begin{cases} \hat{\mu}_n(\alpha)\omega(i) + \hat{\mu}_m(\alpha)\omega(i)u(i) & \text{if } i = j, \\[2mm] \frac{h(j)}{h(i)}\hat{\mu}_m(\alpha)\omega(i)u(j) & \text{if } i \neq j. \end{cases}$$

The equation $\varphi(j) = \sum_i \varphi(i)\mu_\alpha(i,j)$, becomes

$$\frac{\varphi(j)}{h(j)} = \frac{\varphi(j)}{h(j)}\hat{\mu}_n(\alpha)\omega(j) + \sum_i \hat{\mu}_m(\alpha)\omega(i)u(j)\frac{\varphi(i)}{h(i)}.$$

If we now take $\pi = \varphi/h$, we obtain $\pi(j) = \frac{u(j)\hat{\mu}_m(\alpha)\hat{\omega}}{1-\omega(j)\hat{\mu}_n(\alpha)}$, where $\hat{\omega} = \sum_j \omega(j)\pi(j)$ can be interpreted as the mean fitness. In order to be able to compare our expression for the stable genotype distribution with its counterpart in Kingman's work, we replace $\omega(i)$ by the normalized fitness $\omega(i)/\hat{\omega}$. This gives

$$\pi(j) = \frac{u(j)\hat{\mu}_m(\alpha)}{1 - \hat{\omega}^{-1}\omega(j)\hat{\mu}_n(\alpha)}$$

and if we assume that a mutation occurs with probability θ independently of everything else, we get $\hat{\mu}_m(\alpha) = \theta$ and

$$\pi(j) = \frac{u(j)\theta}{1 - \hat{\omega}^{-1}\omega(j)(1-\theta)},$$

i.e. the exact expression obtained in [7] for a completely different model (compare also with some of the formulas in [9]).

To fully appreciate the above result, we give a brief account of Kingman's house-of-cards model to show how different it is from our branching process model. Consider a population which is so large that random fluctuations can be ignored and the gene frequencies can be studied deterministically. Suppose that there are k possible alleles $(A_1, \ldots , A_k)$ and let $p_1, \ldots , p_k$ (such that $\sum_i p_i = 1$ and $p_i \geq 0$, $i = 1, \ldots, k$) denote the corresponding gene frequencies. We will assume that the mutation matrix is given by

$$u_{ij} = \begin{cases} 1 - u - u_i & \text{if } i = j; \\ u_j & \text{otherwise} \end{cases}$$

where $u_i > 0$ and $\sum_i u_i = u < 1$. To obtain the next generation individuals, gametes are drawn from the gamete pool in such a way that an individual is of genotype $A_i A_j$ with probability $p_i p_j$. Such an individual will survive to maturity with probability w_{ij} which is assumed to be multiplicative (i.e. $w_{ij} = w_i w_j$ for some w_i, $i = 1, \ldots, k$). This means that selection operates on gametes rather than on individuals. Using this notation, the gene frequencies, $P_j(t)$, in the th generation can be shown to satisfy the recursion

$$P_j(t + 1) = (1 - u)W_t^{-1}P_j(t)w_j + u_j$$

where $W_t = \sum_{j=1}^{k} P_j(t)w_i$. Using this recursion Kingman shows that

$$P_j(t) \to \sum_{r=0}^{\infty}(1 - u)^r W^{-r}u_j w_j^r = \frac{u_j}{1 - W^{-1}\omega_j(1 - u)}$$

where $W = \lim_{t \to \infty} W_t$.

Another example is provided by Kingman's extension of the house-of-cards model to the case of a continuous genotype space (cf. [6]). Omitting the details, the main result is that the stable genotype distribution, $\pi(\gamma)$, turns out to be

$$\pi(\gamma) = \frac{u(\gamma)\theta}{1 - \hat{\omega}^{-1}\omega(\gamma)(1 - \theta)}.$$

REFERENCES

[1] R. BÜRGER (1986) *On the maintenance of genetic variation: Global analysis of Kimura's continuum-of-alleles model.* J. Math. Biol. 24: 341–351.

[2] P. JAGERS (1989) *General branching processes as Markov fields.* Stochastic Processes and their Applications 32: 183–212.

[3] P. JAGERS AND O. NERMAN (1984) *The growth and composition of branching populations.* Adv. Appl. Prob. 16: 221–256.

[4] P. JAGERS O. NERMAN AND Z. TAIB (1989) *When did Joe's great ... greatgrandfather live? Or: On the time scale of evolution.* IMS Lecture Notes-Monograph Series 18: 118–126.

[5] P. JAGERS (1991) *The growth and stabilisation of populations.* Statsitical Science Vol. 6, No. 3, 269–283.

[6] J.F.C. KINGMAN (1978) *A simple model for the balance between selection and mutation.* J. Appl. Prob. 15, 1–12.

[7] J.F.C. KINGMAN (1980) *Mathematics of Genetic Diversity.* SIAM. Philadelphia, Pennsylvania.

[8] M.E. MOODY (1988) *A branching process model for the evolution of transposable elements.* J. Math. Biol. 26: 347–357.

[9] E.D. ROTHMAN AND N.C. WEBER (1986) *A model for weak selection in the infinite allele framework.* J. Math. Biol. 24: 353–360.

[10] Z. TAIB (1992) *Branching Processes and Neutral Evolution.* Lecture Notes in Biomathematics No 93, Springer Verlag.

[11] Z. TAIB (1993) *A note on modelling the dynamics of budding yeast populations using branching processes.* J. Math. Biol. 31: 805–815.

IMA SUMMER PROGRAMS

SPRINGER LECTURE NOTES FROM THE IMA:

The IMA Volumes in Mathematics and its Applications

Current Volumes: